THE NEW MANUFACTURING VALUE CHAIN

Leveraging AI, Machine Learning, and Advanced Analytics for Success

Arun Gupta, PhD

J.ROSS PUBLISHING

ISBN: 978-1-60427-210-9

e-ISBN: 978-1-60427-865-1

Printed and bound in the U.S.A. Printed on acid-free paper.

10 9 8 7 6 5 4 3 2 1

Library of Congress Cataloging-in-Publication Data can be found in the WAV section of the publisher's website at www.jrosspub.com/wav.

Direct all inquiries to J. Ross Publishing, Inc., 151 N. Nob Hill Rd., Suite 476, Plantation, FL 33324. For EU safety/GPSR concerns, email: gpsr@mare-nostrum.co.uk or contact: Mare Nostrum Group B.V., Mauritskade 21D, 1091 GC Amsterdam, The Netherlands.

Phone: (954) 727-9333
Fax: (561) 892-0700
Web: www.jrosspub.com

This book is dedicated to

Anshu, Avni, Ankit, Prerana, and not to forget Avik, who continues to be my inspiration.

CONTENTS

Foreword ix
Acknowledgments xi
About the Author xiii
WAV™ Page xv

1 Why Focus on Manufacturing? 1

Introduction 1
Manufacturing and Advanced Manufacturing 1
Definition and Concept of the Manufacturing Value Chain 4
Key Components of the Manufacturing Value Chain 6
  Primary Activities: Creating and Delivering Value 6
  Support Activities: Enabling Efficiency and Effectiveness 7
Summary 9

2 The Manufacturing Value Chain 11

Introduction 11
Overview of the Manufacturing Value Chain 15
  Starting the Manufacturing Value Chain—Conceptualization 16
  Procurement in the Manufacturing Value Chain 17
  The Production Stage in the Manufacturing Value Chain 19
  Distributing Products Made in the Manufacturing Value Chain 21
  Sales and Marketing in the Manufacturing Value Chain 22
  After-Sales Service in the Manufacturing Value Chain 24
  Summary—Stages in the Manufacturing Value Chain 25
Industry 4.0 and Advanced Technologies 25
Challenges, Opportunities, and Considerations 29
Success Stories 32
Future Trends 35
Summary and Key Insights 37

3 Leveraging Advanced Analytics and Business Intelligence in Manufacturing 39

Introduction 39
Common Advanced Analytics Algorithms and BI Tools for the Manufacturing Sector 44
Data Collection and Preparation in Manufacturing 46

Applications of Advanced Analytics and BI in Manufacturing 51
Predictive Maintenance 51
Quality Control 54
Supply Chain Optimization 58
Demand Forecasting 61
Predictive Analytics for Production Scheduling 64
Implementing Advanced Analytics Solutions 66
Data Security and Governance 81
Future Trends and Opportunities 88
Summary and Key Insights 89

4 Common Algorithms, Tools, and Frameworks in AI/ML for Manufacturing 91

Machine Learning Algorithms in Manufacturing 92
Supervised Learning Algorithms 92
Unsupervised Learning Algorithms 102
Reinforcement Learning 109
AI Algorithms in Manufacturing 111
The AI Architectures of Intelligent Manufacturing 112
AI Tools and Frameworks in Manufacturing 123
Scikit-Learn 124
TensorFlow 126
Keras 128
PyTorch 130
AutoML Tools 131
Summary and Key Insights 134

5 Product Design and Development in Manufacturing 137

Fundamentals of Product Design and Development 139
Leveraging Advanced Analytics and BI in Product Design and Development 144
Utilizing BI in Product Design and Development 145
Utilizing AI/ML in Product Design and Development 148
Summary and Key Insights 149

6 Leveraging Advanced Analytics, BI, ML, and AI in Procurement 151

Understanding Data Sources and Integration 153
Role of Advanced Analytics and BI in Raw Material Procurement 158
AI and ML Applications in Raw Material Procurement 159
Advanced Analytics, BI, ML, and AI in Cost Optimization 161
Risk Management and Compliance 164
Data Governance and Security 165
Managing Procurement Processes 166
Future Trends and Opportunities 168
Summary and Key Insights 169

7 Incorporating Advanced Analytics, BI, ML, and AI in Manufacturing and Production 171

Understanding Manufacturing and Production Processes.... 173
Understanding Data Sources and Integration 175
Key Algorithm Classes Used in Manufacturing and Production 177
Predictive Maintenance Algorithms 177
Demand Forecasting Algorithms.... 177
Quality Control Algorithms 179
Supply Chain Optimization Algorithms.... 179
Production Scheduling Algorithms.... 179
Anomaly Detection Algorithms.... 179
Energy Optimization Algorithms 179
Application of Advanced Analytics and BI in Manufacturing and Production—Enhancing Operational Visibility and Performance.... 179
Harnessing ML and AI in Manufacturing and Production 181
Advanced Analytics, BI, ML, and AI in Supply Chain Management: Driving Efficiency and Optimization 182
Implementation Challenges and Best Practices.... 184
Future Trends and Outlook.... 186
Summary and Key Insights.... 186

8 Leveraging Advanced Analytics, BI, ML, and AI in Distribution and Logistics 189

Understanding Data Sources and Integration 191
Application of Advanced Analytics and BI in Distribution 196
Harnessing ML and AI for Route Optimization 199
Optimization of Warehouse Operations with AI.... 201
Inventory Management with AI.... 201
Order Picking and Packing with AI.... 201
Warehouse Layout Optimization.... 201
Real-Time Insights and Automation.... 202
Sustainability and Risk Mitigation.... 202
Environmental Sustainability.... 202
Social and Ethical Considerations.... 203
Disruption and Security Risk Mitigation 203
Implementation Challenges and Best Practices.... 203
Future Trends and Outlook.... 205
Advanced Autonomous Systems 205
Intelligent Inventory Management.... 205
Sustainability and Ethical Considerations 205
Human-AI Collaboration 205
Additional Considerations 206
Summary and Key Insights.... 206

9 Advanced Analytics, BI, ML, and AI in Sales & Marketing and After-Sales Service.... 209

Understanding Data Sources and Integration 212
Application of Advanced Analytics and BI in Sales & Marketing and After-Sales Support 216
ML and AI in Sales & Marketing and After-Sales Support.... 217
Predictive Analytics for Sales Forecasting and Service 219
Optimizing Marketing Campaigns and Customer Support with AI.... 221
Sustainability and Ethical Considerations.... 223
Implementation Challenges and Best Practices 225
Future Trends and Outlook.... 225
Summary and Key Insights.... 226

10 Unleashing the Potential of Advanced Analytics, BI, ML, and AI in Manufacturing 229

Summarizing Key Insights 230
Algorithm Usage and Trends 233
Broader Implications 236
Ethical Considerations and Regulatory Challenges.... 236
Sustainability Issues 237
Innovation and Transformation.... 237
The Role of Automation and Smart Factories 237
Predictive Maintenance and Its Impact.... 237
Challenges and Opportunities 238
Future Outlook.... 238
Inspiring Action 238
Call to Action 239

Chapter References.... 241

Chapter 1 241
Chapter 2 242
Chapter 3 243
Chapter 4 245
Chapter 5 245
Chapter 6 246
Chapter 7 247
Chapter 8 248
Chapter 9 249
Chapter 10 250

Index.... 251

FOREWORD

We are in the early stages of a transformational period in our economic and digital age. Many executives today are overwhelmed by the buzzwords surrounding the emerging field of digital transformation. This is particularly true in manufacturing and supply chain fields, where investors and start-ups are throwing around terms such as *artificial intelligence (AI), predictive analytics, machine-based learning, robotic process automation*, and *generative AI* without really understanding to what applications these technologies can truly be applied. Executive decision makers are often reluctant to admit they do not fully understand what these technologies are all about—but know that the problems they are facing require a different set of solutions. The reality of what manufacturing managers do is a far cry from this future world of machines making decisions. The typical workday of an average manufacturing worker is a type of firefighting—putting out one fire after another, often requiring multiple, spur-of-the-moment decisions in response to an ongoing set of problems and constraints. These decisions are often made based on rudimentary insights and gut reactions. An outside observer might get the impression that significant technological know-how lies behind supply chain and manufacturing decision making. However, today, that is not the case—but it is changing.

Most of the activity that occurs behind the scenes at all stages of manufacturing and the supply chain involves human activities, like processing transactions on screens and Excel sheets. Today, there are massive white space opportunities in the daily activities that manufacturers, supply chain planners, distributors, buyers, and contract officers undertake. This is where opportunities lie in the future of manufacturing. By applying technologies like AI, ML (machine learning), advanced analytics, and generative AI, organizations will be able to drive efficiencies in the way we work that will create trillions of dollars in economic value by reducing working capital, increasing agility, and reducing waste, while making better decisions based on data and analytics rather than on gut feelings.

But what will this look like? My colleague, Arun Gupta, has synthesized the opportunities for applying these emerging technologies in a succinct and practical book that captures the current state of unprecedented opportunities to enhance efficiency, productivity, and competitiveness. With a doctoral degree in computer science and engineering, and decades of work experience in the trenches with companies such as Weyerhaeuser, IBM, Caterpillar, and Deloitte, Arun understands the delicate balance that exists between the world of programmers and the world of managers and leaders—and how to create the correct language that allows both parties to collaborate and innovate.

In this book, he explores the various ways in which ML, AI, advanced analytics, and generative AI are leveraged to gain real-time visibility into production metrics, equipment performance, and supply chain dynamics. By leveraging cutting-edge algorithms and tools, manufacturers can optimize processes, improve decision making, and achieve superior performance across the production value chain. This book will show you how.

Rob Handfield, Ph.D.
Bank of America University Distinguished Professor of Supply Chain Management
North Carolina State University

ACKNOWLEDGMENTS

Writing a book is no easy feat, and with this, I have much to be grateful for. I received valuable feedback from several people in my network, most importantly from Prof. Robert Handfield, who provided guidance and encouragement. I also received critical input from my colleagues, students, and clients while building the foundations for this book. This book emerged from an opportunity in the early 2010s at SAS Manufacturing, where I was asked to present to a group of marketing professionals about analytics in manufacturing. Thank you, Mark Demers and Tim Fairchild, for giving me the opportunity to speak and set the foundation for this book. I am also grateful to my peers and leaders at my past places of employment who engaged me in assignments that would lay the groundwork for this book to come to life.

I would also like to acknowledge my family, without whom this book would not have been possible. They not only provided valuable suggestions but continued to encourage me throughout this effort:

Avni Gupta
Ankit Gupta
Anshu Gupta
Prerana Chhabria
Dr. Ajay Gupta
Sanjay Jain
Kokil Jain

ABOUT THE AUTHOR

Arun P. Gupta, Ph.D., holds a bachelor's in Chemical Engineering and a doctorate in Computer Science Engineering from the University of Texas at Arlington, with over 25 years of expertise spanning academia, manufacturing, and management consulting. His work focuses on information technology (IT), analytics, artificial intelligence (AI), and machine learning (ML) across diverse industries.

As an academician, Dr. Gupta developed and taught courses in computer science and the application of IT in supply chain management within both the computer science and business departments. This foundational work laid the groundwork for his role in the manufacturing sector, where, among other roles, he also served as a trusted advisor for a large industrial manufacturing organization. There, he helped guide the company's IT, digital, and AI-powered transformation to enhance manufacturing processes and optimize supply chain operations. It was during this time that Dr. Gupta first explored AI, developing natural language processing (NLP) models to extract actionable insights from documents—a pivotal moment in his AI journey. Transitioning to management consulting, Dr. Gupta architected innovative e-business solutions, led process development initiatives, and directed strategy and business transformation efforts across a wide range of industries. His recent focus has been on leveraging data, analytics, AI, ML, and generative AI to drive digital transformation and business optimization. Along the way, he has trained numerous global practitioners and leaders, sharing his expertise in these cutting-edge technologies.

Dr. Gupta's career is defined by a proven track record of success across various sectors. For the past 25 years, he has also served as a research associate with the Supply Chain Resource Consortium at North Carolina State University, contributing to the development of a widely recognized supply chain maturity model. Today, Dr. Gupta is at the forefront of applying AI, ML, and generative AI to revolutionize business processes, optimize supply chains, and mitigate environmental and social risks for his clients. His thought leadership has earned him invitations to present at major conferences, and he has published several influential conference papers.

This book has free material available for download from the Web Added Value™ resource center at *www.jrosspub.com*

At J. Ross Publishing we are committed to providing today's professional with practical, hands-on tools that enhance the learning experience and give readers an opportunity to apply what they have learned. That is why we offer free ancillary materials available for download on this book and all participating Web Added Value™ publications. These online resources may include interactive versions of material that appears in the book or supplemental templates, worksheets, models, plans, case studies, proposals, spreadsheets and assessment tools, among other things. Whenever you see the WAV™ symbol in any of our publications, it means bonus materials accompany the book and are available from the Web Added Value Download Resource Center at www.jrosspub.com.

Downloads for *The New Manufacturing Value Chain* include various figures from the book to use in your AI implementations.

1

WHY FOCUS ON MANUFACTURING?

"Manufacturing is more than just putting parts together. It's coming up with ideas, testing principles, perfecting the engineering, and completing the final assembly."

—Sir James Dyson, founder of Dyson

INTRODUCTION

In the ever-evolving landscape of modern industry, the manufacturing value chain stands as the cornerstone of global production, orchestrating the conversion of raw materials into products that satisfy global market demands. This intricate network of interconnected processes encompasses everything from product design and raw material sourcing to manufacturing, distribution, and customer delivery. Each phase adds value and shapes the journey from conception to consumption, impacting global dynamics, technology, and sustainability. Understanding this chain is crucial for achieving operational excellence, cost efficiency, and sustained competitiveness in the marketplace.

This chapter explores the core concepts of manufacturing and the manufacturing value chain, delving into the key components and stages from design to delivery. We discuss the challenges and interdependencies across the chain as well as the influence of globalization, technology, risk management, and sustainability considerations. This sets the stage for a broader examination of how manufacturing impacts the world and the insights offered by the rest of the book.

MANUFACTURING AND ADVANCED MANUFACTURING

There are many definitions for manufacturing. Some of these are listed here:

1. **The United States Department of Labor** classifies manufacturing as a subset of the "goods-producing industries supersector."[1]
2. **The North American Industry Classification System Manufacturing** (NAICSM) defines the manufacturing sector as ". . . the mechanical, physical, or chemical transformation of materials, substances, or components into new products."[2] Further, NAICSM defines manufacturing companies as ". . . plants, factories, or mills that characteristically use power-driven machines and materials-handling equipment . . . Manufacturing establishments may process materials or may contract with other establishments to process their materials for them."
3. **The National Institute of Standards and Technology** (NIST) defines manufacturing as the systematic process of converting raw materials, components, or parts into finished

goods that meet a customer's expectations or specifications.[3] This involves a combination of manual and automated labor, machinery, and the application of scientific principles and technology.
4. **The United Nations Industrial Development Organization** defines manufacturing as a set of economic activities that involve the transformation of raw materials and components into finished products for sale. It encompasses a range of processes, from traditional craft production to advanced, high-tech methods, contributing significantly to economic growth and employment.[4]
5. **The World Economic Forum** defines manufacturing as the systematic process of creating, producing, and assembling goods on a large scale, often involving the use of advanced technologies and innovation.[5] It is a key driver of economic development, providing employment opportunities, fostering innovation, and contributing to global trade.

All of the aforementioned definitions have in common the fact that manufacturing produces goods using raw materials and components while relying on labor and machinery and the help of advanced technologies and innovation.

Of the various defined business categories, the *manufacturing sector* is the largest and contains a broad group of sub-sectors, such as industrial manufacturing, transportation equipment, high technology, apparel and footwear, food and beverage, pharmaceuticals, and chemicals. Manufacturing is critical for the economy because it employs a large segment of the workforce. Additionally, products from the manufacturing sector are used by other business categories (e.g., agriculture, mining, trade, services, administration, etc.) in conducting their business.

Manufacturing has been and continues to be the economic driver for countries around the globe. A high-quality manufacturing sector is required to forge a path to development and create wealth for countries. The health of the manufacturing industry drives the health of a country's economy and is critical to economic growth and prosperity. Further, approximately 80% or more of global trade is in the form of goods, while less than 20% is in the form of services, supporting the importance of manufacturing in the global economy.

Global manufacturing is a $16.3 trillion industry and contributes 11.4% of global gross domestic product (GDP). The top three countries in manufacturing are China, the United States, and Germany—making up over 53% of manufacturing. Over 70% of global trade involved products created in manufacturing.

In the mid-2010s, countries like Australia[6] and South Africa[7] saw that their shrinking manufacturing capabilities were having an adverse effect on their respective abilities to conduct research and development (R&D) and, hence, innovation. However, during a similar timeline, the United States also experienced a shrinking manufacturing base, but the level of R&D and innovation did not see a comparable reduction. In the United States, even though the manufacturing industry is just over 11% of GDP,[8,9] it constitutes 54% of the private sector R&D spend.[10] Most U.S. companies have realized that moving their manufacturing overseas for an anticipated lower cost structure has resulted in reduced manufacturing maturity. They have concluded that a large geographic separation between manufacturing and innovation is slowly leading to reduced innovation capability in terms of overall volume, cycle time, cost, and quality, eroding their competitive advantage. With rising labor costs in target offshoring countries, the cheap labor advantage of offshoring has all but vanished.

The extended supply chain, cultural differences, lower productivity, higher transportation costs, lower control over the product, and longer cycle times to introduce new products have further tarnished its image, resulting in many companies *reshoring* their manufacturing. In a global business environment, reshoring their overseas plants makes sense to meet regional demand at competitive prices. For products with increasing global demand, U.S. companies have selected to redistribute their manufacturing capacities by adding capacity in the United States to incorporate reshoring.[11] For example, in 2014, GE reshored its appliance manufacturing. In 2013, Walmart decided to domestically source an additional $50 billion worth of U.S.-made products over the next 10 years to help with reshoring.[12] Caterpillar has reshored production of two product lines from Japan to the United States. Other companies like Ford and GM have also been on the reshoring bandwagon.[13] Other countries have similarly been pushing their own reshoring initiatives.

The manufacturing industry also plays an important role in nonmanufacturing industries. These industries depend on manufactured goods to conduct their core business operations. For example, the retail industry is mainly involved in trading manufactured goods. The healthcare segment extensively uses output from the manufacturing industry in the form of medical equipment and pharmaceuticals. Airlines conduct most of their business using aircraft and interiors from the manufacturing industry. Additionally, improvements in the nonmanufacturing sectors are highly dependent on technological improvements in the manufacturing industry. For example, the tremendous improvements in computing hardware due to innovation in manufacturing have resulted in faster processing and lower costs, improved software, and newer capabilities. These improvements are being utilized worldwide in all aspects of trade in the form of electronic commerce.

> *"It's not the big that eat the small, it's the fast that eat the slow."*
>
> —Jason Jennings, author and business leadership expert

It is well-known that manufacturing creates jobs. Of the three largest manufacturing countries previously listed, China employs over 70 million people, the United States employs over 14 million people, and Germany employs almost 8 million people. In general, studies on economic multipliers have shown that manufacturing tends to have a higher multiplier compared to other sectors.[14] For example, a report by the United States NIST suggested that *manufacturing* has a higher multiplier effect than *services*, meaning that each dollar of final demand in manufacturing has a greater impact on the overall economy. A job added in manufacturing results in about nine nonmanufacturing jobs that support it.

Advanced technology and innovation have resulted in a new term called *advanced manufacturing*, which is now used in the industry. Advanced manufacturing represents a paradigm shift in industrial production, leveraging state-of-the-art technologies like automation, artificial intelligence (AI), machine learning (ML), and data analytics to optimize manufacturing processes. Advanced manufacturing—according to the President's Council of Advisors on Science and Technology Report to the President on Ensuring American Leadership in Advanced Manufacturing[15]—is "a family of activities that (a) depend on the use and coordination of information, automation, computation, software, sensing, and networking and/or (b) make use of cutting-edge materials and emerging capabilities enabled by the physical and biological sciences, for example, nanotechnology, chemistry, and biology. It involves both new ways to manufacture existing products and the manufacture of new products emerging from new advanced technologies" (like additive or 3-D manufacturing).

Advanced manufacturing enhances efficiency, precision, and customization, thereby fostering a more agile and sustainable manufacturing ecosystem. By incorporating industry principles, advanced manufacturing transforms traditional factories into smart, interconnected systems that can adapt to dynamic market demands.[16] As a key driver of innovation, it not only accelerates production but also plays a crucial role in shaping the future of manufacturing by embracing the full potential of modern technologies.[17]

The associated productivity gains have consistently resulted due to advances in technology that, in turn, have driven economic growth and movement toward advanced manufacturing. In the United States, these productivity gains have averaged about 3.4% annually since 2005. Using the United States as the proxy for other countries—along with examples from Australia and South Africa—shows the need for a robust manufacturing sector, especially advanced manufacturing. This is typically reached through reshoring.

Advanced manufacturing and *manufacturing for the grand challenges* initiatives have worked to overcome the perception that manufacturing is mechanical or dirty work. Advanced manufacturing seeks to establish a stronger foundation for innovation—one that is stronger in learning and developing critical skills. Similarly, the National Academy of Engineering, with their global partners, established 14 grand engineering challenges of the twenty-first century.[18] These grand challenges addressed four areas that impact humanity—sustainability, health, vulnerability, and joy of living. A 2013 summit focused on the need to examine the engineering and manufacturing capabilities required to solve the grand challenges in manufacturing.[19] It was recognized that the success and growth of advanced manufacturing require a collaborative effort between governments, industry, and academia with an increased global student base possessing a high interest in technology- and engineering-related fields. The grand challenges mandated new ways of thinking. These efforts have started to invalidate the negative perception associated with manufacturing. The advantages developed countries achieved when they grew through manufacturing are slowly being eroded by developing countries that are focusing on advanced manufacturing through an increased level of innovation.

Like other industries, manufacturing has its challenges as well. Most manufacturing requires physical presence on the manufacturing floor, although that has seen some changes since the COVID-19 pandemic. While productivity and other external factors may have reduced the labor force, a zero-labor manufacturing environment is not possible in the near future, thus making labor shortages one of the biggest challenges. Further, just having warm bodies in the manufacturing environment is insufficient; these people must be skilled in manufacturing, computing, data analysis, safety, etc. Getting the right skills to the workforce and establishing policies to make this happen are additional challenges faced not just by manufacturing but also by other industries and policymakers. Manufacturing facilities have an impact in the areas where they are located and are impacted by legislation established locally, federally, and globally. In addition to legislation, the impact on manufacturing by customer requirements (e.g., environment, social, and governance or diversity, equity, and inclusion) adds to the challenges.

DEFINITION AND CONCEPT OF THE MANUFACTURING VALUE CHAIN

The manufacturing value chain is a comprehensive and intricate framework that encapsulates the entire life cycle of a product. It starts with the ideation of a product and progresses through multiple stages, such as design, procurement, manufacturing, distribution, sales, and service. Each stage plays a crucial role in enhancing the overall value of the final product, emphasizing the necessity for seamlessly integrated and efficient processes. It serves as a strategic model, highlighting the interconnectedness of diverse activities within the manufacturing domain. Understanding the manufacturing value chain provides companies with a strategic advantage. It serves as a roadmap for identifying areas of improvement, fostering innovation, making informed decisions, and emphasizing efficiency, quality, and adaptability. It offers a systematic approach to achieving operational excellence and sustained competitiveness.

The manufacturing value chain journey commences with the conceptualization phase, where innovative ideas are translated into tangible product concepts. This stage emphasizes creativity, innovation, and iterative refinement to meet market demands and includes customer feedback through voice of the customer. Following this, the procurement and sourcing phase involves efficiently acquiring raw materials and components through collaboration with suppliers. Sustainable and ethical sourcing practices have become integral in the contemporary manufacturing landscape.

The heart of the value chain lies in the production or manufacturing phase, where raw materials undergo transformation into finished products. Utilizing advanced algorithms, machinery, automation, and skilled labor, manufacturers can optimize efficiency and quality control. Logistics and distribution dynamics play a crucial role in ensuring timely and cost-effective material and product delivery, aligning with the need for agility in responding to market demands.

In the manufacturing value chain, marketing and sales activities are indispensable components that shape the success and sustainability of the entire process. Manufacturers undertake comprehensive market research to understand customer needs, strategically promote and position their products, and utilize digital channels for enhanced visibility and value proposition. Understanding consumer behavior and market trends is essential for devising effective marketing strategies. Integration of e-commerce facilitates a broader market reach, while customer relationship management systems optimize sales processes. Post-sales support and service, coupled with analytics and key performance indicators, contribute to continuous improvement, ensuring manufacturers remain agile, responsive, and competitive in the evolving market landscape.

Industry 4.0 represents a transformative wave of technological advancements, ushering in an era of smart manufacturing and interconnected production systems. This paradigm shift encompasses a spectrum of cutting-edge technologies, including the Internet of Things, AI, ML, big data analytics, digital twins, and cyber-physical systems. In the context of the manufacturing value chain, Industry 4.0 fosters seamless integration and communication between various stages, optimizing processes, enhancing efficiency, and enabling data-driven decision making.[20,21] Smart factories leverage real-time data from interconnected devices and sensors to facilitate predictive maintenance, improve production planning, and enhance overall operational agility.[22] As a result, Industry 4.0 not only streamlines traditional manufacturing processes but also paves the way for innovative business models and customization. This allows manufacturing to usher in a new era of responsiveness and adaptability in the manufacturing value chain. The holistic approach of Industry 4.0 optimizes individual stages of the manufacturing process as well as fundamentally transforms the entire value chain, aligning it with the demands of the digital age. Figure 1.1 depicts the interconnectedness of manufacturing with other sectors of the economy. Not all sectors are represented but most of the important ones are. This visual representation underscores the integral role manufacturing plays in supporting diverse industries and driving overall economic development.

In essence, the manufacturing value chain underscores the holistic approach to product creation and delivery. It accentuates the synergy between various stages, recognizing the importance

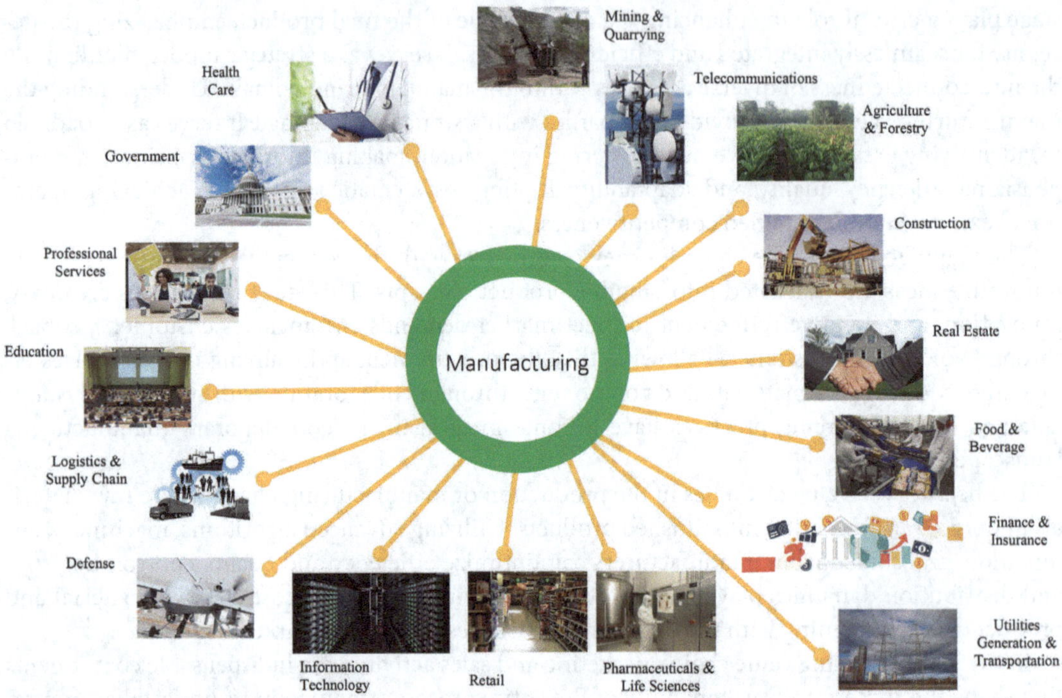

Figure 1.1 Interconnectedness of manufacturing with other sectors of the economy.

of innovation, efficiency, and adaptability in the dynamic market landscape. Navigating this intricate network of processes positions manufacturers to thrive amid technological advancements and evolving consumer preferences.

KEY COMPONENTS OF THE MANUFACTURING VALUE CHAIN

All manufacturing organizations have a value chain. The value chain provides a process flow view of organizations where inputs are transformed into output. Each step in the transformation adds value to the input. The effectiveness of a value chain is governed by the methods in which these processes are linked and the transformations that take place at each step. Since value chains exist to serve the customer, the basic input of a value chain is demand, actual demand (firm orders), or perceived demand (forecasted orders).

The value chain consists of two main sets of activities—primary and supporting. Each of these has multiple components mainly geared to manufacturing but is generic enough to be industry agnostic. The primary set of activities is focused on creating and delivering value through products. The secondary set of activities enables efficiency and effectiveness in the value chain. Figure 1.2 depicts the elements of a value chain. The following paragraphs cover these in a little more detail.

Primary Activities: Creating and Delivering Value

There are five primary activities that directly contribute to the creation and delivery of a product or service. These include:

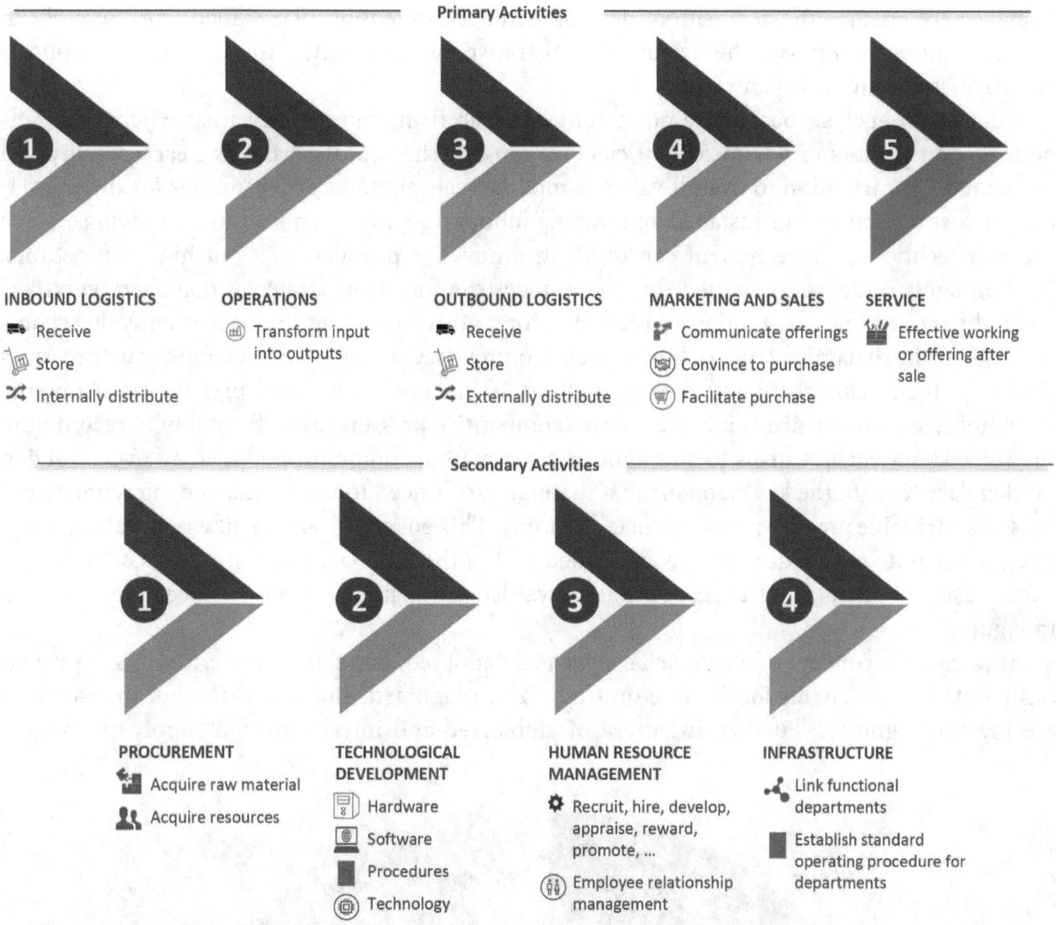

Figure 1.2 Generic representation of a value chain.

1. Inbound logistics—involve the transporting, receiving, storing, and managing of inputs, ensuring a smooth flow of material into the production process
2. Operations (or manufacturing/production)—focus on the transformation of these inputs into finished products or services
3. Outbound logistics—encompass activities related to the storage and distribution of the final product
4. Marketing and sales—involve promoting and selling the product
5. Service—activities that address post-sale customer support, enhancing overall customer satisfaction

Support Activities: Enabling Efficiency and Effectiveness

Support activities, on the other hand, provide the infrastructure and resources necessary to carry out the primary activities efficiently. These include procurement, technology development, human resource management, and firm infrastructure. Procurement involves sourcing materials and negotiating with suppliers to optimize costs and quality. Technology development pertains to research and development efforts that enhance product or process innovation. Human resource management ensures the acquisition, development, and retention of a skilled workforce. Firm infrastructure encompasses the overall administrative and managerial framework that supports all activities within the organization.

Value chain analysis becomes a powerful tool when firms can identify areas where they outperform competitors or where cost efficiencies can be achieved. By dissecting each activity and assessing its contribution to overall value, companies can pinpoint opportunities for differentiation or cost reduction. For instance, optimizing inbound logistics can lead to cost savings, while superior technology development can result in innovative products that outshine competitors. By thoroughly understanding the value chain, firms can formulate strategies that align with their strengths and weaknesses, gaining a sustainable competitive advantage in the market. Value chains account for the dynamic nature of business environments. Companies must continuously reassess and adapt their value chain activities to respond to changes in customer preferences, economic conditions, technological advancements, and competitive pressures. The flexibility to reconfigure the value chain enables firms to stay agile and maintain a competitive edge in an ever-evolving marketplace. One of the key elements that manufacturers need to focus on is the interconnectedness of sustainable practices for a circular economy. This ensures practices like renewable energy integration and waste reduction are embedded within their standard operating procedure to advance sustainability and foster responsible stewardship of natural resources, succinctly captured in Figure 1.3.

Although the concept of a value chain is widely adopted, there are some criticisms. As previously stated, value chains focus more on traditional manufacturing industries, but the elements are industry-agnostic. Further, in an era of globalized and interconnected supply chains, the

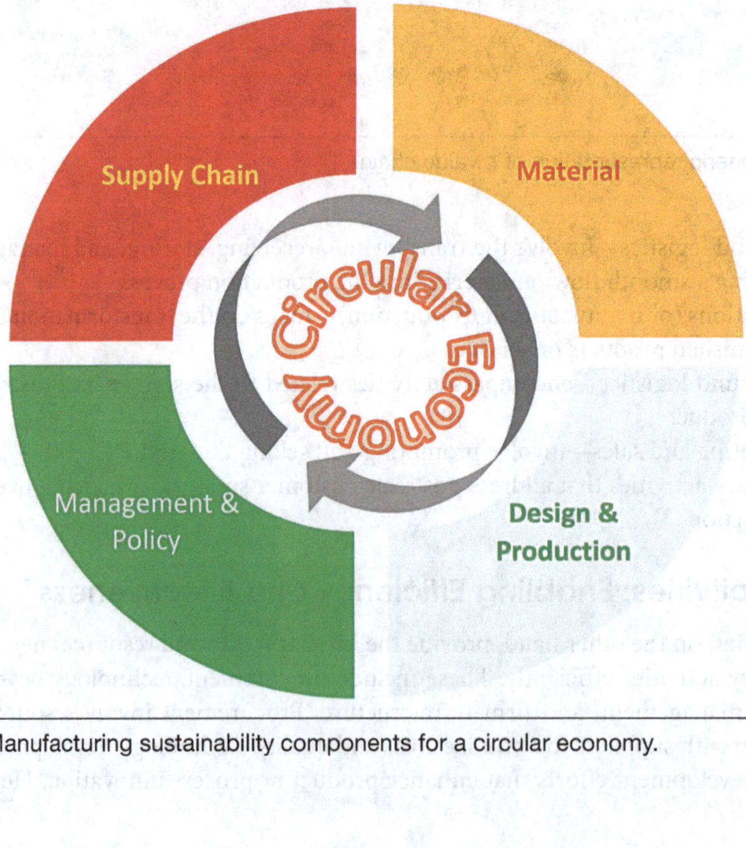

Figure 1.3 Manufacturing sustainability components for a circular economy.

traditional linear representation of the value chain is an oversimplification of the complexities of modern business networks. Despite these critiques, the value chain remains a fundamental tool in strategic management, offering valuable insights into a company's internal processes and avenues for enhancing overall competitiveness.

> *"The essence of strategy is choosing what not to do."*
>
> —Michael Porter, economist, researcher, author, speaker, and teacher

SUMMARY

The manufacturing value chain is a multifaceted system that continues to evolve with technological advancements, globalization, and changing consumer expectations. Recognizing the importance of each stage and adapting to emerging trends is imperative for manufacturers aiming to thrive in a changing, competitive, and dynamic environment. As you go through this book, specifically in chapters on the various phases of the manufacturing value chain (starting in Chapter 5), you will encounter a running vignette for a fictitious manufacturing company, Stellar Electronics. This vignette contains personas involved in the specific value chain phase and potential value added by AI, ML, and advanced analytics solutions. Note that particular metrics can vary based on the context and implementation. You will also realize that choosing not to get on board with advanced analytics, business intelligence, AI, and ML prevents you from building a strong and competitive strategy. Chapter 2 will cover the manufacturing value chain in more detail.

2

THE MANUFACTURING VALUE CHAIN

"Innovation distinguishes between a leader and a follower."
—Steve Jobs, Former CEO, Apple

INTRODUCTION

In the vast landscape of contemporary industry, the manufacturing value chain stands as the backbone of production, distribution, and consumption. It symbolizes a holistic journey laden with challenges, innovations, and transformative technologies. The chain begins with the procurement of raw materials and ends with the delivery of finished goods. All of the other stages—design, production, distribution, and service—play a crucial role in adding value. As a reminder, Figure 1.2 is a visual representation of the primary and secondary activities of the traditional manufacturing value chain, highlighting each stage from raw materials to end products.

Recent years have witnessed the advent of Industry 4.0, marking the integration of advanced technologies, such as advanced analytics, machine learning (ML), artificial intelligence (AI), the Internet of Things (IoT), and smart manufacturing techniques, into the manufacturing value chain. This infusion has sparked a revolution by enhancing precision, agility, and decision-making capabilities across the value chain. This shift is not just a technological upgrade; it represents a classic transformation where machines communicate, adapt, and autonomously optimize processes, leading to unparalleled levels of efficiency.

While delving into the scope of the manufacturing value chain, it becomes evident that the dynamics are not confined to the shop floor alone. They encompass the globalization of supply chains, heightened consumer expectations, and an imperative focus on sustainability. This further amplifies the complexity and intricacies of this ecosystem. Recent global events showcasing supply chain disruptions underscore the need for resilience and agility in the manufacturing value chain. Furthermore, the growing emphasis on sustainability prompts manufacturers to reassess their processes by adopting eco-friendly practices that align with evolving consumer values.

In the ensuing pages, we will navigate the various facets of the manufacturing value chain—from its historical roots to the contemporary landscape shaped by technology and global dynamics. Through an exploration of case studies, emerging trends, and critical analysis, we will unravel the complexities and unveil the opportunities that lie within this integral element of the industrial tapestry. As we embark on this journey, I invite you to envision the manufacturing value chain not merely as a sequence of operations but as a living organism, adapting, evolving, and propelling industries into a future where innovation is the driving force.

To grasp the essence of the manufacturing value chain, a journey back in time to trace its roots through pivotal historical epochs becomes imperative. The inception lies in the throes of the Industrial Revolution, a transformative era that witnessed the metamorphosis of manual labor into mechanized processes. This revolution laid the groundwork for structured manufacturing, with machines powered by steam engines heralding a new era of production. The advent of mass production techniques in the late nineteenth century further solidified the value chain, introducing assembly lines and specialization of labor.

The twentieth century ushered in the second wave of industrialization, marked by automation and the widespread use of electrical energy. This era witnessed the refinement of manufacturing processes, paving the way for enhanced efficiency and increased production capacities. Notable figures, such as Henry Ford, played a pivotal role by introducing assembly-line production that revolutionized the automobile industry and set a precedent for the organization of manufacturing processes.

Fast-forward to the latter half of the twentieth century, where the integration of computers and electronics propelled manufacturing into the realm of precision and control. The rise of computer-aided design (CAD) and computer-aided manufacturing (CAM) systems marked a turning point, allowing for greater accuracy and customization in the production process. This era set the stage for the gradual evolution of manufacturing into a more interconnected and technologically advanced ecosystem.

The present-day landscape of the manufacturing value chain finds itself intricately entwined with the advent of Industry 4.0. This latest industrial revolution—characterized by the integration of advanced technologies such as advanced analytics, ML, AI, IoT, and smart manufacturing—has redefined the manufacturing paradigm. Machines equipped with sensors gather real-time data, enabling predictive maintenance and efficient resource allocation. ML and/or AI algorithms optimize production schedules, and IoT devices enhance communication between various stages. The manufacturing value chain has evolved into a dynamic, interconnected network where digital technologies play a central role in orchestrating operations. Figure 2.1 depicts the history of manufacturing, from steam engines to assembly lines to smart manufacturing to digital Industry 4.0.

The globalization of supply chains has emerged as a pivotal factor influencing the manufacturing value chain. A product's journey now spans continents, necessitating a robust and interconnected network. While this globalization opens avenues for cost-effective production and broader market reach, it also introduces challenges. Recent events, such as the global COVID-19 pandemic, underscored the vulnerability of overly decentralized supply chains. Disruptions in one part of the world can reverberate across the entire value chain, emphasizing the need for resilience and flexibility.

The future of the manufacturing value chain lies in balancing the advantages of globalization with strategies to mitigate risks. It also lies in balancing the digital industry with the physical industry. Figure 2.2 shows the interconnectedness of Industry 4.0 with biological and physical components, allowing the synergies to result in insights into manufacturing needs. This fusion of historical evolution and technological integration underpins the multifaceted landscape of the contemporary manufacturing value chain. Robust risk planning and management are required as supply chains become longer; black swans can show up at any time, but risk planning helps alleviate their significant impacts. This involves regionalizing supply chains, leveraging technology for real-time monitoring, conducting war-game exercises, and embracing agile manufacturing practices. By doing so, manufacturers can navigate the complexities of long supply chains and a

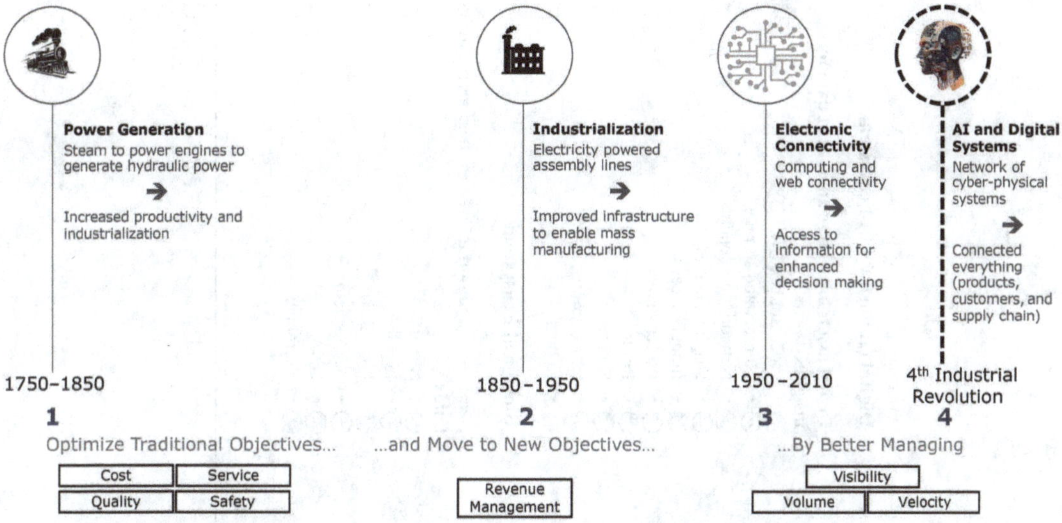

Figure 2.1 Manufacturing history timeline.

globalized world while ensuring stability and adaptability in the face of unforeseen challenges. Figure 2.3 shows the data framework for manufacturing excellence and the related complexities. This framework helps manufacturers enhance and build on capabilities, develop new partnerships among trading partners, and determine the maturity and gaps within their data-centric applications.

> *"Accelerating the transition to a green digitized economy is key to put Europe at the forefront of industrial competitiveness worldwide. As part of the CEO Alliance, we are inviting EU policymakers to pursue an industrial policy that supports digital infrastructure, innovation, and the roll-out of cutting-edge technology, such as AI, to make it happen."*
>
> —Björn Rosengren, ABB CEO

In an era dominated by environmental consciousness, sustainability has become a core consideration in the manufacturing value chain. Consumers increasingly demand eco-friendly products, and regulatory frameworks worldwide are mostly tightening environmental standards. Manufacturers are reevaluating their processes, seeking innovative ways to reduce their ecological footprint.

Sustainable practices encompass various aspects of the value chain, from sourcing raw materials responsibly to optimizing water and energy consumption and reducing waste. Technologies such as IoT and AI play crucial roles in monitoring and improving sustainability metrics. IoT sensors track resource usage in real time, and AI algorithms provide insights to enhance efficiency while minimizing environmental impact. Sustainability practices encompass a holistic approach. They involve not only the environmental impact but also promoting human rights, fair labor practices, community development, and addressing social inequalities. Transparency and accountable management of resources and operations emphasizing ethical business practices and responsible decision making are core to sustainability. Embracing sustainability is not just a moral imperative but also a strategic move to align with evolving consumer expectations and regulatory requirements.

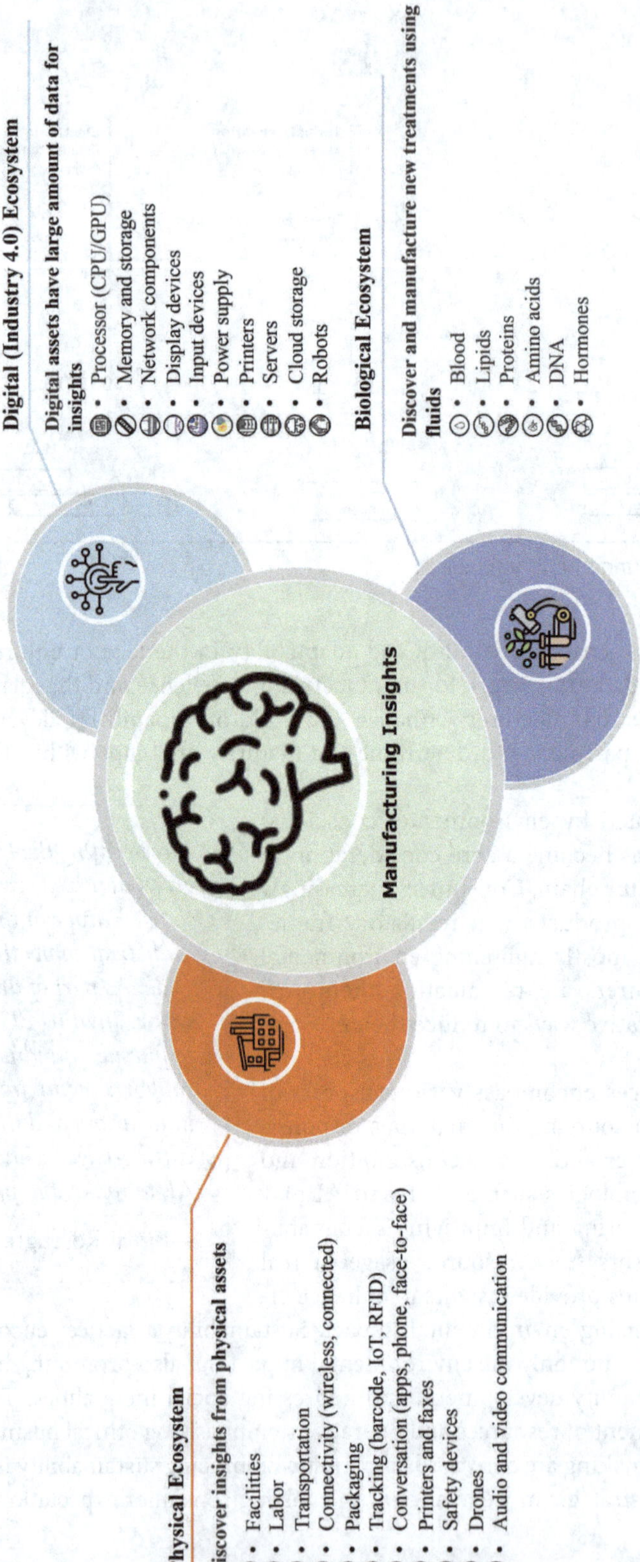

Figure 2.2 Synergies in physical, biological, and digital ecosystems for manufacturing insights.

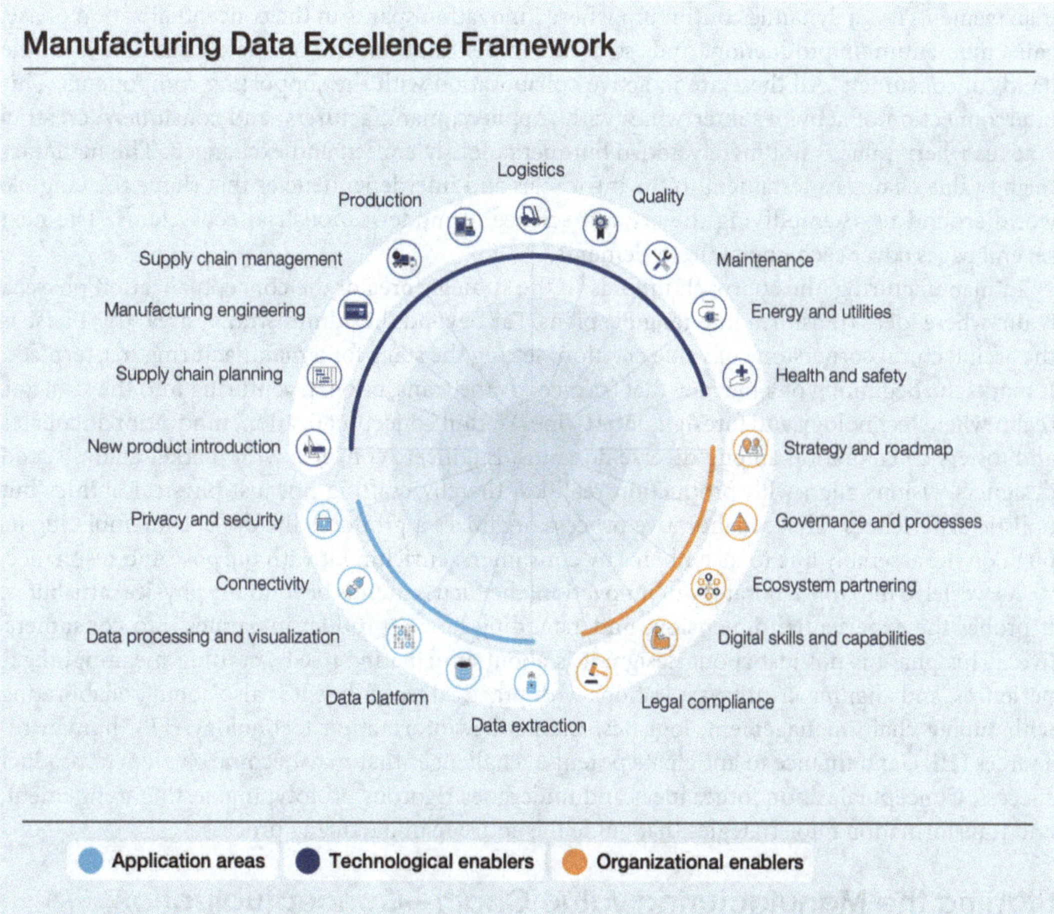

Figure 2.3 Manufacturing data excellence framework. Image from the World Economic Forum.

OVERVIEW OF THE MANUFACTURING VALUE CHAIN

The manufacturing value chain is a complex network of interconnected processes designed to transform raw materials into finished products, delivering value at each stage, as depicted in Figure 1.2. This intricate web of processes, stages, and interactions is the heartbeat of industrial creation, planning all that is required to seamlessly produce goods. In the vast landscape of manufacturing, the value chain serves as the backbone, linking every facet from concept to consumption. In the intricate processes within manufacturing, the initial phase of the manufacturing value chain lays the foundation for the work that follows. This phase marks the commencement of the intricate planning of the value chain. The choices made here reverberate throughout manufacturing, setting the tempo for what follows. This is a chapter where the art of negotiation, the science of analytics, and the ethos of sustainability converge, creating a synchronized set of processes in the manufacturing value chain.

At its essence, the manufacturing value chain encapsulates a product's many stages, encompassing design, production, distribution, and service. It also includes the supporting components of procurement, technological development, human resource management, and infrastructure management. It is a dynamic continuum where innovation sparks in the conceptualization phase, gains momentum in production, finds structure in distribution, and ultimately culminates in the hands of consumers. All these are in active collaboration with the supporting components. This interconnection of activities intertwines with suppliers, manufacturers, and consumers, creating a nexus where value is not merely added but meticulously crafted and exchanged. The manufacturing value chain is a testament to the intricacies and interdependencies that shape the tangible world around us, exemplifying the art and science of modern industrial ecosystems. The next several pages cover each one of these elements.

In manufacturing, the approach unfolds in the strategic area of the conceptualization phase, a realm where ideas transform into tangible plans. Far beyond the administrative area, this phase is the architectural cornerstone of value creation, setting the stage for a manufacturing masterpiece. It marks the beginning of a journey that transcends the transactional, venturing into the strategic realm where technology and foresight intertwine. Within conceptualization, innovation incubates and fosters collaboration among diverse domains. Engineers converse with market analysts, and designers' visions align with production realities, thereby crafting not just physical entities but holistic experiences. This collaborative process organizes a product's life cycle, contemplating its birth on the assembly line to its embrace by consumers, enriching it with purpose and resonance.

As we delve into this laboratory of innovation, the focus extends beyond the physical attributes. It probes the experiential dimension, understanding how a product integrates into consumers' lives. This phase is not just about design; it is about anticipating needs, crafting meaningful experiences, and aligning creative aspirations with practical feasibility. It is also about collaborating with supply chain management, logistics, marketing, information technology (IT), human resources (HR), and finance to anticipate potential challenges that may become barriers to product success. Conceptualization forges ideas and undergoes rigorous prototyping, testing, refinement, and transformation into strategies that guide the entire manufacturing process.

Starting the Manufacturing Value Chain—Conceptualization

Beyond envisioning products, the conceptualization phase becomes a compass to navigate trends, understand market dynamics, and align strategies with the pulse of consumer landscapes. Conceptualization stands as the incubator of innovation and the architect of product ideation. This pivotal stage transcends the mundane, propelling the manufacturing value chain into the realm of visionary design and strategic planning. Conceptualization is not merely about envisioning products. It is a profound exploration of the *art of the possible*, a canvas where ideas come to life and take the shape of a tangible manufacturing reality. Hence, it emerges as the visionary force shaping the trajectory of the manufacturing value chain, its significance echoing far beyond the initial strokes, resonating in every subsequent stage of manufacturing.

As the conceptualization phase unfolds, it becomes a conduit for strategic decision making. It involves not just envisioning the product but mapping out the entire journey from raw materials to end-user experience. This foresight enables manufacturers to align their strategies with market demands, optimize production processes, and lay the groundwork for a resilient and adaptive manufacturing value chain. In essence, conceptualization is the spark that ignites the fire of innovation, propelling the manufacturing value chain into a future defined by visionary design and strategic acumen.

During conceptualization, multiple teams are actively collaborating with the engineers. The finance team is ensuring that the product cost is something that is feasible in the market. The sales team is looking at how to position it in the market and at what cost. The HR team is figuring out whether there are enough human resources with the right skill set. If not, how soon can they find such people and onboard them near the manufacturing location(s). The IT team is ensuring that the data and IT tools required by the new product are available and that the new product does not generate so much data as to break the IT infrastructure. The supply chain team, especially procurement, is actively working on identifying potential glitches in getting components to the manufacturing floor. Professionals in this phase, armed with a palette of negotiation skills and market insights, craft the foundation upon which the entire manufacturing edifice will rest. Transportation and logistics focus on the smooth flow of raw materials to the line side or finished products to the distribution centers. The choices made here ripple through the subsequent stages, shaping the success of the final product. If the design team does not actively collaborate, it can result in expensive workarounds.

One of the key facets of this phase is the exploration of emerging technologies and their potential integration into the manufacturing process. From AI to advanced robotics, conceptualization is the playground where these innovations are assessed for their transformative impact. The goal is not just to create products that fulfill immediate needs but to anticipate future trends and position the manufacturing value chain at the forefront of industry evolution.

In the modern-day manufacturing narrative, this phase has surpassed its transactional roots. It has evolved into a strategic maneuver where technology and data analytics wield influence. Organizations are harnessing the power of predictive analytics to foresee market trends, optimize inventory levels, and preemptively address potential disruptions. This metamorphosis brings agility to manufacturing, allowing organizations to not only respond to market dynamics but to anticipate and shape them.

> As an example, designers at a large manufacturing company were working on an upgrade to a product. The previous version was so large that two would barely fit on a truck. Not collaborating with logistics, the new product ended up being an inch wider than the old version. This product would not fit on a truck in the same layout as before. The new layout had to be positioned differently—resulting in only one product fitting on a truck and doubling the product's outbound logistics cost. The finance and the sales teams balked at the added cost as unsustainable in the market. This upgrade was eventually scrapped after the initial run of products, causing the company to incur a large financial loss.

Procurement in the Manufacturing Value Chain

The procurement phase follows the conceptualization phase. Procurement involves strategic decision making with respect to sourcing raw materials, components, crucial production services, and supplier engagements when setting the foundation for product quality and cost-effectiveness. Manufacturers carefully identify and secure suppliers for the necessary resources, negotiate with them, and ensure a reliable supply chain. Moving beyond the conventional notions of transactional negotiations, modern procurement is a strategic enabler that shapes the very fabric of a company's success. As such, this stage sets the foundation for successful manufacturing.

In this transformation, *sustainability* emerges not merely as a buzzword but as a guiding principle that will steer organizations toward responsible sourcing and ethical practices. Modern consumers demand more than just products; they seek ethically sourced materials, eco-friendly practices, and a commitment to social responsibility. Sustainability, ethical sourcing, and social responsibility become integral elements of procurement negotiations, reflecting a broader shift in consumer expectations. As depicted in Figure 2.4, sustainability is a multidimensional issue that starts with lobbyists pressuring the government to help their cause, whether for or against sustainability. Then, multiple interrelated areas act in concert to negatively impact the climate and the earth. Organizations, recognizing the shift in the narrative, redefine their strategies to align with sustainable, ethical sourcing and social responsibility as integral elements, reflecting a broader shift in consumer expectations. It is a strategic pivot that adds layers to the narrative and resonates with a conscientious consumer base.

At the forefront of this transformation is the art of negotiation where professionals—armed with a palette of negotiation and financial skills and market insights—navigate a complex landscape of contracts, pricing structures, and delivery timelines. The choices made here ripple through the subsequent stages, shaping the destiny of the final product. This phase has transcended its transactional roots into long-term strategic thinking, where technology and data analytics wield influence. Organizations are harnessing the power of predictive analytics to foresee market trends, optimize inventory levels, and preemptively address potential disruptions. This change brings agility to manufacturing, allowing organizations to not only respond to market dynamics but to anticipate and shape them.

Boeing and Apple illustrate contrasting approaches to procurement, with significant differences in outcomes and reputational impact. Boeing's 737 MAX crisis highlights the severe consequences of hasty and flawed procurement processes, which contributed to design oversights and safety failures. In this case, a missing component on a 737 MAX led to a midair incident, sparking lawsuits, federal investigations, and executive changes. Boeing's focus on rapid production compromised quality control and supplier oversight, revealing how procurement decisions can critically affect safety and brand reputation when not aligned with rigorous standards.

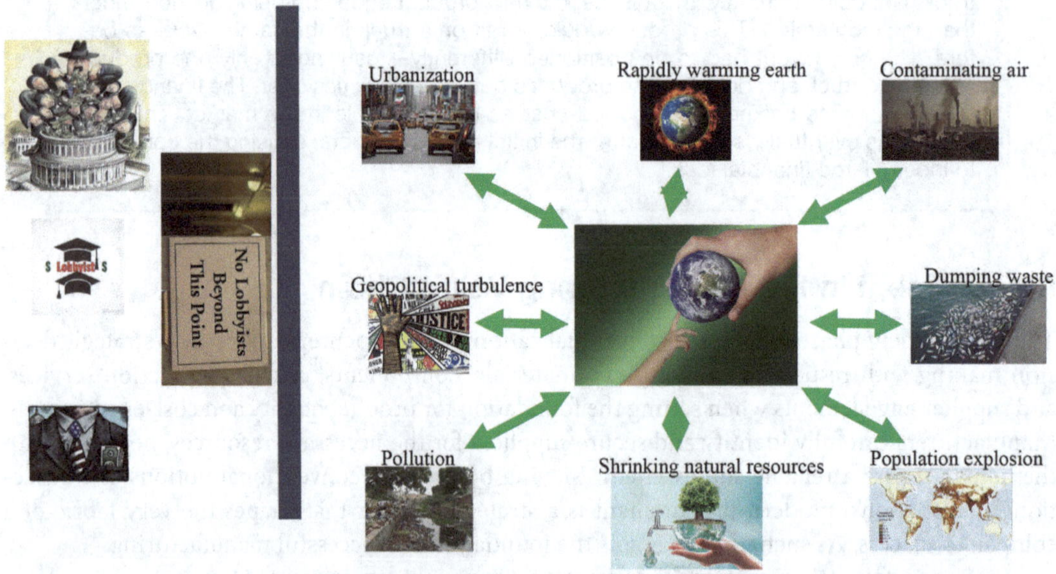

Figure 2.4 Some of the many dimensions impacting sustainability.

In contrast, Apple demonstrates how robust, strategically managed procurement processes can enhance market competitiveness and brand perception. Apple's sustainable supply chain practices emphasize procurement excellence, focusing on responsible sourcing, renewable energy, and ethical manufacturing.[1] Proactive procurement practices, whether aimed at securing critical resources or enhancing sustainability, can drive operational success, improve customer satisfaction, and build resilience, underscoring the benefits of procurement as a strategic, value-driven function.

Sustainable procurement also extends to the technology sector. Companies sourcing minerals for electronic components, such as cobalt and lithium, are under increasing scrutiny to ensure responsible mining practices. Technology giants are proactively engaging with suppliers to trace the origin of raw materials and address other concerns related to environmental impact and human rights violations.

In the context of Industry 4.0, procurement leverages technologies like blockchain and AI to enhance transparency and efficiency. Blockchain, with its decentralized and tamper-resistant nature, finds application in supply chain traceability. For instance, the food industry utilizes blockchain to provide consumers with real-time visibility into the journey of products from farm to table, ensuring food safety and authenticity.[2] Other food products (like seafood) and luxury items (like diamonds) are also using blockchain in procurement to ensure legitimate catch or mining and prevent blood diamonds (from Angola, Ivory Coast, Sierra Leone, Liberia, Guinea, and Guinea-Bissau) from entering the legal supply chain.

AI, on the other hand, transforms procurement by automating routine tasks, analyzing vast datasets for insights, and predicting market trends. An illustrative example is the use of AI-powered demand forecasting in retail. By analyzing historical data, market trends, and external factors, retailers can optimize inventory levels, minimize stockouts, and enhance overall supply chain efficiency.

Strategic partnerships emerge as a cornerstone of modern procurement practices. Beyond transactional relationships, organizations are fostering collaborations with suppliers to drive innovation. Joint research and development initiatives, shared sustainability goals, and collaborative problem-solving become integral aspects of these partnerships. An exemplary case is seen in the automotive industry, where manufacturers collaborate with suppliers to co-create advanced materials, contributing to the development of fuel-efficient and environmentally friendly vehicles. But beware, cutting corners, unrealistic timelines, and cost constraints can have a negative effect, even within these partnerships.

Modern procurement is a strategic powerhouse that extends far beyond traditional transactional realms. It navigates the delicate balance between economic imperatives and ethical considerations, leveraging technology, sustainability, and collaborative partnerships to sculpt a narrative of responsible and forward-thinking procurement. The stories of successful procurement strategies are not just tales of cost savings; they are accounts of resilience, innovation, and social responsibility that are symbolic of the evolving attitude in contemporary manufacturing.

The Production Stage in the Manufacturing Value Chain

The production stage in manufacturing is the heart of the entire value chain. This is where raw materials undergo a metamorphosis into finished goods through a meticulously arranged set of processes utilizing cutting-edge machinery, innovative technology, and skilled labor. It is a realm where efficiency, precision, and innovation converge to shape the tangible output of manufacturing. Quality control measures are implemented to ensure the final product meets stringent standards. Production is the dynamic landscape where organizations leverage advanced manufacturing technologies such as robotics, automation, and the Industrial Internet of Things (IIoT) to enhance operational efficiency, increase productivity, and optimize production costs.

In recent years, the advent of smart manufacturing has revolutionized traditional production paradigms, creating a connected and intelligent ecosystem. This transformative approach facilitates real-time monitoring and control, predictive maintenance, and adaptive production scheduling. Global industry leaders like Siemens and General Electric have embraced *smart* manufacturing practices, optimized their production lines, and dynamically responded to shifting market demands. This evolution reflects a broader trend in the industry toward embracing digital transformation to remain competitive and resilient.

A notable trend in the production stage is the emergence of mass customization, allowing manufacturers to tailor products to individual customer needs while maintaining the efficiency of mass production. Companies like Adidas have successfully implemented additive manufacturing (3-D printing) to produce customized shoe midsoles. They use customers' individual foot scans, demonstrating the potential of technology in achieving both personalization and efficiency.[3]

The rise of sustainable manufacturing practices is another significant trend in the production stage. Companies are increasingly integrating environmentally friendly processes and materials, aligning with the global drive toward a circular economy. Patagonia, renowned for its commitment to sustainable manufacturing, utilizes recycled materials and promotes *repair and reuse* to minimize its environmental impact.

Efficient supply chain collaboration plays a pivotal role in the production stage of manufacturing. It acts as the backbone to ensure a seamless and synchronized flow of materials and components. In today's globalized manufacturing landscape, where supply chains often span across multiple regions and involve various stakeholders, collaboration becomes indispensable. Timely and coordinated efforts among suppliers, manufacturers, and distributors are essential to maintain operational efficiency and meet production schedules. Automotive giant Toyota exemplifies this with its renowned Toyota Production System with a lean production system that minimizes waste and optimizes efficiency through just-in-time manufacturing and continuous improvement practices.

One primary reason for the criticality of efficient supply chain collaboration in the production stage lies in the intricate interdependencies within the manufacturing process. Components and materials have specific timelines and sequences in the production line, and any disruption or delay in the supply chain can have cascading effects on the entire production cycle. For instance, if a critical component is delayed in reaching the production floor, it can lead to idle machinery and labor, increased lead times, and potential production bottlenecks. Collaborative relationships help in aligning these timelines, ensuring that all elements of the supply chain work in harmony.

Moreover, effective collaboration enables better risk management and resilience in the face of uncertainties. The modern manufacturing environment is susceptible to various disruptions, such as geopolitical events, natural disasters, or unexpected changes in demand. By fostering close ties with suppliers and other partners, manufacturers can proactively address potential challenges. For example, having alternative suppliers or contingency plans in place allows companies to navigate disruptions more smoothly, minimizing the impact on production schedules and overall business continuity. Additionally, emphasis on lean manufacturing principles, just-in-time production, and inventory optimization underscores the need for efficient supply chain collaboration. Close coordination helps achieve lean objectives by streamlining processes, reducing waste, and maintaining optimal inventory levels. This collaborative approach allows for the synchronization of production cycles with demand patterns, preventing overstock or stockouts and enhancing cost-effectiveness. Sometimes this does not work as expected.

> Maintaining alternative suppliers and robust contingency plans is essential for companies to navigate supply disruptions, ensuring production schedules and business continuity remain intact. However, even with precautions, unforeseen issues can arise. In one instance, a major manufacturer relied on a simple yet crucial bolt, costing less than five dollars, to join two large product components in the first stage of assembly. When an unapproved change in the bolt's metallurgy led to failures, the resulting downtime not only disrupted production but also caused products to break down at customer sites. The manufacturer faced significant warranty costs and potential liability risks—though, fortunately, no injuries occurred due to these malfunctions.
>
> To address the issue, the manufacturer switched to a new supplier with strict guidelines to enforce proper metallurgy standards. However, as production demands grew, the supplier struggled to keep up with the pace, leading to a stockout of this critical bolt and a complete halt in the assembly line. To prevent future disruptions, the manufacturer and supplier collaborated closely to develop an effective inventory strategy, securing buffer stock both at the supplier's facilities and within the manufacturer's own operations. This case highlights the importance of rigorous quality assurance in supplier management and proactive supply planning to prevent single points of failure in the production process.

While the production stage presents opportunities for innovation and efficiency, it also poses challenges:

- The need for skilled labor in the face of evolving technologies
- Constant pressure to balance cost-effectiveness with sustainability
- Constantly evolving government regulations across the globe
- Risks related to weather or pandemics

However, organizations that plan for and navigate these challenges effectively gain a competitive edge in the dynamic landscape of modern manufacturing.

Distributing Products Made in the Manufacturing Value Chain

The distribution stage in the manufacturing value chain is a critical link that bridges production and consumption and is where finished products make the transition from production facilities to consumers. This multifaceted phase involves a sophisticated network of logistics, transportation, and warehousing activities to ensure timely and efficient delivery. In today's era of e-commerce dominance and global supply chains, the need for swift and reliable delivery methods is crucial. Real-time tracking and efficient inventory control through the distribution stage have become more intricate and interconnected than ever before.

At the heart of the distribution stage lies the optimization of logistics and transportation. Companies strive to strike the right balance between cost-effectiveness and swift delivery, often leveraging advanced technologies such as route optimization algorithms and real-time tracking systems. Real-time data analytics have become a prerequisite in optimizing distribution networks. Through advanced analytics tools, companies gain actionable insights into inventory levels, demand patterns, and consumer behavior. This data-driven approach allows for dynamic decision making, helping companies adjust their distribution strategies promptly. For instance, predictive analytics can anticipate fluctuations in demand, enabling companies to adjust inventory levels and distribution routes proactively. Major logistics providers such as FedEx[4] and UPS[5] exemplify this approach, adopting cutting-edge technologies to enhance their distribution networks and offering customers precise tracking information along with reliable delivery services.

Within the distribution stage of manufacturing, the advent of technology has been a transformative force, reshaping traditional paradigms and introducing unprecedented efficiencies. Warehousing and inventory management technologies have become integral components of the distribution stage, influencing the overall supply chain's agility. The rise of automated warehouses equipped with robotics and AI has revolutionized how companies handle, store, and retrieve inventory. Automated order fulfillment systems, powered by robotics and AI, have emerged as stalwarts in streamlining warehouse operations. E-commerce giants like Amazon, Alibaba, and JD.com have embraced robotic warehouses where machines efficiently pick, pack, and dispatch orders, thus minimizing errors and human intervention, significantly expediting the distribution process, and enhancing overall operational efficiency.

The last-mile delivery segment, marking the final leg of distribution, has witnessed a paradigm shift with the rise of crowdsourced and gig economy-based delivery models. Companies like Uber Eats and DoorDash leverage independent contractors for last-mile deliveries, fostering flexibility and scalability in responding to fluctuating demand. This decentralized approach has proven particularly effective in urban areas with high population density.

Omnichannel retailing has introduced an additional layer of complexity to the distribution stage. Companies must now seamlessly integrate their multiple online channels and offline/brick-and-mortar channels, ensuring a unified customer experience. COVID-19 has motivated retail giants like Walmart, Target, and several grocery stores to embrace omnichannel strategies, offering customers the flexibility to order online and receive products through various channels, including in-store pickup and home delivery.

Furthermore, sustainability has become a focal point in the distribution stage, with an increasing emphasis on eco-friendly packaging, reduced carbon emissions, and ethical sourcing. Companies are actively exploring greener transportation options, such as electric vehicles and alternative fuels, to align with environmental goals. In addition, logistics and distribution companies are working to optimize their back-haul to minimize empty loads (in the industry, sometimes called *carrying sailboat fuel*). IKEA, for instance, has committed to making all of its last-mile deliveries emission-free by 2030.[6] The collaboration between Mercedes-Benz and logistics company DB Schenker to implement a carbon-neutral transportation solution exemplifies a commitment to eco-friendly distribution practices.[7]

Strategic partnerships and collaborations play a pivotal role in optimizing the distribution stage. Companies often join forces with third-party logistics providers or leverage shared distribution networks to enhance efficiency. An example is the collaboration between Starbucks and Alibaba's Cainiao Network to establish a new retail distribution network in China that seamlessly integrates online and offline channels.[8]

As the distribution stage continues to evolve, integration with other emerging technologies, such as blockchain, holds promise for enhancing transparency and traceability. Blockchain's decentralized and immutable ledger can authenticate the provenance of products, providing consumers with detailed information about the journey of a product from manufacturing to delivery. This transparency fosters trust and aligns with the growing demand for ethically sourced and sustainable products.

Sales and Marketing in the Manufacturing Value Chain

Sales and marketing in manufacturing are undergoing a profound transformation, driven by the integration of advanced analytics, business intelligence (BI), AI, and ML technologies. These technologies enable manufacturers to achieve unprecedented levels of customer engagement, targeted marketing, and sales optimization. As manufacturers adapt to the digital era, leveraging these advanced tools provides a competitive advantage that allows companies to anticipate customer needs, tailor their strategies, and maximize revenue potential. Here, we explore how these technologies are reshaping sales and marketing in the manufacturing sector by empowering companies to innovate, streamline operations, and ultimately drive sustainable growth.

Manufacturers are increasingly leveraging advanced analytics and BI tools to gain actionable insights into customer behavior, market trends, and sales performance. By analyzing large datasets from various sources—including customer interactions, social media, and market data—companies can identify patterns and trends to tailor their marketing strategies and optimize sales processes. Predictive analytics models help forecast sales trends and identify potential opportunities, empowering manufacturers to optimize inventory management, production planning, and resource allocation to meet customer needs and maximize revenue.

For instance, AI-driven customer segmentation allows manufacturers to target specific customer groups with personalized marketing campaigns, resulting in higher engagement and conversion rates. Advanced analytics tools also enable manufacturers to assess the effectiveness of their marketing efforts, allowing for data-driven decision making and continuous optimization. For example, Emerson Electric Company uses AI-powered algorithms to analyze customer data and develop targeted content for different personas. This targeted approach enhances customer experiences and builds stronger relationships, resulting in higher retention rates and customer loyalty.[9]

The future of sales and marketing in manufacturing is also being shaped by emerging technologies like deep learning and natural language processing (NLP). Deep learning algorithms can analyze complex datasets, such as images and audio, to provide deeper insights into customer preferences and market trends. This capability can be used to create more personalized marketing campaigns and optimize customer support interactions. NLP algorithms, on the other hand, enable manufacturers to leverage unstructured data, such as customer reviews and social media comments, to better understand customer sentiment and preferences. This information can be used to refine product offerings and improve customer experiences. Companies like Honeywell have invested in AI-powered sentiment analysis to gauge customer feedback from various channels, such as social media and customer reviews. By understanding customer sentiments, Honeywell can tailor its marketing strategies to align with customer expectations and preferences.[10]

Virtual reality (VR) and augmented reality (AR) technologies are beginning to play a role in marketing by providing immersive and interactive experiences to customers. These technologies can be used to showcase products in virtual showrooms, allowing customers to visualize and interact with products before purchasing. Lockheed Martin has used AR to enhance its marketing and customer presentations. By using AR, Lockheed can provide potential clients with an interactive demonstration of its aerospace products, allowing customers to explore the technology and features in detail, as well as in pilot training. This level of engagement can help to build stronger relationships with customers and improve the chances of successful sales.[11] Similarly, Siemens provides its customers the rFAT (Remote Factory Acceptance Test) service to leverage VR and offer customers the ability to perform a factory acceptance test remotely without the customer being physically present. By providing a behind-the-scenes look at its operations, Siemens builds trust with customers and helps them understand the value of its products and services.[12]

Overall, the integration of advanced analytics, BI, AI, and ML technologies in sales and marketing is empowering manufacturers to make data-driven decisions, optimize operations, and deliver exceptional customer experiences. As technology evolves, it becomes mandatory that manufacturers stay abreast of emerging trends and invest in these advanced tools to maintain a competitive edge in the dynamic manufacturing landscape.

After-Sales Service in the Manufacturing Value Chain

Beyond the sale, *after-sales* service is critical for customer satisfaction. It includes providing warranty services, maintenance, repairs, and support throughout the product's life cycle. Exceptional after-sales service contributes significantly to a positive brand image, fosters brand loyalty, and enhances long-term customer relationships, thereby fostering and enhancing the overall customer experience. Organizations refine their products and services by leveraging data analytics and customer feedback, contributing to continuous improvement. This phase extends far beyond mere transactional activities, evolving into a strategic opportunity to showcase commitment, reliability, and responsiveness.

In the realm of after-sales service within manufacturing, the digital era has ushered in a transformative shift, prominently marked by the incorporation of AR and VR. A prime example is Microsoft's HoloLens, which revolutionizes remote assistance. Technicians equipped with HoloLens can guide customers through troubleshooting processes with unprecedented precision, enhancing overall service experiences.[13] IBM Watson has advertised its capabilities to predict the failure of an elevator and has provided capabilities for remote support.[14] The goal of these examples is to enhance the after-sales service experience, minimizing customer downtime and optimizing service efficiency.

Customer centricity lies at the core of successful after-sales service strategies. Proactive monitoring of product performance, facilitated by IoT, allows companies to anticipate potential issues before they escalate. For example, automotive manufacturers like Tesla utilize IoT sensors to gather real-time data from vehicles, enabling predictive maintenance and reducing the likelihood of unexpected breakdowns.

The integration of AI and ML in after-sales service has revolutionized predictive analytics. By analyzing historical data, these technologies harness the power of IoT sensors to monitor product performance in real time, forecast equipment failures, recommend preventive measures, and optimize spare parts inventory. Industrial manufacturers like John Deere and Caterpillar employ predictive analytics to anticipate maintenance needs, ensuring equipment reliability and minimizing downtime for customers.

In the realm of consumer electronics, subscription-based after-sales services have gained prominence. Apple's AppleCare program exemplifies this trend, offering customers extended warranty coverage, priority technical support, and even accidental damage protection. Beyond traditional warranty services, these models ensure ongoing customer engagement and loyalty by providing ongoing value beyond the initial purchase.

Supply chain resilience plays a key role in after-sales service, particularly concerning genuine spare parts availability. Proactive monitoring of product performance and agile supply chain strategies contribute to minimizing downtime. Leveraging blockchain technology enhances traceability and transparency in spare parts management, mitigating the risk of counterfeit components infiltrating the supply chain. This ensures the authenticity of spare parts and facilitates streamlined processes, reducing lead times for repairs.

In the automotive sector, the shift toward electric vehicles (EV) has prompted manufacturers to reimagine after-sales service. Tesla's over-the-air (OTA) software updates exemplify how digitalization allows manufacturers to remotely enhance vehicle performance, fix software glitches, and introduce new features, minimizing the need for customers to visit service centers. This approach illustrates technological advancement and signifies a shift toward sustainable and customer-centric practices. Strategic partnerships and collaborations are becoming increasingly prevalent in the after-sales service domain. Original equipment manufacturers (OEMs) are collaborating with specialized service providers to showcase a holistic approach to service delivery. Companies like Siemens collaborate with third-party service providers to deliver predictive maintenance solutions, leveraging expertise across domains for holistic service offerings. The adoption of circular economy principles is reshaping after-sales service strategies. Remanufacturing, refurbishing, and recycling products contribute to sustainability goals while creating additional revenue streams. Xerox's Green World Alliance exemplifies a commitment to environmental responsibility by reusing and recycling print cartridges, thus reducing waste and conserving resources.[15]

The after-sales service stage of manufacturing is evolving into a dynamic arena where technological innovations, customer centricity, and sustainability converge. The seamless integration of digital technologies, predictive analytics, and strategic partnerships is redefining service excellence and establishing a paradigm where customer satisfaction transcends the transaction. Incorporating the aforementioned nuances into the manufacturing narrative offers manufacturers a more immersive understanding of the evolving landscape of after-sales service.

Summary—Stages in the Manufacturing Value Chain

The manufacturing value chain is a dynamic and evolving framework where each stage is intricately connected. From conceptualization to procurement to after-sales service, organizations leverage technological advancements to streamline operations, increase productivity, and meet the ever-changing demands of the market. This continual adaptation and integration of innovative technologies define the modern manufacturing landscape. As several earlier examples show, each stage of the manufacturing value chain adds value to the final product. Procurement ensures the use of high-quality inputs, fostering product reliability. Production leverages innovation and efficiency to meet or exceed customer expectations. Distribution ensures prompt delivery, reducing lead times and enhancing customer satisfaction. Sales and after-sales service builds brand loyalty by addressing customer needs beyond the initial purchase.

INDUSTRY 4.0 AND ADVANCED TECHNOLOGIES

In the realm of Industry 4.0, the manufacturing value chain undergoes a profound transformation characterized by the integration of advanced technologies that redefine efficiency, flexibility, and innovation. At its core, Industry 4.0, characterized by the fusion of physical and digital systems (see Figure 2.2), emphasizes the importance of data analytics, IoT, AI, and robotics. Industry 4.0 represents a convergence of digital technologies, data analytics, and smart manufacturing to create a connected and intelligent ecosystem. The integration of IoT is a prerequisite in this transformation, where sensors and devices intercommunicate to collect and share real-time data across the value chain. These technologies provide unprecedented insights, optimize decision making, and facilitate predictive maintenance.

Central to Industry 4.0 is the concept of the *smart factory*, where automation and data exchange coordinate manufacturing processes (see Figure 2.5). This figure also captures Industry 4.0 concepts, including 5G connections within the manufacturing factory.

Robots play a big role in Industry 4.0 as well. Figure 2.6 shows a prime example of the implementation of collaborative robots—also known as cobots—alongside human workers. These cobots, equipped with sensors and ML capabilities, optimize tasks requiring precision and repetitive

Figure 2.5 Robots and humans collaborating on a manufacturing floor.[16]

Figure 2.6 Symbiotic interaction between cobots and humans in manufacturing.[17]

actions. BMW has embraced Industry 4.0 principles by transforming its production facilities into smart factories. The integration of IoT-enabled sensors and collaborative robots has streamlined manufacturing processes, enhancing efficiency and flexibility. BMW's adoption of cobots in assembly lines has not only improved precision but also provided a safer working environment for employees. BMW, with NVIDIA, is now working toward a concept it calls *iFactory* in its goal of virtual production.

Another major component of Industry 4.0 is the use of digital twins, basically, digital models of physical objects. Digital twins use real-time data from physical objects to simulate the operations of the digital twin and compare them to the performance of the physical object. Figure 2.7 shows BMW's smart manufacturing architecture along with a representation of the BMW automobile digital twin. The architecture depicts the key components—IIoT (with its multiple elements), an open collaboration (between BMW and business partners) and manufacturing (including open-source components, hardened IoT) platform, and the platform housing the solution (Microsoft Azure).

The predictive power of AI and ML takes center stage in Industry 4.0, particularly in predictive maintenance (PdM). Through AI algorithms analyzing equipment data, organizations can foresee potential failures and schedule maintenance proactively, minimizing downtime. Figure 2.8 shows a typical automated manufacturing process workflow. Such automation minimizes downtime and reduces production losses.

Consider, for example, General Electric's SmartSignal predictive maintenance software. It utilizes AI for predictive maintenance across its diverse range of industrial equipment. Through the

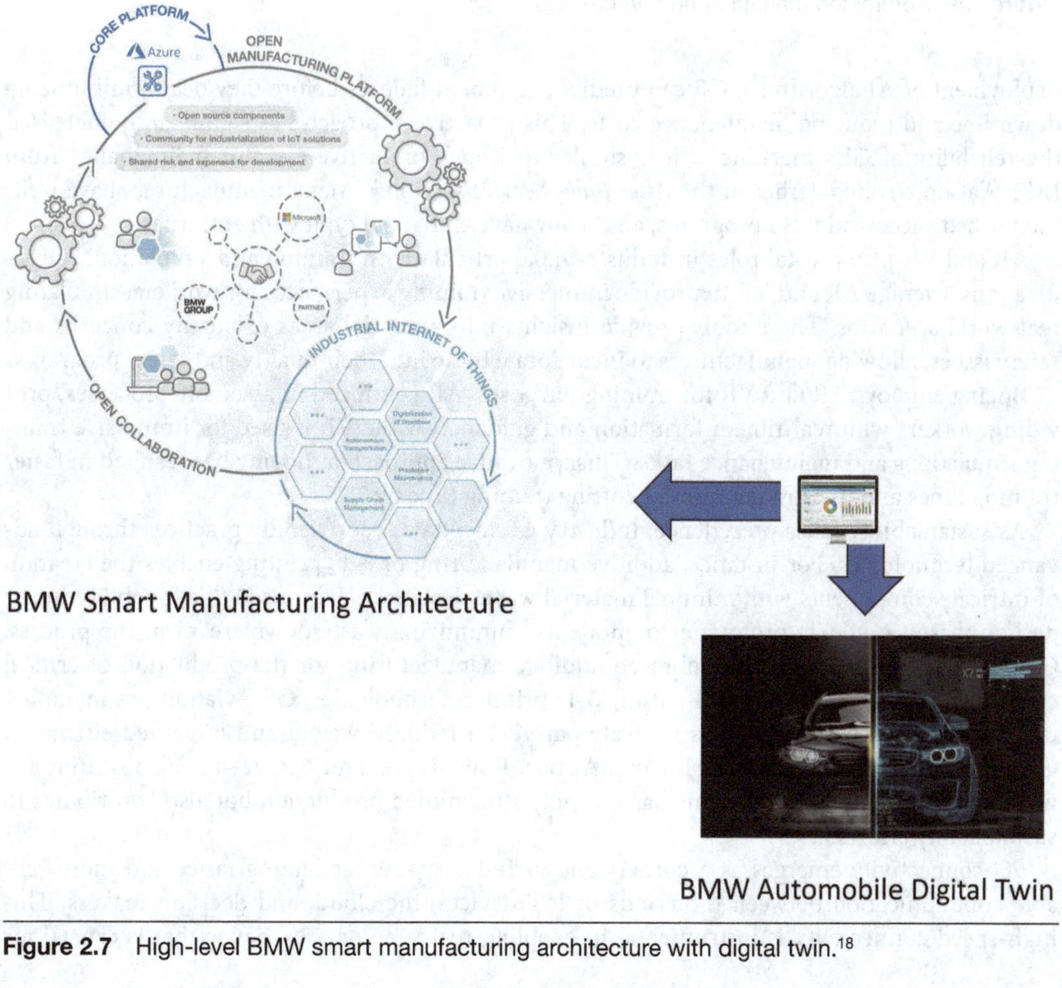

Figure 2.7 High-level BMW smart manufacturing architecture with digital twin.[18]

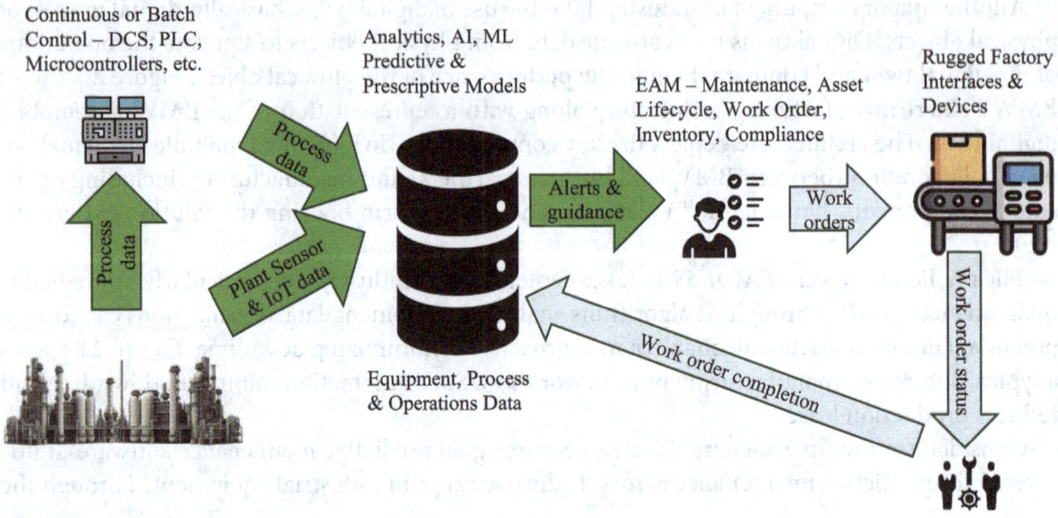

Figure 2.8 Automated manufacturing processes.[20]

deployment of AI algorithms, GE can predict equipment failures before they occur, minimizing downtime and reducing maintenance costs. This proactive approach has significantly increased the reliability of GE's machinery. It is similar to Kone's predictive elevator maintenance from IBM Watson covered earlier in the *After-Sales Service* section.[19] Many manufacturers have well-established successful PdM programs, and many have embarked on it with enthusiasm.

AR and VR play pivotal roles in Industry 4.0, particularly in training and simulations. Organizations leverage AR and VR to provide immersive training experiences for workers, simulating real-world scenarios. These tools provide insights into potential areas of quality concerns and safety issues, allowing manufacturers to incorporate these into their quality and safety programs.

Boeing employs AR and VR for training purposes. AR is utilized in assembly processes, providing workers with real-time information and guidance, while VR is used for immersive training simulations and maintenance tasks. This innovative approach at Boeing has resulted in faster training times and improved accuracy, cutting training time by 75%.[21]

As sustainability takes precedence, Industry 4.0 facilitates eco-friendly practices through advanced technologies. For instance, additive manufacturing or 3-D printing enables the creation of intricate components with minimal material waste. Engineers have used this capability in the design phase to quickly prototype components, minimizing waste downstream in the process. General Electric Aviation has embraced additive manufacturing for the production of critical components in aircraft engines. By using 3-D printing technologies, GE Aviation has manufactured large metal parts as well as intricate parts with reduced weight and improved efficiency. Other large manufacturers—like Boeing, Airbus, Rolls Royce, and NASA—are also shifting toward additive manufacturing. This has not only streamlined production but also contributes to sustainability efforts.

5G connectivity emerges as a cornerstone in Industry 4.0, facilitating faster and more reliable communication between thousands of IoT devices, the cloud, and decision makers. This high-speed connectivity is instrumental in enabling real-time data exchange that is crucial for responsive decision making. The incorporation of 5G connectivity, exemplified by Ericsson's initiatives in manufacturing, is a transformative force driving Industry 4.0. This capability enables instant decision making and enhances coordination in manufacturing processes. Ericsson's 5G solutions create a robust network infrastructure supporting diverse devices, fostering a smart, interconnected ecosystem on the factory floor. Additionally, the scalability and flexibility afforded by 5G contribute to efficient production processes, adapting swiftly to changing demands. The remote accessibility enabled by 5G extends operational efficiency beyond the factory, allowing for centralized control, maintenance, and monitoring, thereby shaping a new era of smart and adaptable manufacturing.[22,23,24]

Industry 4.0, propelled by advanced technologies, reshapes the manufacturing value chain into a dynamic and interconnected ecosystem. The integration of IoT, AI, robots, AR, VR, additive manufacturing, and 5G creates a landscape where efficiency, sustainability, and innovation converge. As this transformative journey unfolds, it presents a vivid portrayal of the Industry 4.0 landscape within the manufacturing sector.

CHALLENGES, OPPORTUNITIES, AND CONSIDERATIONS

The manufacturing value chain, while a bedrock of global industry, encounters a plethora of challenges, opportunities, and considerations that shape the trajectory of businesses and economies. Traversing this complex terrain demands a deep understanding of the industry's nuances. Here, we explore and delve into the intricate facets that define manufacturing.

One significant challenge lies in the inherent volatility of global supply chains, highlighted prominently by disruptions like the COVID-19 pandemic and the Russia-Ukraine and Israel-Hamas wars. The disruptive impact of such events has underscored vulnerabilities, disrupting the seamless flow of materials and components across countries and industries. The automotive sector recently faced a critical semiconductor shortage, sending shockwaves throughout the global production network. These events highlighted the importance of building resilience and agility into supply chain strategies as well as establishing robust risk management. Other, more periodic events, such as hurricanes or winter snowstorms, are predictable but can still wreak havoc in the supply chain.

The integration of sustainability into manufacturing processes adds another layer of complexity that emerges as a pressing challenge and opportunity. Striking a balance between economic viability and environmentally conscious practices confronts manufacturers. The shift toward circular economy models and reduced environmental impact is an opportunity for companies to align with evolving consumer expectations. The implementation of green practices not only mitigates environmental risks but also serves as a market differentiator, contributing to long-term viability. While implementing sustainable practices presents challenges, for example, higher initial costs and intricate supply chain adjustments, it also opens avenues for innovation and positions companies as responsible stewards of the environment. Transitioning to sustainability and a circular economy model represents a notable opportunity for manufacturing. Embracing sustainable practices can lead to reduced waste and heightened resource efficiency.

Leading companies like Unilever and Tesla have successfully integrated sustainable practices into their value chains. Unilever's commitment to the UN's Sustainable Development Goals and Tesla's pursuit of carbon neutrality showcase a paradigm shift toward eco-friendly operations.

Further, Tesla, at the forefront of sustainable manufacturing, has grappled with sourcing ethically sound raw materials amid a dynamic market. Major corporations are actively adopting circular packaging models to minimize their environmental footprint.[25]

While the benefits of IoT, AI, and data analytics are evident, the shift requires significant investments and a fundamental cultural transformation within organizations. Smart factories embracing these technologies may encounter resistance from traditional manufacturing setups. The integration of these technologies necessitates substantial investments and poses challenges related to data security, interoperability, and the upskilling of the workforce. Incorporating the consideration of the ethical use of technology in education, especially in the context of AI and automation, is paramount.

Organizations embracing smart manufacturing, like Siemens with its Digital Enterprise Suite, find opportunities in increased efficiency, predictive maintenance, and data-driven decision making. Ensuring trustworthy and responsible AI practices, as demonstrated by Microsoft's AI principles, is essential to prevent unintended consequences and to build trust among stakeholders.[26] Ethical considerations take center stage in the integration of AI into manufacturing processes. It extends to areas like data privacy, where companies like Apple have implemented stringent measures to protect user information. Ensuring the responsible use of AI to avoid bias and uphold ethical standards is vital. Companies like Google emphasize ethical AI principles in their manufacturing practices, setting a precedent for responsible technology adoption.[27]

The rapid adoption of advanced technologies demands upskilling the workforce, creating a mismatch between the skills available and those needed in the evolving manufacturing landscape. Various initiatives aim to bridge this gap through collaborative education-industry partnerships. Companies like Caterpillar, General Electric, and Siemens have addressed this challenge through extensive training programs and partnerships with educational institutions. Bridging this gap not only ensures a qualified workforce but also supports innovation and competitiveness in the long run. Unfortunately, leaders must realize that this is not a one-time effort, but requires continuous training within the workforce. Regulatory compliance remains a critical consideration, particularly in industries with stringent standards. Adhering to diverse regulatory requirements is essential for maintaining operational integrity. Pharmaceutical manufacturers, for instance, must navigate complex Good Manufacturing Practice (GMP) standards to ensure compliance. These principles and processes must be an integral part of training programs.

Collaboration and strategic alliances are powerful tools for fostering innovation and resilience. In the face of global challenges, forming partnerships can pave the way for knowledge sharing and collaborative problem-solving. Initiatives like the World Economic Forum's *Global Lighthouse Network* for the future of advanced manufacturing and production foster collective problem-solving and the sharing of best practices. It exemplifies the impact of global collaboration on manufacturing excellence.[28]

The manufacturing value chain is undergoing a transformative phase with challenges acting as catalysts for innovation and opportunities paving the way for sustainable growth. Companies that strategically navigate these dynamics stand poised to shape the future of manufacturing. Figure 2.9 provides a glimpse of the challenges posed by this transformation and the multiple dimensions with complex dependencies that need to be addressed.

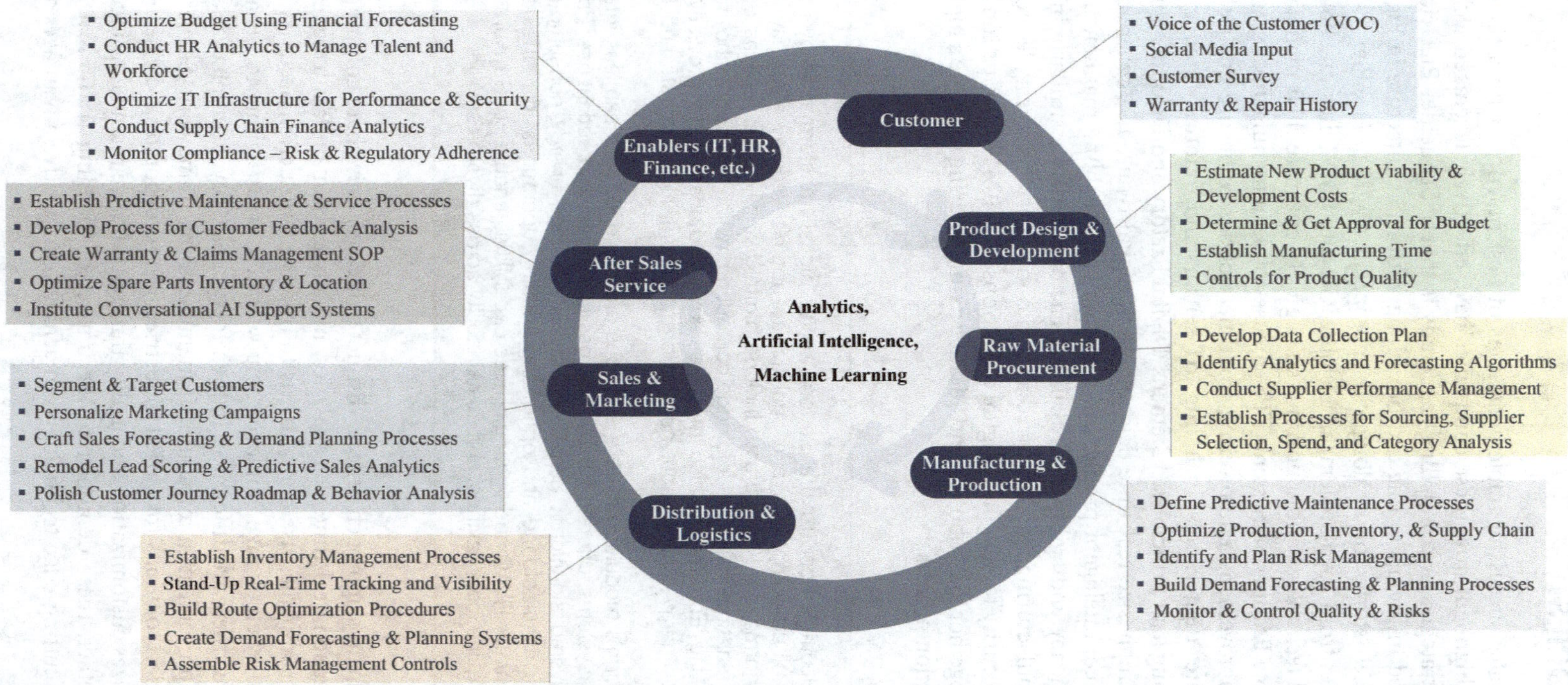

Figure 2.9 Interconnected opportunities and complex dependencies in manufacturing.

SUCCESS STORIES

This section illustrates successful implementations of best practices within the manufacturing value chain. They highlight how these companies have effectively leveraged technology or adopted innovative strategies. A compelling case study is that of Schneider Electric's EcoStruxure platform highlighted in the upcoming text box along with its architecture in Figure 2.10. Summaries of other success stories follow along with appropriate diagrams (see Figures 2.11 and 2.12).

Other case studies from Toyota, Siemens, Deloitte, GE, and BMW illustrate how these companies have leveraged technology and innovative strategies to optimize their manufacturing operations. Toyota's Toyota Production System (TPS) focuses on lean manufacturing, waste reduction, and continuous improvement, empowering frontline workers to ensure quality and efficiency. Toyota also integrates digital tools, including IoT sensors and robotics, to enhance production flexibility by adopting lean principles.[31] Siemens' Digital Factory and Deloitte's Smart Factory offer solutions that emphasize real-time data analytics, digital twins, and collaborative robots to create interconnected, intelligent production environments, enabling manufacturers to achieve high levels of efficiency and agility. GE's Brilliant Factory Initiative similarly integrates digital twins, additive manufacturing, and advanced analytics for smarter, more agile production, while prioritizing cybersecurity to safeguard digital operations. The BMW Group's Industry 4.0 transformation brings together IoT, robotics, and additive manufacturing to create a highly flexible and customized production process.

> Schneider Electric's EcoStruxure platform drives digital transformation by using IoT sensors, AI, cloud computing, and analytics to optimize energy efficiency and operational performance. Key features include digital twins—virtual replicas of assets that provide real-time insights—and edge computing, which reduces latency for faster decision making. EcoStruxure also integrates advanced analytics for predictive maintenance and energy optimization, helping organizations reduce costs and improve sustainability.
>
> The platform's modular, scalable solutions are designed for seamless integration with existing systems, offering flexibility across industries. In practical use, EcoStruxure has delivered significant benefits, such as reducing energy consumption and downtime in manufacturing and optimizing building operations in real estate with smart systems for HVAC and lighting. These capabilities enable EcoStruxure to drive productivity gains, cost savings, and sustainability across diverse sectors.

While each company's approach reflects distinct priorities, they share commonalities in their embrace of digitalization, real-time data analytics, and advanced manufacturing technologies. All use cases for these companies use digital twins to improve production accuracy and predictive maintenance, and they emphasize employee empowerment or upskilling to sustain these transformations. However, there are differences in their focus areas: Toyota emphasizes lean principles and continuous improvement in a traditional manufacturing framework, Siemens and Deloitte focus on digitalization solutions for varied industries, GE incorporates cybersecurity rigorously, and BMW emphasizes customization and agility through additive manufacturing and workforce development. Together, these demonstrate the impact of blending traditional manufacturing best practices with cutting-edge technologies to achieve competitive advantages across different sectors.

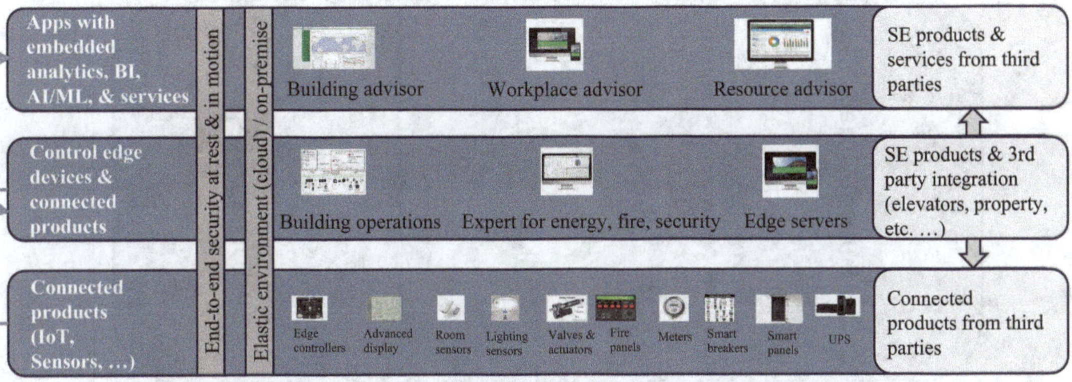

Figure 2.10 Schneider Electric (SE) EcoStruxure Power Architecture.[29]

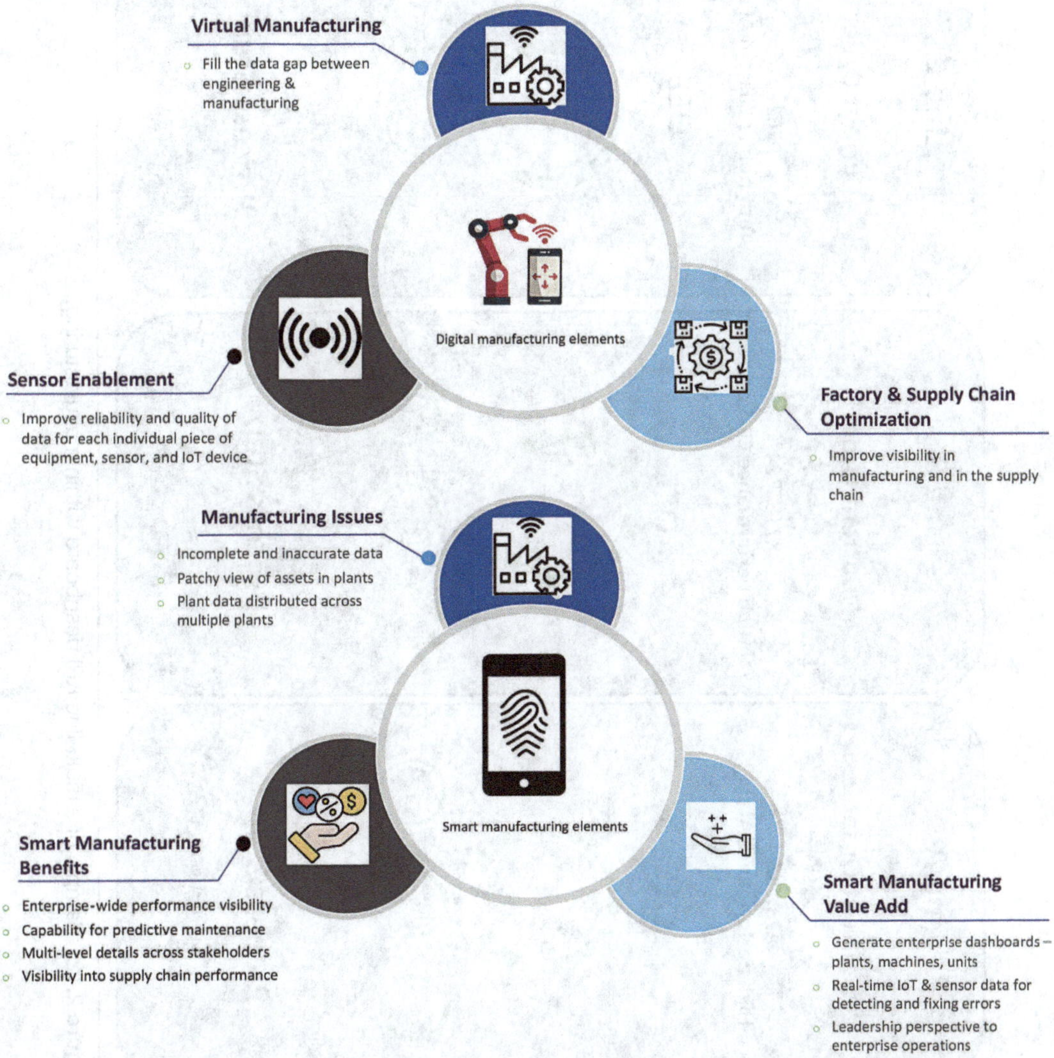

Figure 2.11 GE's Dimensions of the Brilliant Factory Initiative.[30]

Figure 2.12 Sample of a multi-line KPI dashboard within manufacturing.

A sample of visualization giving real-time performance within a smart factory environment through key performance indicators (KPIs) is given in Figure 2.12. As depicted here, such a dashboard (also known as an ANDON board) captures overall equipment effectiveness along with key metrics like quality, performance, and availability across multiple lines. It also shows details of what is on the line (product and description) as well as process steps for each line. Manufacturers can use similar dashboards specific to their needs.

FUTURE TRENDS

Future trends in the manufacturing industry are poised to reshape the landscape, driven by rapid advancements in technology, shifting consumer preferences, and the imperative for sustainable practices. One key trend is the increasing integration of automation and robotics into manufacturing processes, representing a fundamental shift in how goods are produced. This drives efficiency, cost reduction, and productivity gains while improving safety. Automation technologies, such as AI-driven robotics, cobots, and autonomous systems, are revolutionizing traditional manufacturing operations. They perform repetitive tasks with speed and precision, enabling faster production cycles, higher precision, and improved quality control.

These technologies not only improve efficiency and productivity but also enhance flexibility and agility in response to changing market demands. For example, AI-powered robots can perform complex tasks with speed and precision, leading to shorter production cycles and higher-quality output. The use of digital twins, virtual simulation, and predictive analytics enables manufacturers to optimize operations, predict maintenance needs, and identify opportunities for process improvement in real time. As a result, manufacturers achieve greater cost savings, minimize downtime, and maintain a competitive edge in today's fast-paced market.

The increasing focus on sustainability practices and environmental stewardship is driving significant changes in the manufacturing industry. With growing concerns about climate change and resource depletion, manufacturers are under pressure to adopt more sustainable practices across their value chains. This includes reducing carbon emissions, minimizing waste generation, and embracing circular economic principles to promote resource efficiency and recycling. Companies are investing in renewable energy sources like solar and wind power to reduce their carbon footprint and lower energy costs. By embracing sustainability, manufacturers can not only meet regulatory requirements and consumer expectations but also drive innovation and long-term business success.

Innovative technologies like 3-D printing (additive manufacturing) are enabling more sustainable production methods by reducing material waste and energy consumption. Additive manufacturing continues to disrupt traditional manufacturing processes by enabling rapid prototyping, customization, and on-demand production. With advancements in materials and printing technologies, manufacturers can now produce complex geometries, lightweight structures, and functional components with high precision and efficiency. Additive manufacturing offers cost savings, reduced lead times, and design flexibility, making it an increasingly attractive option for various industries, including aerospace, automotive, and healthcare. The materials used to create objects, referred to as filaments, come in various types, including thermoplastics, metals, ceramics, and composite materials. The choice of filament depends on factors such as the desired properties of the final object, the printing method, and the 3-D printer's capabilities. As the need for various types of filaments increases, expect more novel filaments to emerge in the market.

Evolving consumer demands and digitalization continue to drive improvements and transform the manufacturing industry. Today's consumers expect greater customization, convenience, and speed in their product experiences, prompting manufacturers to adopt more flexible and agile production processes. Digital technologies like the IoT, AR, and blockchain are playing crucial roles in enabling this transformation. IoT-connected devices provide valuable data insights into product usage patterns and performance metrics, allowing manufacturers to optimize production and enhance product quality. AR technologies are being used to improve training and maintenance processes, which will reduce downtime and improve workforce productivity. Blockchain is revolutionizing supply chain management by providing transparent and secure transactions, enhancing traceability and authenticity throughout the manufacturing value chain. By embracing digitalization and meeting growing consumer demands, manufacturers stay ahead of the competition and drive growth in the digital age. As this technology gets entrenched in manufacturing, many more use cases and improvements will result.

The IoT is driving the convergence of physical and digital worlds in manufacturing, enabling real-time monitoring, predictive maintenance, and intelligent decision making. Connected sensors, actuators, and devices collect data from machines, equipment, and production processes, providing valuable insights into performance, utilization, and energy consumption. Industrial connectivity solutions, like IIoT platforms and edge computing, enable seamless integration of data across the manufacturing value chain, thus facilitating optimization, automation, and innovation. By harnessing the power of IoT and IIoT, manufacturers can enhance visibility, agility, and resilience in their operations, paving the way for Industry 4.0 transformations.

The development of advanced materials and nanotechnology is also changing manufacturing processes and product design. These materials offer superior strength, durability, and performance characteristics compared to traditional materials. Nanotechnology, in particular, enables precise control at the molecular level, leading to innovations in areas such as lightweight composites, self-healing materials, and nano-scale sensors. Manufacturers are leveraging these advancements to create products with enhanced functionality, efficiency, and sustainability.

Finally, with the proliferation of digital technologies and interconnected systems in manufacturing, cybersecurity and data privacy have emerged as critical concerns. Manufacturers continue to invest in robust cybersecurity measures to protect their networks, systems, and sensitive data from cyber threats, ransomware attacks, and data breaches. Implementing encryption, access controls, and network segmentation helps safeguard critical assets and intellectual property. Moreover, compliance with data privacy regulations, like the European Union's General Data Protection Regulation and the California Consumer Privacy Act, is essential to maintaining trust with customers and stakeholders.

These multifaceted trends that are shaping the future of the manufacturing industry cover automation and robotics integration, sustainability practices, environmental stewardship, evolving consumer demands, and digitalization. These trends also encompass advancements in materials science; IoT, AI, and ML technologies; digital twins; virtual simulation; and cybersecurity. As manufacturers embrace digitalization and adopt innovative technologies, they are well-positioned to capitalize on new opportunities, address emerging challenges, and shape the future of manufacturing.

SUMMARY AND KEY INSIGHTS

Throughout this comprehensive exploration of the manufacturing value chain—spanning its diverse stages from conceptualization and procurement to production, distribution, and after-sales service—we unraveled the intricate web of processes, technologies, and stakeholders shaping the global manufacturing landscape. The conceptualization phase, often the inception point of transformative ideas, serves as the foundational bedrock where innovative concepts materialize into tangible products. This stage is characterized by rigorous market research, strategic planning, and forward-thinking ideation, laying the groundwork for subsequent stages in the value-chain journey.

Indeed, while the conceptualization phase sets the initial trajectory, other stages of the manufacturing value chain play equally crucial roles in realizing the vision and delivering value to customers. Procurement, for instance, stands as the gateway to securing the necessary resources and materials, underpinning the foundation upon which the entire manufacturing process hinges. Strategic sourcing, supplier relationship management, and risk mitigation strategies are vital elements in ensuring a resilient and efficient supply chain, essential for uninterrupted production and the timely delivery of goods.

The production stage represents the heart of manufacturing operations, where raw materials are transformed into finished products. By leveraging advanced technologies such as robotics, automation, and advanced analytics, manufacturers can optimize production processes, enhance product quality, and drive operational efficiency. Real-time monitoring and predictive maintenance solutions further bolster production uptime, reduce downtime, and minimize costly disruptions.

Distribution marks the pivotal juncture where products reach their intended destinations, bridging the gap between production facilities and end consumers. Efficient logistics management, inventory optimization, and last-mile delivery solutions are paramount in ensuring timely and cost-effective distribution which are essential to meeting customer expectations for speed, reliability, and transparency. Technologies such as RFID, GPS tracking, drone delivery, and route optimization algorithms empower manufacturers to optimize transportation networks, streamline warehousing operations, and deliver superior customer experiences.

Finally, after-sales service represents a critical component of the manufacturing value chain, extending beyond the point of sale to nurture long-term customer relationships and drive post-purchase satisfaction. Offering responsive technical support, proactive maintenance services, and product customization options enhances customer loyalty and fosters repeat business. Leveraging IoT-enabled devices and remote diagnostics capabilities enables manufacturers to offer predictive maintenance solutions, anticipate customer needs, and deliver personalized service experiences, further differentiating their offerings in a competitive marketplace.

We have witnessed how industry frontrunners like BMW, Siemens, and Schneider Electric are harnessing cutting-edge technologies to optimize their operations, drive innovation, and stay ahead in an increasingly competitive market. These cases underscore the pivotal role of efficient supply chain collaboration, strategic procurement, and agile production processes in achieving operational excellence and fostering customer satisfaction and loyalty. BMW's Industry 4.0 transformation, for instance, epitomizes the seamless integration of automation, IoT, and data analytics to streamline production processes, elevate product quality, and swiftly respond to dynamic market demands. Similarly, Schneider Electric's EcoStruxure platform exemplifies the power of digitalization and connectivity in enabling smart, sustainable manufacturing practices, driving efficiency gains and resource optimization.

As we contemplate the horizon of manufacturing's future, a myriad of emerging trends promises to reshape the industry landscape. From advancements in additive manufacturing and sustainable practices to evolving consumer preferences, these trends are poised to revolutionize traditional manufacturing paradigms. The integration of AI, IoT, and advanced analytics will further augment efficiency, flexibility, and customization capabilities within manufacturing, unlocking new realms of productivity and innovation. Embracing sustainable practices, such as waste reduction, energy efficiency, and circular economy principles, aligns with environmental imperatives and drives cost savings and regulatory compliance, positioning companies as responsible corporate citizens. The power of servitization models, where manufacturers offer services alongside products, will bring about a paradigm shift in business models, fostering deeper customer engagement and loyalty.

In summation, the manufacturing value chain remains essential to global economic progress and technological advancement. Its seamless operation is instrumental in delivering high-quality products to discerning consumers, driving economic prosperity, and addressing pressing societal challenges. As we chart the path forward, continuous research and innovation into areas such as sustainability, digitalization, human-centric manufacturing, regulatory compliance, ethical considerations, and supply chain resilience will be imperative for ensuring the sustained success and resilience of manufacturing. The convergence of technology, sustainability, and customer-centricity will undoubtedly shape the future trajectory of manufacturing, presenting both formidable challenges and boundless opportunities for industry stakeholders to explore and seize.

Some of the key takeaways from this chapter include:

1. The manufacturing value chain is a complex network of interconnected processes designed to transform raw materials into finished products, delivering value at each stage.
2. As manufacturers embrace digitalization and adopt innovative technologies, they are well-positioned to capitalize on new opportunities, address emerging challenges, and shape the future of manufacturing.
3. The conceptualization phase, often the inception point of transformative ideas, serves as the foundational bedrock where innovative concepts materialize into tangible products. This stage is characterized by rigorous market research, strategic planning, and forward-thinking ideation, laying the groundwork for subsequent stages in the value chain journey.
4. While the conceptualization phase sets the initial trajectory, other stages of the value chain journey are equally important in ensuring the successful launch and commercialization of new products and services.

3

LEVERAGING ADVANCED ANALYTICS AND BUSINESS INTELLIGENCE IN MANUFACTURING

"Analytics isn't a magic idea that will deliver results sometime in the future. It's real, and it's working, here and now."

—David Longridge, Former VP of Sales and Marketing, Boeing Global Services

"The most dangerous kind of waste is the waste we do not recognize."

—Shigeo Shingo, industrial engineer and Toyota production system expert

INTRODUCTION

In modern manufacturing, advanced analytics and business intelligence (BI) have risen to prominence as essential tools for driving operational excellence, enhancing decision-making processes, and gaining a competitive edge in the market. Advanced analytics refers to the utilization of sophisticated algorithms and statistical techniques to analyze vast amounts of data, extract meaningful insights, and predict future trends or outcomes. Conversely, BI encompasses a broader set of methodologies, technologies, and processes aimed at transforming raw data into actionable intelligence that is presented visually and can inform strategic business decisions.

The significance of advanced analytics and BI in the manufacturing sector cannot be overstated. Manufacturers are constantly inundated with a deluge of data from various sources, including sensors, machines, production lines, and supply chain systems. Harnessing this data effectively can unlock valuable insights that drive operational improvements, cost reductions, and revenue growth. Advanced analytics enables manufacturers to move beyond traditional descriptive analytics and delve into predictive and prescriptive analytics (see Figure 3.1), allowing them to anticipate potential issues, optimize processes, and capitalize on emerging opportunities in real time.

Advanced analytics refers to the process of analyzing large and complex data sets using sophisticated algorithms to uncover deeper insights, patterns, and trends that are not readily apparent through traditional analysis methods. With the prevalence of streaming real-time data from the cloud and thousands of Internet of Things (IoT) devices, advanced analytics become crucial in gaining insights from this data as well. Unlike descriptive analytics, which focuses on summarizing historical data as a snapshot of past performance, advanced analytics gets into predictive and prescriptive analytics. This facilitates forecasting future outcomes and trends, recommending

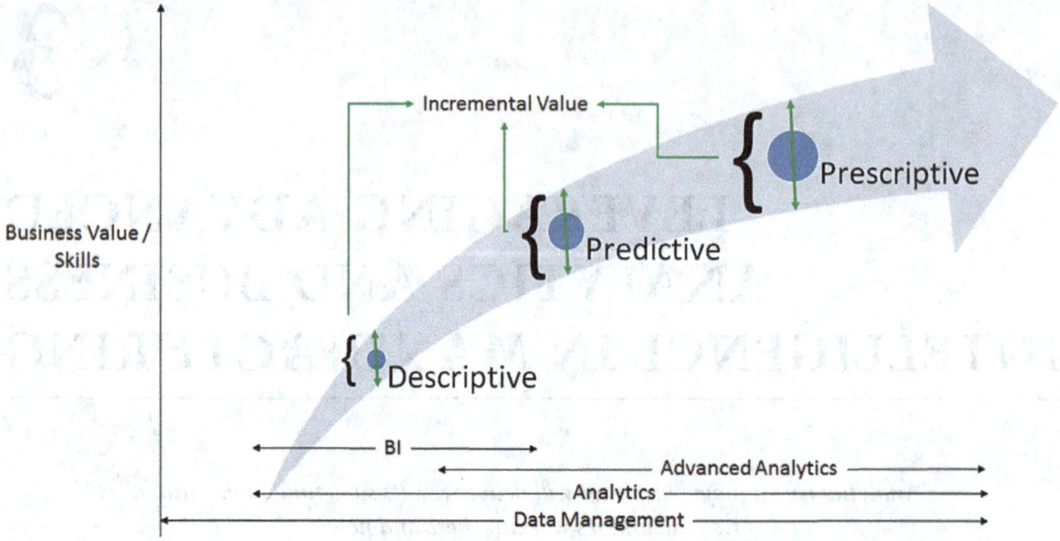

Figure 3.1 The analytics continuum.

optimal courses of action, identifying emerging opportunities, and mitigating potential risks. This includes techniques such as predictive modeling, data mining, and optimization algorithms. Advanced analytics serves as a powerful tool for driving data-driven decision making and optimizing business processes across various domains, including manufacturing.

> *"In God we trust, all others must bring data."*
>
> —W. Edwards Deming, renowned statistician and quality guru

In manufacturing, advanced analytics is pivotal in improving operational efficiency, enhancing product quality, and optimizing supply chain management. For example, predictive maintenance leverages advanced analytics algorithms to anticipate equipment failures before they occur, which enables proactive maintenance interventions that minimize unplanned downtime and reduce maintenance costs. Similarly, predictive analytics can be applied to demand forecasting, allowing manufacturers to anticipate shifts in customer demand and optimize production schedules accordingly. By harnessing the power of advanced analytics, manufacturing organizations can unlock new insights, drive innovation, and gain a competitive edge in today's rapidly evolving market landscape.

BI, on the other hand, encompasses a set of methodologies, processes, architectures, technologies, and tools designed to transform raw data into actionable insights that support decision making within organizations. At its core, BI involves the collection, integration, analysis, and visualization of data from multiple sources to provide stakeholders with a comprehensive view of business performance and trends. Through the use of sophisticated dashboards, reports, interactive visualizations, and data analytics tools, BI enables users to explore data, uncover patterns, and derive meaningful insights that inform strategic, operational, and tactical decisions. By leveraging BI capabilities, manufacturing

> *"The goal of business intelligence is to turn data into information, and information into insights."*
>
> —Carly Fiorina, former CEO, Hewlett-Packard

companies can gain a deeper understanding of their operations, identify areas for improvement, and drive continuous optimization and innovation.

In the manufacturing context, BI applications can range from monitoring key performance indicators (KPIs) on production efficiency and inventory levels to analyzing customer feedback and market trends. For instance, real-time monitoring dashboards provide plant managers with immediate insights into production metrics, allowing them to identify bottlenecks and inefficiencies in the manufacturing process. Additionally, BI tools can integrate data from various internal sources, including IoT sensors and enterprise resource planning (ERP) systems, to provide a holistic view of the manufacturing value chain. It can also source and integrate data from external sources like free government data and paid data from data providers. By empowering decision makers with timely and accurate information, BI enables manufacturing organizations to make informed decisions, streamline operations, and stay agile in an increasingly competitive environment. Figure 3.2 captures a sample dashboard used mostly in supply chain management, depicting KPIs for a given time period (one quarter here). However, similar dashboards can also be used at the management and leadership level.

The manufacturing sector is characterized by complex processes like supply chains, diverse production processes, and large volumes of data generated at various stages of operations. Advanced analytics and BI are crucial in harnessing this data to optimize processes, improve product quality, and drive innovation. Manufacturers can minimize downtime, reduce costs, and enhance overall efficiency by leveraging advanced analytics techniques such as predictive maintenance, anomaly detection, and demand forecasting. Similarly, BI tools enable executives and decision makers to gain real-time visibility into operational performance, identify areas for improvement, and make informed decisions to achieve strategic objectives.

Within the manufacturing sector, advanced analytics and BI are pivotal in driving innovation, optimizing processes, and improving overall performance. By leveraging advanced analytics techniques such as predictive modeling, manufacturers can gain deeper visibility into their operations, identify patterns or anomalies, and make data-driven decisions to enhance efficiency and quality. For instance, predictive maintenance models can analyze equipment sensor data to predict potential failures before they occur, enabling proactive maintenance and minimizing costly downtime. Similarly, BI tools empower decision makers at all levels of the organization with actionable insights derived from data. From shop floor supervisors to C-suite executives, stakeholders can access intuitive dashboards, reports, and visualizations that provide real-time visibility into KPIs, production metrics, and supply chain performance. This enables timely decision making, fosters collaboration across departments, and drives continuous improvement initiatives.

Advanced analytics and BI can optimize multiple stages of the manufacturing value chain. This includes product design, production planning, inventory management, quality control, and supply chain optimization. For example, in product design, manufacturers can use predictive analytics to simulate different design scenarios, perform virtual testing to identify potential issues, and optimize product specifications. This helps in managing cost, manufacturability, and other characteristics to iterate product designs to enhance performance, reduce costs, and accelerate time-to-market. In production planning, advanced analytics techniques can optimize production schedules, allocate resources efficiently, and minimize lead times. For example, automotive companies use simulation software along with wind tunnel experiments to optimize vehicle aerodynamics, improve fuel efficiency, and enhance safety features before physical prototypes are built. Similarly, BI dashboards can provide real-time visibility into inventory levels, production status, and supplier performance, which enables proactive decision making to minimize stockouts or delays.

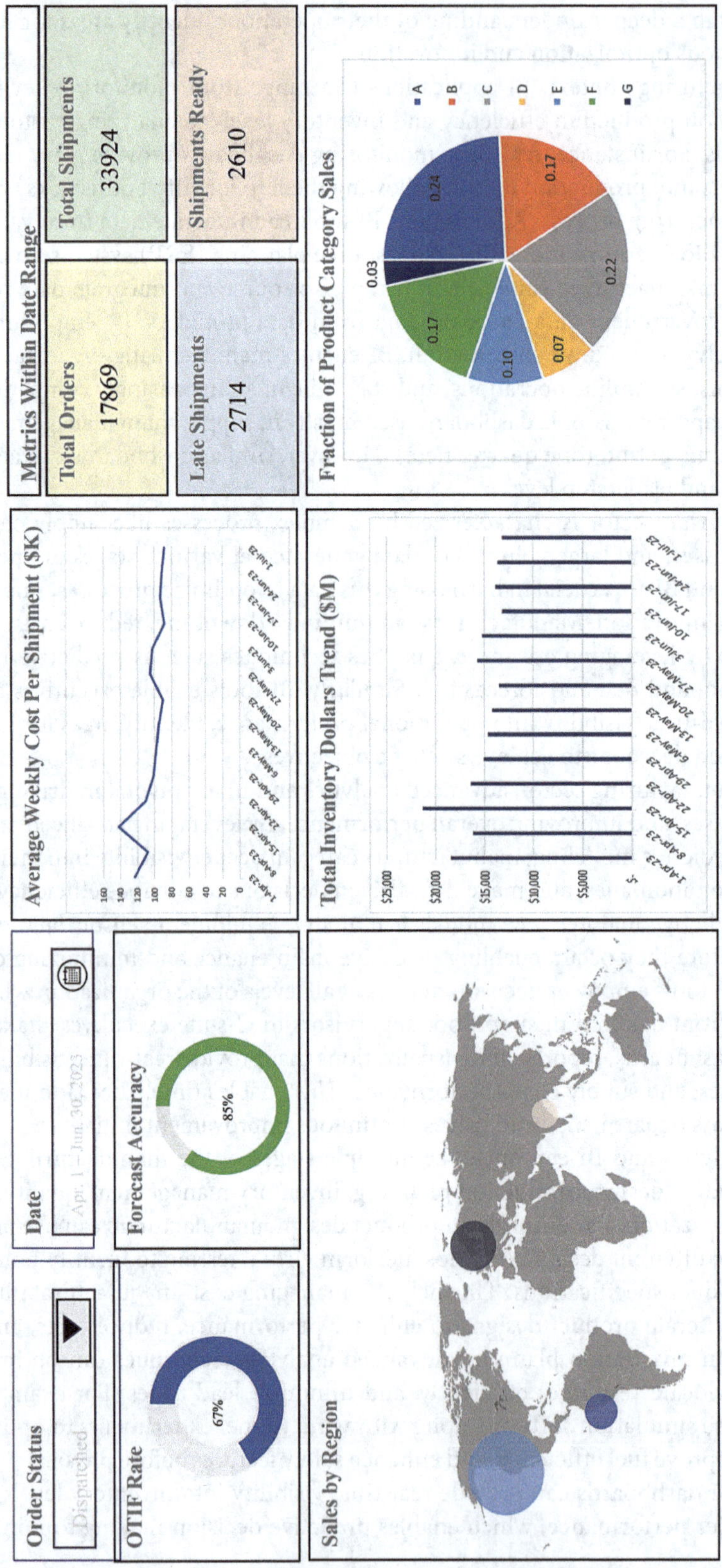

Figure 3.2 Sample supply chain dashboard.

In production planning and scheduling, advanced analytics algorithms are used to analyze historical production data, demand forecasts, and supply chain constraints to generate optimized production schedules that balance production capacity, inventory levels, and customer demand. By applying advanced analytics algorithms to historical data, manufacturers can identify patterns, trends, and correlations. This results in more accurate demand forecasts, thereby enabling better inventory management, reduced stockouts, and improved customer service levels.

The following relevant case studies demonstrate the use of advanced analytics in manufacturing. They start with a summary of a case from SKF and then continue with summaries of other cases. These cases show that just building an advanced analytics solution is not sufficient; a lot of thought and effort needs to go into its deployment to achieve success.

One exemplary case of predictive maintenance implementation is SKF's Rotating Equipment Performance (REP) Center. The Center has demonstrated the transformative potential of predictive maintenance (PdM) in manufacturing. The REP Center operates as a hub for remote monitoring and PdM services, offering manufacturers (SKF's customers) comprehensive solutions to optimize the performance and reliability of industrial rotating equipment. Leveraging advanced analytics, IoT technologies, and decades of expertise in bearing and rotating machinery, SKF delivers actionable insights and prescriptive recommendations to clients worldwide.

At the heart of the REP Center's operations is its ability to remotely monitor critical machinery in real time, continuously collecting data on equipment health and performance through the deployment of a network of IoT sensors and condition monitoring systems. SKF captures a wealth of operational data, including temperature, vibration, lubrication, and energy consumption metrics, enabling detailed analysis and early identification of potential equipment failures. This extensive data collection allows SKF's experts to offer a range of PdM services tailored to meet the unique needs of its clients. These services encompass condition-based monitoring, predictive and prescriptive analytics, and performance optimization, providing manufacturers with actionable insights to enhance asset reliability and operational efficiency. Through PdM strategies, SKF's REP Center has helped numerous clients achieve significant cost savings, reduce maintenance expenditures, and extend the lifespan of critical machinery, positioning the company at the forefront of innovation and excellence in PdM solutions amidst the ongoing digital transformation of the manufacturing industry.

PdM and advanced analytics-driven quality control (QC) are transforming manufacturing and related industries. Companies like Bosch, BMW, GE Aviation, Rolls-Royce, Schindler, and Union Pacific have implemented PdM solutions using IoT sensors and advanced analytics to monitor and predict equipment performance. BMW's PdM system, developed in partnership with Bosch, uses IoT sensors to detect potential equipment failures, thus reducing disruptions, enhancing operational efficiency, and improving on-time delivery.[1] Similarly, GE Aviation and Rolls-Royce employ IoT and predictive algorithms to maintain jet engines, resulting in increased asset uptime and reduced maintenance costs. Schindler has applied PdM to monitor elevators, improving equipment reliability and safety, while Union Pacific uses sensors to build predictive models for its locomotive fleet, enhancing fuel efficiency and operational performance. These PdM initiatives underscore the value of real-time monitoring and proactive maintenance across diverse sectors.

Companies like Ford, Siemens, Dow, and BASF have implemented advanced analytics and BI initiatives to enhance QC processes. Ford and Siemens use analytics tools to gain real-time insights into manufacturing workflows, allowing for proactive detection and resolution of quality issues. This has led to significant improvements in product quality, reduced defect rates, and enhanced customer satisfaction. Dow and BASF have leveraged cloud-based analytics platforms to optimize QC in their chemical production processes, using predictive models to proactively identify and mitigate quality deviations. These analytics-driven QC initiatives have enabled these companies to optimize production costs, improve profitability, and uphold high product quality standards.

Supply chain and demand forecasting are other use cases where companies have achieved significant success (or failure). Sophisticated models enabled Procter & Gamble (P&G) to optimize inventory levels, streamline production planning, and minimize stockouts and excess inventory. This helped P&G to improve their product demand forecasts.[2] However, Ford encountered problems in applying advanced analytics to optimize its supply chain operations with siloed data and difficult integration with global suppliers. The lack of standardization and interoperability was the primary culprit that hindered the effectiveness of Ford's advanced analytics initiatives.[3]

Together, these stories highlight the role of PdM, analytics-based QC, and demand forecasting in driving operational efficiency, reliability, and customer satisfaction in manufacturing and beyond.

The rest of this chapter follows a specific script. We start with some standard algorithms, followed by data collection and preparation. As previously mentioned, data and analysis are crucial to the success of advanced analytics initiatives, which leads to the correct and usable BI solutions. Then, we will cover applications of advanced analytics and BI in manufacturing, followed by the implementation of advanced analytics solutions. Next, because of the vast amounts of proprietary data that are generated in these deployments, we will cover data security and governance. Last but not least, before we conclude this chapter, we will cover future trends, challenges, and opportunities.

COMMON ADVANCED ANALYTICS ALGORITHMS AND BI TOOLS FOR THE MANUFACTURING SECTOR

Before exploring the specific techniques and methodologies employed in advanced analytics and BI in manufacturing, it is essential to understand their foundational roles in enabling data-driven decision making and strategic insights. Advanced analytics encompasses a variety of sophisticated techniques and algorithms that allow manufacturers to extract valuable insights from large and complex data sets, enabling predictive capabilities and proactive decision making. BI, on the other hand, encompasses a set of processes, technologies, and tools that empower manufacturers to transform raw data into actionable information, facilitating quick and informed decision making at all levels of the organization. By leveraging advanced analytics and BI solutions, manufacturers can gain valuable insights into their KPIs, monitor operational metrics in real time, and identify trends and patterns to optimize processes and drive business growth. With robust infrastructures in place, manufacturers can unlock the full potential of their data assets and gain a competitive edge in today's dynamic marketplace. Now, let's explore some of the core components and methodologies of advanced analytics and BI in manufacturing.

> *"Information is the oil of the 21st century, and analytics is the combustion engine."*
>
> —Peter Sondergaard, SVP and global head of research at Gartner, Inc.

We will start with two of the most common statistical algorithms: time series analysis and regression analysis. Both of these algorithms are used for prediction and inference. Time series analysis focuses on understanding and forecasting temporal patterns in data, while regression analysis examines the relationship between variables and makes predictions based on this relationship. Time series analysis analyzes historical temporal data to identify patterns, trends, and seasonal variations. By analyzing past patterns and extrapolating them into the future, organizations can make informed decisions on the parameters of interest. Regression models, on the other hand, identify relationships between a dependent variable and one or more independent variables and not necessarily on temporal data. The goal is to quantify the strength and direction of the relationship between independent and dependent variables. Methods like moving averages, exponential smoothing, autoregressive integrated moving averages, Holt-Winters, and Box-Jenkins models are extensions of time series analysis. These models are mostly used in demand forecasting and use weighted models with more recent observations getting higher weights. Decomposition models, also used in demand forecasting, decompose observations by a dimension of interest (e.g., seasonality, whether actual season, or months, or weeks).

In manufacturing, optimization algorithms are mostly used in supply chain management but can also be used in other manufacturing areas like production scheduling. Methods like linear, integer, and dynamic programming are used to optimize simple to complex problems. Optimization algorithms use mathematical techniques to maximize or minimize an objective function that is subject to constraints, including those on decision variables. Other optimizations, such as network optimization or inventory optimization, are specialized algorithms mostly used in supply chain processes. They determine the optimal allocation of resources, inventory levels, production schedules, and distribution routes to minimize costs and maximize efficiency. Heuristics algorithms can also be used to quickly find approximate solutions to complex optimization problems. Another dimension in optimization algorithms is simulation modeling, where algorithms are used to simulate and optimize processes, mostly in the supply chain and production process.

Algorithms that cover maintenance and reliability within manufacturing have matured over the years, and advanced analytics have started playing a key role in this industry. There are many algorithms here, each with a specific role. Many are based on statistical analysis—like statistical process control (SPC), Six Sigma Methodology, and Control Plan Development. Others are more focused—like failure mode and effects analysis (FMEA), which identifies potential failure modes within equipment and evaluates their potential impact on operations. By analyzing failure modes, their causes, and associated effects, organizations can prioritize maintenance activities and develop preventive strategies to mitigate. Root cause analysis (RCA) is used to investigate the underlying causes of failures or malfunctions. This helps manufacturers implement corrective actions and preventive measures to address recurring issues and enhance equipment reliability. Techniques for reliability analysis are commonly used to predict the time until a particular failure event occurs.

Manufacturers employ various BI tools to enhance operational efficiency and decision making across different areas such as predictive maintenance, quality control, supply chain optimization, demand forecasting, and production scheduling. Most BI products are commercial, with some focusing only on one area of the business while others focus across areas. Common tools for dashboarding and report generation include Microsoft Power BI, Tableau, Microstrategy, SAS Visual Analytics, and Qlik Sense. Some products focusing on supply chain management include Oracle Supply Chain Management Cloud, SAP Integrated Business Planning, JDA Software, and Kinaxis' RapidResponse. Finally, Siemens Opcenter, Plex Systems, and Dassault Systèmes DELMIA Ortems concentrate on scheduling production activities. Manufacturers can develop their own reporting tools for free using tools within libraries in popular programming languages like Python.

Python developers can use libraries like pandas, matplotlib, and seaborn, and new libraries are emerging frequently as well.

That is just a sampling of the algorithms and tools available to manufacturers to conduct their business effectively. Obviously, there are many advanced analytics techniques and BI tools available to the practitioner.

Transitioning from discussing tools and algorithms to data collection and preparation involves shifting focus from the methods used for analysis to the crucial steps involved in acquiring and preparing the data for analysis. While advanced analytics algorithms and BI tools play a pivotal role in extracting insights from data, the effectiveness of these analyses relies heavily on the quality, relevance, timeliness, and availability of the underlying data. Therefore, attention must be given to meticulously collecting data from various sources, ensuring its accuracy and completeness, and preprocessing it to make it suitable for analysis.

DATA COLLECTION AND PREPARATION IN MANUFACTURING

In modern manufacturing, data is generated from various sources, providing valuable insights to improve operational efficiency and inform decision-making processes. One primary data source is sensors that are embedded within machinery and equipment to monitor parameters such as temperature, pressure, vibration, and energy consumption. For example, in automotive manufacturing, sensors are used to monitor the performance of robotic assembly lines, detect equipment failures, and optimize production processes in real time. Similarly, IoT devices are another source of data. IoT devices are interconnected devices that communicate with each other and with central systems over the Internet. In manufacturing, IoT devices are deployed across the production floor to capture data from equipment, vehicles, and even products themselves. For instance, in a smart factory setting, IoT sensors embedded in machinery can collect data on machine uptime, production throughput, and quality metrics, enabling manufacturers to identify bottlenecks, optimize production schedules, and improve overall equipment effectiveness.

ERP systems serve as centralized repositories of operational data, integrating information from various departments such as production, procurement, inventory management, and sales. By consolidating data from different sources into a single platform, ERP systems provide a holistic view of manufacturing operations, facilitating data-driven decision making and resource allocation. Additionally, manufacturing execution systems (MES) capture detailed production data in real time, tracking work orders, production schedules, and quality metrics to optimize manufacturing processes and ensure compliance with industry standards. Many ERP systems, together with MES, capture quality, financial, and supply chain data as well.

Financial and operational data encompass information related to manufacturing costs, revenue, profitability, and operational performance metrics across multiple product lines. This includes data from financial statements, cost accounting systems, production reports, and KPIs. By analyzing this data, manufacturers can assess the financial health of their operations, identify cost-saving opportunities, optimize resource allocation, and build strategies around product lines. For example, analyzing production yield data and equipment utilization rates can help manufacturers identify inefficiencies in production processes, minimize waste, and improve overall operational efficiency.

Supply chain data encompass information related to the flow of materials, components, and finished products throughout the supply chain network. This includes data on supplier performance, consumption and inventory levels, lead times, transportation routes, and logistics costs.

By analyzing supply chain data, manufacturers can identify opportunities to optimize sourcing strategies and reduce inventory carrying costs and shrinkage to enhance overall supply chain efficiency. For example, real-time visibility into supplier transit, lead times, and inventory levels enables manufacturers to proactively address potential supply shortages or disruptions, minimizing production downtime and ensuring continuity of operations. An example of this using actual data from a large manufacturing company for about 260,000 data points for almost 5,500 distinct parts comprising over 42.2 million distinct pieces for a calendar year is shown in Figure 3.3. As you can see, there is a significant deviation from the expected ± 1-day variation, mainly on the side of higher inventory (+1 side), indicating excess inventory.

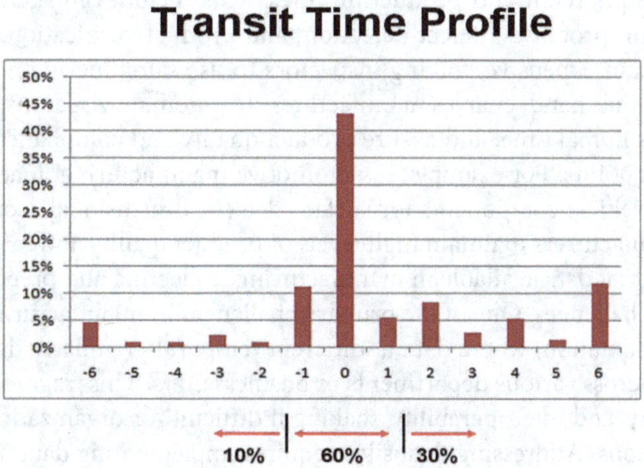

Variation from normal (± 1 day)
Mean: -0.06 days
Std Dev.: 7.56 days
Profit Impact: ~ $2.2 Million

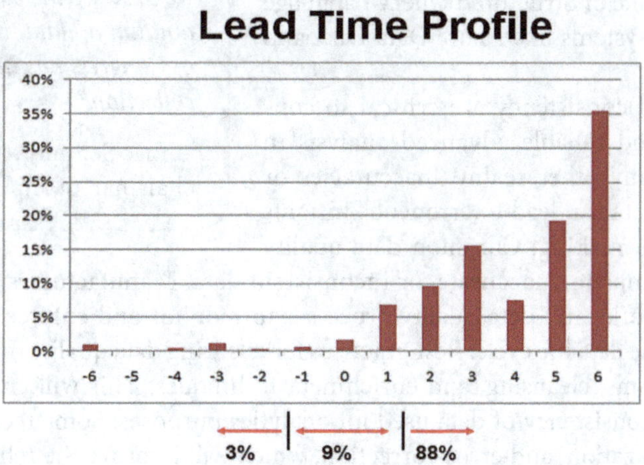

Variation from normal (± 1 day)
Mean: 5.47 days
Std Dev.: 10.82 days
Profit Impact: ~ $21 Million

Figure 3.3 Actual lead-time insights from a manufacturing company.

Customer feedback and market data provide valuable insights into customer preferences, market trends, and competitive dynamics, helping manufacturers align their products and services with customer needs and market demand. This includes data from customer surveys, online reviews, social media interactions, and market research reports. By analyzing customer feedback and market data, manufacturers can identify emerging trends, anticipate shifts in consumer behavior, and develop innovative products and services that resonate with target audiences. For instance, a manufacturer may use sentiment analysis techniques to analyze customer reviews and social media conversations, gaining actionable insights into product performance, brand perception, and areas for improvement.

Machine vision systems capture and analyze visual data from cameras and sensors installed on manufacturing equipment and production lines. This includes image data generated during quality inspection processes, defect detection, and product verification. Manual inspection is time-consuming and expensive, forcing inspectors to use sampling methods. Machine vision technology, on the other hand, enables manufacturers to automate visual inspection tasks, detect defects or anomalies in real time, and ensure product quality and compliance with specifications for many more data points. For example, in automotive manufacturing, machine vision systems are used to inspect vehicle components for surface defects, dimensional accuracy, and assembly errors, helping manufacturers maintain high levels of product quality and safety.

Despite the wealth of data available in manufacturing, collecting and preparing data for analysis presents several challenges. One of the primary challenges in manufacturing is the presence of data silos and fragmentation, where data at different temporal granularity is stored in disparate systems or formats across various departments or business units. This fragmentation inhibits data visibility, accessibility, and interoperability, making it difficult for organizations to gain a holistic view of their operations. Addressing data silos requires implementing data integration strategies that enable seamless communication and interoperability between different systems and platforms. Manufacturers can use integration tools like Informatica, Qlik, Tibco, and Alteryx. These can be used in conjunction with cloud-based data warehouse systems like Snowflake or Structured Query Language Server or data lake systems like Azure Data Lake and Databricks.

> *"Every day, three times per second, we produce the equivalent of the amount of data that the Library of Congress has in its entire print collection."*
>
> —Eric Schmidt, former executive chairman of Alphabet Inc.

Data quality and consistency are critical to conducting accurate and reliable advanced analysis in manufacturing or any other realm. Inaccuracies or inconsistencies in data can lead to erroneous insights and flawed decision making. Common data quality issues include incomplete, inaccurate, or inconsistent data. Manufacturers must establish data governance frameworks and data quality processes to monitor and enforce data integrity standards throughout the data life cycle. Best practices for ensuring data quality include data profiling, validation mechanisms, cleansing, and enrichment techniques. This will enhance the accuracy, completeness, and consistency of data used for analytics purposes. Some techniques include deduplication, normalization, and error correction, which will improve the reliability and usability of manufacturing data.

Data security and privacy concerns are paramount in manufacturing, given the sensitive nature of manufacturing data and the potential impact of data breaches or unauthorized access. Manufacturers must implement robust cybersecurity measures to safeguard confidential information, intellectual property (IP), and sensitive operational data from internal and external threats. This includes implementing access controls, encryption, authentication mechanisms, and intrusion detection systems to mitigate security risks and ensure compliance with regulatory requirements such as the General Data Protection Regulation (GDPR) and the California Consumer Privacy Act (CCPA). This also includes establishing checks and balances for data being stored on the public cloud. Case in point, as recently as May 2023, Samsung discovered an accidental leak of sensitive internal code on the ChatGPT platform. Samsung has since banned the use of generative artificial intelligence (AI) tools. Amazon, JPMorgan Chase, Bank of America, Citigroup, and other banks have followed suit. However, generative AI is not bad as long as use cases are judiciously identified, such as when writing and testing code.[4]

Integrating data from heterogeneous sources and systems is a complex challenge in manufacturing, particularly with the proliferation of IoT devices, sensors, and legacy information technology (IT) systems. Adding to the complexity is the granularity at which different data items are collected. Some are sub-second, others at second frequency, yet others at minutes and hours frequency. Data that is slow to change, like the level of ore mined or the height of a chip pile in a paper mill, is even collected at day or week frequency. Achieving seamless data integration and interoperability requires standardized data formats, protocols, and interfaces that facilitate the exchange and integration of data across disparate systems and platforms. Manufacturers can leverage technologies such as application programming interfaces (APIs); extract, transform, load (ETL); and extract, load, transform (ELT) tools, along with cloud storage. They can also use middleware solutions to streamline data integration processes and enable real-time data exchange between different applications and databases.

Effective data governance is essential for ensuring regulatory compliance and accountability in manufacturing. Manufacturers must establish a governance board to define and manage policies, processes, and controls. They govern data usage, access, and retention in accordance with industry regulations and data privacy laws. This includes implementing data classification schemes, access controls, audit trails, and data stewardship roles to ensure data accountability and traceability throughout the organization. Compliance with regulations such as GDPR, CCPA, and the International Standards Organization (ISO) requires ongoing monitoring, auditing, and documentation of data management practices to mitigate legal and reputational risks associated with noncompliance. By establishing robust data governance frameworks and investing in data quality management processes, manufacturers can unlock the full potential of their data assets and derive maximum value from advanced analytics and BI solutions.

As manufacturing operations scale and evolve, scalability and seamless integration of data collection and preparation processes become increasingly important. With new technology, manufacturers face the scalability issue from a data and user count perspective. Manufacturers must invest in flexible and interoperable technologies that accommodate growing data volumes and diverse data sources. This enables a system or solution to accommodate the growth of users, data, and systems put in place to handle increased analytical workloads. All these must be managed without compromising performance or reliability.

Manufacturers must invest in scalable analytics platforms, infrastructure, and architectures that can adapt and expand in response to evolving business needs and market dynamics. This includes flexible cloud-based analytics solutions, distributed computing frameworks, and modular architectures. These enable seamless scaling of on-demand computing resources, storage capacity, and analytical capabilities. By embracing scalable analytics solutions, manufacturers can future-proof their operations and accelerate their journey toward data-driven decision making and innovation.

Integration, on the other hand, is crucial for unlocking the full value of data by aggregating, harmonizing, and analyzing information from disparate sources and systems across the manufacturing value chain. Integration involves connecting and unifying data from sensors, IoT devices, machinery, production systems, enterprise applications, and external sources to create a harmonized and unified data ecosystem for analysis and insights generation. Achieving seamless integration requires interoperability standards, data integration technologies, and data management practices. This facilitates data exchange, transformation, and synchronization across heterogeneous environments. Manufacturers must adopt an integrated approach to data architecture and governance that prioritizes data interoperability, consistency, and accessibility across the organization. This includes implementing data integration platforms, middleware solutions, and API frameworks that enable real-time data exchange, orchestration, and synchronization between different systems and applications. By embracing integration, manufacturers can break down data silos, enhance collaboration, and derive actionable insights from a unified view of their operations. This enables them to optimize processes, improve decision making, and drive competitive advantage in today's digital manufacturing landscape.

The skills and talent gap poses a significant challenge for manufacturers aiming to effectively leverage advanced analytics and BI. As manufacturing becomes increasingly digitized and data-driven, there is a growing demand for individuals with expertise in data analysis, statistics, advanced analytics, and data visualization, among other technical skills. However, many manufacturers struggle to find and retain talent with the necessary skills and experience to drive their analytics initiatives forward. This skills gap is exacerbated by the rapid pace of technological change and the evolving nature of analytics tools and techniques, which require continuous learning and upskilling among workforce members; it is not a set-it-and-forget-it philosophy! Building a skilled workforce capable of effectively managing and analyzing manufacturing data is essential for successful advanced analytics and BI initiatives. Training programs and talent development initiatives can help bridge the skills gap and empower employees to leverage data-driven insights to propel innovation and improve business outcomes. Data and enterprise architects play a crucial role in the success of advanced analytics and BI in a manufacturing company.

Addressing the skills and talent gap requires a multifaceted approach that encompasses recruitment, training, and development strategies. Manufacturers must invest in attracting top talent—collaborating with educational institutions, industry associations, and training providers to develop skills needed for analytics roles in manufacturing. Additionally, manufacturers can foster a culture of continuous learning and knowledge-sharing within their organizations by providing access to online courses, workshops, and mentorship programs. A last but cost-prohibitive solution is to get external help. There are hundreds of consulting companies charging from $20 per hour (for off-shore resources) to over $500 per hour (for on-shore resources). Many of these firms are helpful and try to build a good relationship with clients, while many others serve only as transactional partners. Like any other relationship, due diligence regarding any external parties you engage with is critical, not just the first time but with each additional engagement.

Further, manufacturers must address the cultural barriers and mindset shifts associated with adopting advanced analytics and BI. This includes promoting data literacy and analytical thinking among employees at all levels of the organization, encouraging cross-functional collaboration, and fostering a data-driven decision-making culture. Being disciplined about data from *the cradle to the grave* and treating it as an asset impacting virtually everything on financial statements is crucial. By empowering employees with the necessary skills and knowledge to harness the power of data, manufacturers can overcome the skills and talent gap and unlock the full potential of advanced analytics and BI to drive innovation, efficiency, and competitiveness in the manufacturing sector.

APPLICATIONS OF ADVANCED ANALYTICS AND BI IN MANUFACTURING

By now, it should be clear that in today's rapidly evolving manufacturing landscape, advanced analytics and BI have emerged as indispensable tools. They have driven operational excellence, optimized processes, and gained actionable insights from vast volumes of data. From predictive maintenance to demand forecasting, manufacturers across various industries have harnessed the power of advanced analytics and BI to enhance decision making, reduce costs, and improve overall efficiency. This section explores some key applications of advanced analytics and BI in different stages of the manufacturing value chain, shedding light on real-world examples and highlighting the transformative impact of these technologies.

> *"The importance of data and analytics is going to continue to grow."*
>
> —Mary Barra, CEO of General Motors

At each stage of the manufacturing value chain, from production to service, advanced analytics and BI offer unique opportunities for optimization and improvement. As referenced earlier, predictive maintenance leverages data analytics to anticipate equipment failures and schedule proactive maintenance activities to minimize unplanned downtime and maximize asset utilization. QC processes benefit from advanced analytics by identifying defects and deviations in real time, ensuring adherence to stringent quality standards. Further, supply chain optimization relies on BI to streamline inventory management, reduce lead times, and enhance supplier collaboration, driving greater efficiency across the entire supply chain network. Remember, these are only some key examples—there are many more use cases, as many use cases as you can envision. This section presents the *art of the possible* and will help you determine the potential use cases at your facility.

Numerous companies have successfully implemented advanced analytics and BI solutions to achieve tangible benefits in their manufacturing operations. As mentioned earlier, Bosch implemented predictive maintenance algorithms and achieved improved operational efficiency, increased equipment reliability, and enhanced overall productivity. Tesla utilized advanced analytics techniques for quality control, but the complex nature of manufacturing processes and the variability of input factors posed challenges to this initiative. These examples underline the transformative potential of advanced analytics and BI—but manufacturers need to proceed with caution.

Despite the promise of advanced analytics and BI, manufacturers face several challenges in implementing these technologies effectively. Data integration from disparate sources, ensuring data quality and reliability, and addressing cybersecurity concerns are among the key challenges in leveraging advanced analytics and BI. Additionally, the shortage of skilled data scientists and the need for effective organizational change management (OCM) pose significant hurdles to successful adoption. Overcoming these challenges requires a holistic approach that encompasses technological innovation, organizational readiness, and strategic planning.

Predictive Maintenance

PdM is a cornerstone of modern manufacturing that utilizes advanced analytics and BI to forecast equipment failures before they occur, enabling proactive maintenance actions and minimizing unplanned downtime. By harnessing the power of data analytics, manufacturers can optimize maintenance schedules, reduce maintenance costs, and extend the lifespan of critical assets.

PdM is based on sophisticated advanced analytics algorithms (like time series analysis, statistical analysis, FMEA, and root cause analysis) capable of analyzing vast amounts of data in real time. These algorithms enable the analysis of large datasets, including sensor and IoT readings, historical maintenance records, and equipment performance metrics. The algorithms use this data to detect subtle changes in equipment performance patterns, allowing for the early identification of potential faults or anomalies. For example, a leading industrial manufacturer deployed advanced analytics models to monitor the health of the equipment manufacturing factories, resulting in close to a 20% reduction in unscheduled maintenance events and significant cost savings.

The PdM market is growing significantly. Many vendors are involved in deploying different types of solutions for PdM—failure prediction, anomaly detection, and remaining useful life. This is concisely captured in Figure 3.4 (IoT Analytics Research 2023-Predictive Maintenance Market Report 2023–2028).

One of the key advantages of advanced analytics algorithms is their ability to adapt and improve over time through continuous learning. As new data becomes available, PdM models can be updated and refined, leading to more accurate predictions and better-informed decision making. For example, in the chemical industry, manufacturers have used advanced analytics algorithms to analyze process control equipment data and predict component failures by detecting potential upcoming malfunctions in pump or temperature sensors, based on real-time performance metrics. By continuously refining these models, manufacturers optimize maintenance schedules and minimize unplanned downtime, ultimately enhancing operational efficiency and customer satisfaction.

Further, advanced analytics algorithms enable manufacturers to move beyond traditional time-based maintenance schedules and adopt a condition-based approach. By monitoring equipment health (such as pump and temperature sensors) in real time and analyzing the current state of

Figure 3.4 Predictive maintenance market, providers, and solutions.[5]

critical assets, PdM models can recommend maintenance actions precisely when they are needed rather than relying on predetermined schedules. This proactive approach reduces maintenance costs and extends equipment lifespan, improving overall asset reliability and performance.

Integration with sensors, IoT, and condition monitoring systems is paramount in enabling PdM strategies within the manufacturing sector. Sensors and IoT embedded within critical equipment and machinery continuously capture and transmit real-time data on various operational parameters such as temperature, vibration, pressure, and energy consumption. These sensors act as the frontline data collectors, providing valuable insights into the health and performance of industrial assets. Alcoa, one of the world's largest players in the aluminum industry, has implemented IoT-enabled sensors across its facilities to monitor various parameters in mining, refining, smelting, fabrication, and recycling. This has enabled them to implement their PdM initiatives, proactive process optimization, and enhanced safety in chemical manufacturing operations.[6] They have achieved almost a 20% reduction in equipment downtime and increased operational efficiency and asset reliability by approximately 30%.[7]

As mentioned earlier, an exemplary case of predictive maintenance implementation is SKF's REP Center. The Center has demonstrated the transformative potential of PdM in manufacturing. The REP Center operates as a hub for remote monitoring and PdM services, offering manufacturers (SKF's customers) comprehensive solutions to optimize the performance and reliability of industrial rotating equipment. Leveraging advanced analytics, IoT technologies, and decades of expertise in bearing and rotating machinery, SKF delivers actionable insights and prescriptive recommendations to clients worldwide.[8]

Another case study is that of BMW implementing PdM solutions in its manufacturing plants. By installing IoT sensors on critical machinery, such as robotic assembly arms and conveyor belts, BMW collects real-time manufacturing data. This data is then analyzed using advanced analytics algorithms developed by Bosch. The algorithms detect anomalies and patterns indicative of impending equipment failures, enabling BMW to schedule maintenance proactively. As a result, BMW has reduced more than 500 minutes of disruption annually for its plant in Regensburg; other plants are seeing similar or better results. This has led to significant cost savings, enhanced worker safety, and increased operational efficiency.[9]

Using advanced analytics and IoT sensors, GE Aviation implemented a PdM solution for its aircraft engines. By collecting real-time data on aircraft engine performance, temperature, air flow rate, and vibration, GE Aviation can detect potential failures before they occur. Advanced analytics algorithms analyze this data to predict maintenance needs and schedule proactive repairs. As a result, GE Aviation has improved aircraft reliability through its engines, reduced unplanned downtime, and minimized maintenance costs. Success metrics include increased asset uptime, reduced maintenance labor hours, and enhanced flight safety.[10]

Similarly, Rolls-Royce implemented PdM solutions for its jet engines using advanced analytics and machine learning (ML) algorithms. By leveraging IoT sensors embedded in engines, Rolls-Royce collects real-time performance data and monitors key operating parameters such as temperature and fuel consumption. Advanced analytics models analyze this data to predict component wear and potential failures, enabling proactive maintenance actions. As a result, Rolls-Royce has reported a significant reduction in unplanned downtime. This improvement allowed for better asset utilization and minimized disruptions to production processes. Success metrics include increased engine lifespan, minimized downtime, and optimized maintenance schedules.

Schindler Elevator Corporation implemented PdM solutions for its elevator and escalator systems using IoT sensors and analytics.[11] By equipping elevators with sensors that monitor factors such as vibration and door operations, Schindler collected real-time data on equipment performance. PdM algorithms analyzed this data to detect anomalies and predict potential failures. As a result, Schindler increased elevator uptime and improved passenger safety.[12]

The final case study—though not from the world of manufacturing—is from Union Pacific Railroad. It built sophisticated models for its locomotive fleet using data from sensors in locomotives and multiple track detectors. These captured data like temperature, vibration, and sound to feed into the models and build a holistic view of the locomotive including predicting potential failures. These predictions were used to schedule proactive maintenance. As a result, Union Pacific has used these predictions to build a five-year window into their future, thereby reducing unplanned downtime, optimizing fuel costs, and improving on-time performance.[13]

Each case study presented here demonstrates the adoption and success of PdM solutions across various industries, highlighting common themes and approaches. Each case study showcases the integration of advanced analytics and IoT sensors to collect real-time data on equipment health and performance. By leveraging this data, manufacturers can predict potential failures and schedule proactive maintenance, leading to reduced downtime, optimized maintenance schedules, and enhanced operational efficiency. Success metrics such as increased equipment uptime, minimized maintenance costs, and improved reliability underscore the tangible benefits of PdM strategies. Furthermore, these case studies emphasize the importance of continuous monitoring, real-time data, advanced analytics, and proactive interventions to maximize the effectiveness of PdM initiatives. These companies are among the many that lead the way in PdM innovation. Manufacturers worldwide are poised to leverage these solutions to drive cost savings, improve asset reliability, and thrive in an increasingly competitive market landscape.

Quality Control

QC is the cornerstone of manufacturing operations, encompassing a range of processes aimed at upholding product standards and ensuring customer satisfaction. It involves meticulously scrutinizing raw materials, production processes, work in process, and finished goods to detect and rectify deviations from predefined quality parameters. In today's dynamic manufacturing landscape, with a strong need for precision and consistency, quality control has evolved significantly. With advancements in technology and the introduction of sophisticated analytics tools, manufacturers now have unprecedented capabilities to monitor, analyze, and optimize quality control processes in real time. Some algorithms and tools that play in the QC arena include SPC, FMEA, Six Sigma, and RCA.

QC plays a pivotal role in driving business success and upholding brand reputation. Beyond regulatory compliance, robust QC practices instill customer confidence, fostering loyalty and brand advocacy. By acting as a safeguard against costly defects, product recalls, warranty costs, and reputational damage, QC ensures the financial viability and sustainability of manufacturing enterprises. Moreover, in industries where safety and reliability are paramount, such as aerospace, automotive, and healthcare, stringent QC measures are indispensable to ensure compliance with standards and regulatory mandates. As manufacturers strive for excellence and differentiation in competitive markets, QC emerges as a strategic imperative, driving continuous improvement and innovation across manufacturing.

QC in manufacturing relies on a diverse array of data sources to ensure product quality and consistency. One primary source of data is the production line sensors that continuously monitor various parameters such as temperature, pressure, and speed. Similar to PdM, these sensors and IoT devices provide real-time data on manufacturing processes, enabling manufacturers to detect deviations from desired specifications and take corrective actions promptly. Another crucial data source for quality control is the integration of MES and supervisory control and data acquisition (SCADA) systems. MES captures detailed information about production activities, including each machine's status, production rate, and material usage. On the other hand, SCADA systems monitor and control industrial processes in real time, collecting data from various sensors and equipment. By integrating MES and SCADA systems, manufacturers gain high visibility into production operations, allowing them to analyze data trends, identify quality issues, and implement preventive measures to ensure product quality. Figure 3.5 shows a high-level quality management process using enterprise data, building an AI model, and using this AI model to classify anomalies.

Additionally, data from historical quality records, inspection reports, and customer feedback serve as valuable inputs for QC, enabling manufacturers to improve processes and enhance product quality continuously. Integrating these diverse data sources empowers manufacturers to adopt proactive QC strategies and uphold high standards throughout the manufacturing value chain. Moreover, data from customer feedback and complaints serve as valuable inputs for QC, allowing manufacturers to address customer concerns and continuously improve product quality and product changes. Integrating these diverse data sources enables manufacturers to adopt a data-driven approach to QC, fostering a culture of continuous improvement and innovation in manufacturing processes.

QC in manufacturing faces several challenges and requires adherence to best practices to ensure product quality and consistency. One significant challenge is the complexity of modern manufacturing processes, which involve numerous variables and interconnected systems. Managing and analyzing vast amounts of data generated from sensors, IoT devices, MES, and ERP systems can be daunting, leading to difficulties in identifying and addressing quality issues effectively. To overcome this challenge, manufacturers must invest in advanced analytics capabilities and data management systems that can process and analyze diverse datasets in real time. Implementing robust data visualization tools also enables manufacturers to gain actionable insights from complex data sets, facilitating informed decision making and optimizing QC measures proactively.

Another challenge in QC is maintaining consistency and accuracy across different production facilities and supply chain partners. Variations in equipment, processes, components, and personnel can lead to inconsistencies in product quality, affecting customer satisfaction and brand reputation. To mitigate this challenge, manufacturers must establish standardized data and QC protocols and procedures across all facilities and implement stringent quality assurance measures throughout the supply chain. Collaborating closely with suppliers and partners to align quality standards and monitor performance ensures consistency and adherence to quality specifications. Additionally, investing in employee training and certification programs helps build a skilled workforce capable of executing QC processes. By enhancing employees' understanding of quality standards and procedures, manufacturers can foster a culture of accountability and continuous improvement. Additionally, providing employees with access to advanced tools and technologies enables them to streamline QC processes and make data-driven decisions. By addressing these challenges and adopting best practices in QC, manufacturers can enhance product quality, reduce defects, and maintain a competitive edge in the market.

Looking at some case studies focused on QC, we start with the Ford Motor Company. Ford implemented an advanced analytics and BI quality initiative to enhance its QC processes across its manufacturing operations. Leveraging advanced analytics tools and techniques, Ford aimed to gain deeper insights into its production processes, identify potential defects or inefficiencies, and

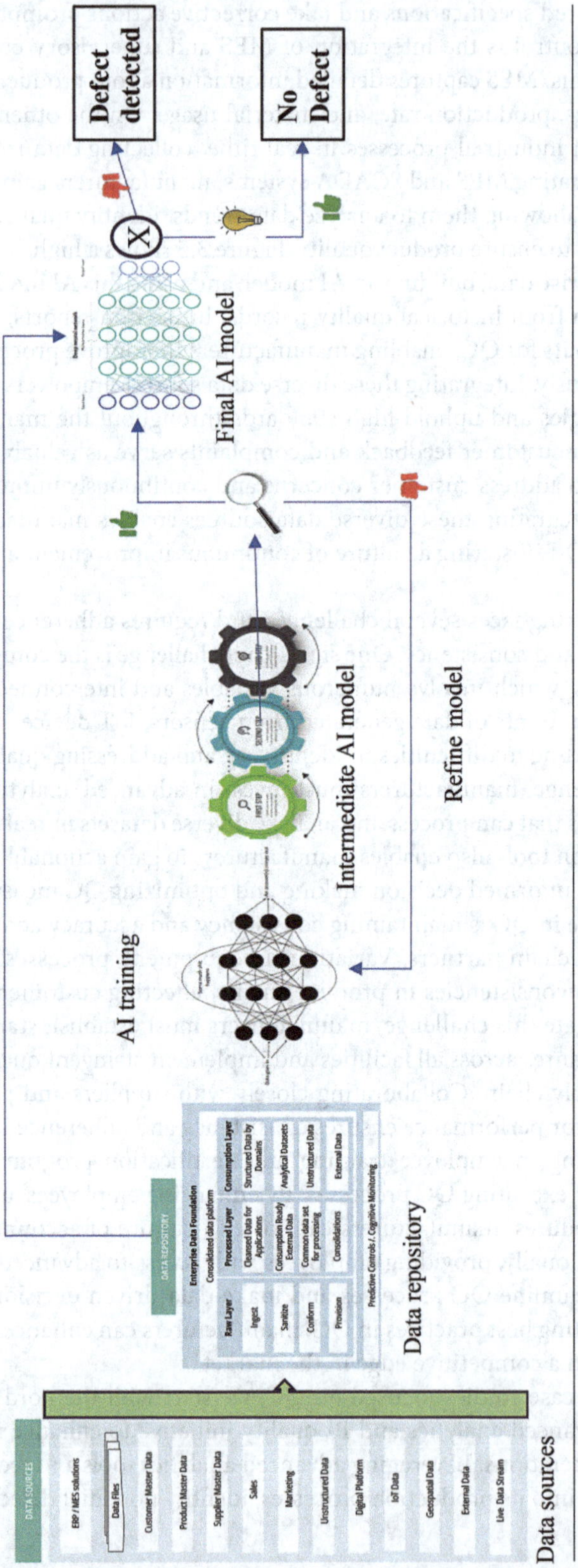

Figure 3.5 Typical quality check process flow.

optimize QC measures. Ford could monitor KPIs in real time and proactively address deviations from quality standards. Through this initiative, Ford achieved significant improvements in its QC practices, leading to enhanced product quality and operational efficiency. By leveraging advanced analytics, Ford could detect and address quality issues more swiftly, reducing the likelihood of defects reaching customers. This proactive approach not only minimized the risk of recalls and warranty claims but also contributed to higher customer satisfaction and brand loyalty. Additionally, the insights gained from advanced analytics enabled Ford to optimize its manufacturing processes, resulting in reduced production costs and improved overall profitability. Overall, Ford's investment in advanced analytics and BI for QC underscored its commitment to delivering superior products and driving continuous improvement across its manufacturing operations.

Siemens is a global leader in manufacturing technology and has implemented an advanced analytics and BI program to enhance QC across its operations. By leveraging sophisticated data analytics tools, Siemens aimed to optimize product quality, streamline manufacturing processes, and reduce defects. Through this initiative, Siemens gained valuable insights into its production workflows, allowing it to identify patterns, trends, and anomalies in manufacturing data. This proactive approach enabled Siemens to detect quality issues in real time, facilitating timely interventions to prevent defects and minimize production downtime. One notable outcome of Siemens' advanced analytics and BI quality control initiative was a significant improvement in product quality metrics. Siemens could identify factors contributing to quality variations and take corrective actions promptly. This reduced defects and rework, leading to enhanced customer satisfaction and increased operational efficiency.

Next, we move on to several chemical companies, starting with the Dow Chemical Company. Dow implemented an advanced analytics- and BI-based QC initiative using a cloud solution provider. The initiative aimed to streamline its quality processes and drive operational efficiency across its manufacturing facilities worldwide. Leveraging the analytics capabilities within the cloud solution, Dow gained real-time insights into its data, enabling proactive monitoring of quality metrics and identification of potential issues. Using predictive analytics algorithms, Dow optimized its QC processes, resulting in a notable reduction in defects and nonconformities in its product lines. For instance, by analyzing historical production data and correlating it with quality outcomes, Dow identified key factors contributing to quality deviations and implemented targeted interventions to address them. As a result, Dow experienced enhanced product quality and reliability, leading to improved customer satisfaction and operational performance.

BASF Chemicals developed a platform to initiate an advanced analytics and BI QC initiative to optimize its manufacturing processes. By implementing these analytics solutions, BASF aimed to enhance its QC mechanisms and drive operational excellence. Leveraging the solution, BASF gained real-time insights into production data, enabling proactive identification of quality issues and rapid decision making to mitigate risks. The implementation of these advanced analytics and BI solutions resulted in tangible improvements. By harnessing predictive modeling of chemical reactions and process parameters, BASF identified potential quality issues before they occurred, allowing for preemptive corrective actions. As a result, BASF witnessed a notable increase in product quality and consistency, leading to higher customer satisfaction and retention rates.

The case studies presented here highlight the commonalities and successes of implementing advanced analytics and BI initiatives for QC across various industries, including automotive and chemicals. In each case, companies such as Ford, Siemens, Dow, and BASF Chemicals leveraged sophisticated data analytics tools and platforms to optimize their QC processes and drive operational excellence.

When transitioning from QC to supply chain optimization, it becomes evident that the principles of advanced analytics and BI are equally applicable and beneficial in optimizing manufacturing processes beyond the production floor. Just as companies have used these technologies to enhance product quality and operational efficiency, they can leverage them to streamline supply chain operations by improving inventory management and enhancing demand forecasting accuracy. By harnessing data analytics, companies gain valuable insights into supply chain dynamics, identify areas for improvement, and make data-driven decisions to optimize inventory levels, reduce lead times, and enhance overall supply chain resilience and responsiveness. Thus, the transition from QC to supply chain optimization is a natural evolution in the application of advanced analytics and BI within manufacturing, demonstrating the versatility and value in driving continuous improvement across the entire value chain.

Supply Chain Optimization

Supply chain optimization has become increasingly critical for companies across various industries to meet evolving consumer demands, reduce costs, and enhance operational efficiency. Advanced analytics and BI have emerged as indispensable tools in manufacturing and have revolutionized the way companies manage their operations and optimize their supply chains. By leveraging advanced analytics algorithms in predictive modeling and prescriptive analytics, organizations can gain actionable insights from vast volumes of supply chain data—from procurement and production to distribution and after-sales service. This helps optimize inventory management, enhance demand forecasting accuracy, streamline logistics operations, and mitigate supply chain risks.

In the realm of supply chain optimization, data serves as the lifeblood of decision-making processes. In addition to the traditional sources of data, such as MES and ERP systems and sensor networks, manufacturers are increasingly tapping into a diverse array of emerging data sources to gain deeper insights into their supply chain. One such source is social media data, which provides valuable information about customer sentiment, market trends, and competitor activities. By analyzing social media feeds, manufacturers can identify emerging consumer preferences, anticipate shifts in demand, and tailor their product offerings accordingly. Similarly, web data from online forums, review sites, and industry blogs can offer valuable insights into customer feedback, product performance, and market perception, enabling manufacturers to fine-tune their strategies and improve their competitive positioning.

Another promising data source for manufacturers is geospatial data, which encompasses information about geographic locations, spatial relationships, and physical features of the environment. By integrating geospatial data (including geofencing data) with other sources, such as IoT sensors and satellite imagery, manufacturers can gain valuable insights into supply chain logistics, asset tracking, and site selection. For example, analyzing this data helps manufacturers optimize transportation routes, identify optimal locations for new facilities, and monitor environmental factors that may impact production processes. Leveraging geospatial analytics, manufacturers can enhance efficiency, reduce costs, and mitigate risks across their operations. Manufacturers also harness data sources like warehouse management systems (WMS) and transportation management systems (TMS), along with external sources such as supplier databases and market intelligence. These diverse data streams provide valuable insights into various aspects of the supply chain, from demand patterns and inventory levels to transportation routes and supplier performance.

Despite the immense potential of advanced analytics and BI in manufacturing, organizations face significant challenges in harnessing the full value of these technologies. One key hurdle is the complexity of integrating disparate data sources and systems, which leads to data silos and inconsistencies hindering effective analysis. Ensuring data quality and accuracy remains a persistent challenge as manufacturers grapple with the sheer volume and variety of data generated across their operations. Also, the shortage of skilled data scientists and analysts capable of interpreting and deriving actionable insights from complex datasets highlights the importance of investing in talent development and training initiatives.

To overcome these challenges and maximize the impact of advanced analytics and BI in supply chain optimization, manufacturers must adopt a strategic and holistic approach to data management and analytics. This includes implementing robust data governance frameworks to ensure data integrity, standardization, and security throughout the supply chain. Organizations should prioritize building a data-driven culture that promotes collaboration, transparency, and accountability across all levels of the organization. Investing in user-friendly analytics tools and platforms that empower business users to access and analyze data independently enhances the democratization of data-driven decision making. It also drives greater adoption of advanced analytics capabilities. Additionally, fostering partnerships and collaboration with external experts and technology providers can help manufacturers stay abreast of emerging trends and best practices in advanced analytics and BI. This enables them to continuously innovate and evolve their analytics capabilities to meet changing business needs.

Although there are many techniques, the most common ones are predictive modeling, optimization, and prescriptive analytics. These techniques empower organizations to extract actionable insights from supply chain data and drive continuous improvement. Forecasting demand patterns, anticipating market trends, and optimizing inventory levels to meet customer demands while minimizing excess inventory costs are all enabled by predictive analytics. Optimization algorithms are used to optimize transportation routes, warehouse layouts, and production schedules in order to minimize costs and maximize operational efficiency across the supply chain network. For example, a consumer goods company may use predictive analytics to predict seasonal and weather demand fluctuations for its products. This allows it to adjust production schedules and inventory levels accordingly to meet customer demand while minimizing excess inventory costs. Prescriptive analytics has demonstrated an immense potential for supply chain optimization. Unlike descriptive and predictive analytics, which focus on analyzing past data and predicting future outcomes, prescriptive analytics goes a step further by recommending actions to optimize decision making in real time (see Figure 3.1). For instance, an automotive manufacturer may use prescriptive analytics to determine the most cost-effective transportation routes for its raw materials, considering factors such as costs, lead times, and inventory levels. By providing actionable insights and recommendations, prescriptive analytics empowers manufacturers to make informed decisions that drive efficiency and agility across the supply chain.

We will now discuss some case studies on supply chain optimization, covering five cases across different industry segments. Vast amounts of data are available for supply chain optimization. Examples of this data include customer data, including demographics, behavior and feedback, browsing history, social media interactions, sales data, real-time market data, market trends, competitor pricing, demand versus inventory data for dynamic pricing, weather patterns, and data from manufacturing plants, distribution centers, and suppliers. That is just a sample of supply chain data available and used by the companies in these case studies.

We start off with Amazon, which, in addition to its innovative use of robotics and automation, has leveraged advanced analytics and BI to optimize its supply chain operations. Through the use of sophisticated algorithms and predictive analytics models, Amazon can anticipate and forecast customer demand with remarkable accuracy, enabling it to optimize inventory levels and minimize stockouts. This granular understanding of customer preferences ensures that popular items are readily available while avoiding overstocking less in-demand products. Furthermore, Amazon's use of advanced analytics extends to dynamic pricing optimization. Amazon uses demand versus inventory to adjust prices dynamically to maximize revenue and profitability. Through these advanced analytics-driven pricing strategies, Amazon can achieve optimal pricing decisions that balance revenue generation with customer satisfaction.

Johnson & Johnson (J&J) enhanced its ability to meet customer demand while reducing costs through advanced analytics. One key area of focus was inventory management, where advanced analytics techniques were applied to analyze data and forecast future demand more accurately. Monitoring real-time inventory levels and demand fluctuations enabled J&J to respond swiftly to market dynamics and maintain high service levels. This allowed the company to optimize inventory levels, reduce stockouts, and meet customer demand more effectively while ensuring the availability of critical medications to patients worldwide. J&J also used advanced analytics to enhance its demand forecasting capabilities, enabling it to anticipate shifts in market demand and adjust production accordingly. As a result, J&J significantly improved its supply chain performance, including higher on-time delivery rates and reduced inventory holding costs.

In the footwear and apparel industry, Nike is a prime example of a company using advanced analytics and BI. Nike learned from a speed bump in 2000 and refined its strategy, which now exemplifies the transformative impact of advanced analytics and BI on supply chain optimization. Leveraging sophisticated data analytics techniques such as predictive analytics, Nike has revolutionized its supply chain operations to achieve greater agility, efficiency, and customer-centricity. Nike can anticipate consumer demand with remarkable accuracy, enabling the company to optimize inventory levels, streamline production processes, and minimize stockouts. This proactive approach allows Nike to align its production schedules and inventory management strategies with evolving customer preferences, thereby enhancing overall operational performance and customer satisfaction. Nike's strategic adoption of BI tools, specifically interactive dashboards, has facilitated enhanced visibility and transparency across its supply chain network, providing actionable insights into key performance metrics, operational trends, and supply chain risks. Thus, Nike can rapidly respond to market dynamics, mitigate supply chain disruptions, and deliver superior value to its customers.[14]

Coca-Cola's journey toward supply chain optimization began with a comprehensive assessment of its operations and processes. Recognizing the need for greater efficiency and agility, the company embarked on a digital transformation initiative aimed at leveraging predictive analytics and BI tools. Coca-Cola gained valuable insights into demand patterns, production cycles, and distribution channels that empowered decision makers to identify bottlenecks, streamline workflows, and optimize resource allocation across the supply chain. Coca-Cola used its data-driven approach to anticipate market trends and consumer preferences accurately. It was able to develop more accurate demand forecasts at both macro and micro levels. This enabled fewer stockouts, reduced excess inventory costs, and improved overall customer satisfaction.

Finally, moving to consumer electronics, Samsung deployed predictive analytics and BI tools to optimize its supply chain operations. Samsung gained real-time visibility into its supply chain processes, identifying inefficiencies, bottlenecks, and areas for improvement. It also applied advanced optimization algorithms to optimize inventory and transportation routes, minimize transportation costs, and reduce lead times. As a result of these initiatives, Samsung achieved significant improvements in supply chain efficiency, cost savings, and customer satisfaction.

Across these five case studies on supply chain optimization, several common themes emerge. These commonalities highlight the transformative impact of advanced analytics and BI on improving operational efficiency and customer satisfaction. One such theme is the proactive use of predictive analytics to anticipate and respond to changes in demand patterns that drive supply chains. Companies like Amazon, Nike, and Coca-Cola leverage sophisticated algorithms to forecast customer demand accurately, enabling them to optimize inventory levels, streamline production processes, and minimize stockouts. This proactive approach enhances operational agility and ensures that products are readily available to meet customer needs, ultimately driving higher satisfaction levels and brand loyalty. Organizations like J&J, Nike, and Samsung gain actionable insights into key performance metrics, operational trends, and supply chain risks by harnessing the power of BI tools and analytics dashboards. Additionally, these companies emphasize the importance of data-driven decision making and real-time visibility into supply chain processes. This has helped them identify inefficiencies, mitigate risks, and make informed decisions to optimize their supply chain processes, resulting in improved efficiency, cost savings, and customer satisfaction. Overall, these case studies accentuate the critical role of advanced analytics and BI in driving continuous improvement and innovation within supply chain operations across diverse industry segments.

We now transition to *demand forecasting* since it is a key element of supply chain optimization. Companies recognize the pivotal role of advanced analytics and BI in anticipating market trends and customer preferences. Building on the insights gained from the previously discussed supply chain optimization case studies, organizations leverage predictive analytics models to forecast future demand with remarkable accuracy. By integrating data from various sources—including sales transactions, social media interactions, and market trends—companies can develop more accurate demand forecasts at both macro and micro levels. This proactive approach enables them to align production schedules, inventory levels, and distribution strategies more effectively, thereby minimizing stockouts, reducing excess inventory costs, and improving overall customer satisfaction. Further, the strategic adoption of BI tools empowers decision makers with actionable insights into demand patterns, enabling them to make informed decisions and capitalize on emerging opportunities in the market landscape. As companies continue to invest in advanced analytics and BI, the synergy between supply chain optimization and demand forecasting becomes increasingly vital in driving operational excellence and sustaining a competitive advantage in dynamic market environments.

Demand Forecasting

Demand forecasting is a critical feature of manufacturing, where the ability to accurately predict customer demand drives downstream operational efficiency and customer satisfaction. By harnessing advanced analytics—primarily predictive analytics—and BI techniques, companies leverage vast datasets to anticipate future demand patterns. For example, Apple utilizes sophisticated demand forecasting systems driven by advanced analytics and ML algorithms. By analyzing historical sales data and online browsing behavior, Apple can precisely anticipate demand for its range of products across different regions and market segments. This allows the company to optimize inventory levels and production schedules, ensuring products are available when customers need them. Similarly, Tesla utilizes advanced analytics to forecast demand for its innovative electric vehicle models. By integrating data from various sources, Tesla can anticipate shifts in consumer preferences and adjust production accordingly. This proactive approach enables Tesla to minimize inventory costs, reduce lead times, and maintain high levels of customer satisfaction.

However, despite the benefits, demand forecasting presents challenges. Ensuring the accuracy and reliability of data inputs is crucial since inaccuracies can lead to erroneous forecasts and the disruption of supply chain operations. Additionally, the dynamic nature of consumer behavior and market trends requires continuous refinement of forecasting models to remain relevant. External factors such as economic fluctuations and regulatory changes further complicate forecasting efforts, introducing uncertainties that may impact demand forecasts. By addressing these challenges, companies can unlock the full potential of demand forecasting, driving operational efficiency and strategic decision making in manufacturing. Through continuous innovation and adaptation, manufacturers can stay ahead of market dynamics, seize new opportunities, and achieve sustainable growth in an increasingly competitive landscape.

Having explored the foundational aspects of demand forecasting and its significance in optimizing supply chain operations, it is essential to delve into some real-world case studies that highlight the practical applications and benefits of advanced analytics and BI in this domain. These case studies exemplify how leading companies across various industries have leveraged data-driven approaches to enhance their demand forecasting accuracy, streamline production processes, and achieve greater operational efficiency. We also gain valuable insights into the diverse strategies and technologies employed by organizations to anticipate market demand, optimize inventory management, and drive business growth. These results have not been in a vacuum; companies have to get data to build their advanced algorithms. Data includes but is not restricted to historical sales data, market trends, trading partners data, social media platforms, and external factors like micro and macroeconomic indicators, consumer behavior, market trends, seasonality, promotional activities, and emerging market trends.

P&G has successfully leveraged advanced analytics for demand forecasting.[15] By integrating data from various sources and employing sophisticated algorithms, P&G can dynamically forecast demand and align its production schedules accordingly. The production schedules impact inventory levels and distribution strategies, ensuring product availability while minimizing excess inventory costs and stockouts. P&G continues its relentless pursuit of improving forecast accuracy, a critical metric in gauging the efficacy of its advanced analytics initiatives. By continuously refining its forecasting algorithms and leveraging real-time data insights, P&G has achieved improved forecast accuracy. This reduces the variance between actual and forecasted sales volumes. The company's data-driven approach not only enhances operational efficiency but also fosters stronger relationships with retailers and suppliers because P&G can better anticipate and meet their demands, driving mutual growth and success.

PepsiCo, Inc., a global leader in the food and beverage industry, has embraced advanced analytics and BI to revolutionize its demand forecasting strategies. Through predictive analytics models, the company can forecast demand for its diverse range of products with exceptional accuracy, allowing it to optimize inventory levels and distribution channels. By employing these sophisticated algorithms, PepsiCo dynamically adjusts its production schedules and inventory management strategies, ensuring that popular items remain readily available while minimizing excess stock. Like P&G, demand variance is one of the key metrics used to evaluate the success of PepsiCo's demand forecasting initiatives. PepsiCo has reduced its demand variance, enhancing operational efficiency and customer satisfaction. By embracing advanced analytics and BI, PepsiCo has significantly improved forecast accuracy, resulting in better inventory management and reduced carrying costs.

Nestlé S.A., another food and beverage company, has strategically employed advanced analytics and BI to revolutionize its demand forecasting capabilities. Recognizing the importance of predicting consumer preferences accurately, Nestlé has embraced cutting-edge data analytics tools to analyze vast troves of data. As a result, Nestlé can forecast demand with unprecedented precision, allowing it to adjust production schedules dynamically for optimal inventory levels and minimize stockout risk. In one notable instance, by analyzing seasonality data, such as during holidays or summer months, Nestlé was able to forecast future demand patterns. This allowed Nestlé to optimize its inventory management strategies, stocking popular items in anticipation of increased demand and preventing overstocking of less in-demand products. As a result, Nestlé achieved enhanced operational efficiency and customer satisfaction while minimizing excess inventory costs. Forecast bias, the deviation between forecasted and actual demand, is the critical metric it uses for forecasting. Nestlé's commitment to leveraging data-driven insights extends beyond demand forecasting to include dynamic pricing optimization, which it has successfully used to maximize revenue and profitability.

Next, we check out Kimberly-Clark Corporation, a leading provider of personal care and hygiene products. Kimberly-Clark is a beacon of innovation in the consumer goods industry, employing advanced analytics and BI to elevate its demand forecasting capabilities. Through the meticulous fine-tuning of its predictive analytics models and the relentless pursuit of accuracy and efficiency, Kimberly-Clark has dynamically adjusted production schedules, aligning with evolving consumer preferences, seasonal fluctuations, and promotional activities. This ensures optimal inventory and meeting customer demand with precision, enhancing overall operational efficiency and customer satisfaction. Kimberly-Clark continuously refines its predictive models using demand volatility to reduce forecast variability. This has enabled Kimberly-Clark to optimize inventory levels, minimize excess stock, and mitigate the risk of stockouts, resulting in streamlined supply chain operations.

Johnson Controls, a global leader in smart building technologies, implemented advanced analytics and BI tools to enhance its demand forecasting capabilities. Johnson Controls developed sophisticated predictive models to forecast demand for its HVAC systems and building automation solutions. It found success with the SARIMA model, which accounted for seasonal fluctuations and trends in demand. Johnson Controls achieved remarkable accuracy in predicting future demand patterns. This led to significant improvements in operational efficiency and customer satisfaction. By leveraging predictive models and sophisticated algorithms, the company gained valuable insights into market dynamics and customer preferences, allowing for proactive planning and resource allocation. It was also able to minimize stockouts, reduce excess inventory, and optimize its supply chain operations, ultimately driving business growth and competitive advantage in the smart building industry. In July 2020, Johnson Controls launched its Blue Digital Twin platform, offering customers a digital replica of physical buildings.[16] This innovative solution enabled it to harness real-time data from building systems and occupants, further enhancing its understanding of customer needs and preferences. By integrating advanced analytics capabilities into its OpenBlue platform, Johnson Controls has revolutionized the way buildings are managed and maintained, driving efficiencies and delivering superior value to its clients.

The final case study is on the Lockheed Martin Corporation. Lockheed, a leading aerospace and defense company, utilized advanced analytics and BI solutions to improve demand forecasting for its military aircraft and defense systems. Lockheed developed sophisticated predictive models to forecast demand for its products, including fighter jets, missiles, and radar systems. It has also optimized its supply chain operations in the aerospace and defense industry. Lockheed utilized advanced analytics to enhance collaboration and communication across its supply chain network. It shares demand forecasts and production plans with suppliers and subcontractors to ensure alignment throughout the supply chain for efficient resource allocation and inventory management. This collaborative approach improves responsiveness to customer demand, reduces lead times, and enhances overall supply chain resilience. A key metric for forecasting employed by

Lockheed is the weighted moving average (WMA) method, which allows for accurate predictions by assigning greater significance to recent sales data. Additionally, as a defense contractor, Lockheed incorporated war gaming, scenario analysis, and risk modeling to account for uncertainties.

In examining these case studies, a common thread emerges—the strategic utilization of advanced analytics and BI to optimize demand forecasting processes. These industry giants have demonstrated a commitment to leveraging data-driven approaches, sophisticated predictive models, and real-time insights to anticipate market demand with exceptional accuracy. By harnessing the power of predictive analytics, these companies dynamically adjust production schedules, optimize inventory levels, and align distribution strategies to meet consumer demand effectively. Notably, key performance metrics such as forecast accuracy, demand variance, and sales volume variance have shown consistent improvement, reflecting the tangible benefits of advanced analytics in demand forecasting.

Transitioning to the next part on production scheduling, it is evident that these technologies are interconnected pillars of modern manufacturing operations and optimization. While demand forecasting focuses on anticipating future consumer needs, production scheduling entails the strategic allocation of resources, optimizing production processes, and coordinating activities to meet demand efficiently. By seamlessly integrating predictive analytics and BI into production scheduling processes, companies can achieve greater agility, responsiveness, and cost-effectiveness in managing their manufacturing operations. In the following section, we will delve into the use of predictive analytics and BI in production scheduling and review some case studies that showcase how companies optimize production scheduling, minimize cycle and takt times, and enhance overall operational efficiency.

Predictive Analytics for Production Scheduling

Advanced analytics and BI play a crucial role in optimizing production scheduling within manufacturing. By leveraging historical production data, demand forecasts, market trends, and real-time insights, companies can enhance efficiency, reduce lead times, and minimize production costs.

Challenges in implementing advanced analytics and BI for production scheduling include data silos, legacy systems integration, talent availability, and the complexity of the manufacturing processes themselves. Companies often struggle to aggregate data from disparate sources and harmonize it for analysis. Ensuring data accuracy and reliability is essential for making informed production scheduling decisions. A continuing challenge is data granularity, where data comes at different time periods—seconds, minutes, hours, days, weeks, etc. The challenge is to effectively marry this data into inputs that are usable for analytics. To address this challenge, companies have invested in data integration platforms to facilitate seamless data exchange between systems and departments. Implementing data governance frameworks and quality control processes has also helped maintain data integrity and consistency. Further, adopting advanced analytics tools with predictive modeling capabilities enables companies to forecast demand more accurately and optimize production schedules accordingly. Companies are investing in training their resources to ensure they have a good pipeline of talent.

The advent of advanced analytics and BI technologies has heralded a new era in production scheduling, offering manufacturers unprecedented insights and capabilities to optimize their operations. By harnessing the power of data-driven decision making, companies can streamline production processes, improve resource allocation, and enhance overall efficiency. As we explore the case studies ahead, we witness firsthand how industry leaders leverage these technologies and data-driven approaches to overcome challenges, drive innovation, and achieve excellence in production scheduling. Companies must procure high-quality data to *feed* the advanced algorithms. Data includes but is not restricted to real-time data from manufacturing facilities and processes, production data, supply chain operations, trading partners data, market demand forecasts, industry needs, seasonality, and other external data.

Let's delve into these compelling case studies to understand how some companies have successfully leveraged advanced analytics for production scheduling. A global leader in specialty chemicals and advanced materials implemented advanced analytics and BI solutions to optimize its production scheduling processes and enhance operational efficiency. By leveraging data, the manufacturer enhanced its ability to schedule production activities efficiently. Through the deployment of predictive analytics models, it forecasted demand for various chemical products, allowing proactive adjustments to production schedules and resource allocation. A significant outcome of the implementation was the reduction in changeover times. By analyzing and optimizing production sequencing, the company has achieved an almost 30% reduction in changeover times between product runs. This improvement has enabled the company to increase throughput, reduce idle time between production runs, and enhance overall manufacturing efficiency. Additionally, it has seen about a 25% decrease in production times, allowing it to respond swiftly to shifts in demand and customer preferences.

Another leading global producer of chemicals and polymers employed advanced analytics and BI to optimize production scheduling. By harnessing data with advanced analytics, it enhanced production efficiency and reduced lead times. Through predictive analytics models, it forecasted demand for chemical products with exceptional accuracy, enabling proactive planning and resource allocation. Leveraging advanced analytics for production scheduling optimization, it reduced changeover time by almost 23% and increased throughput by nearly 20%. In one instance, it implemented advanced analytics solutions to optimize production scheduling for one of its most popular hydrocarbon products. By analyzing the data, it identified opportunities to build a new hydrocarbon product with the goal of streamlining production schedules and improving resource utilization. These initiatives reduced production cycle time by nearly 25% and improved cost savings.

A global leader in industrial technology optimized production schedules to enhance operational efficiency and meet customer demand effectively. Through the integration of sophisticated predictive models, it forecasted demand patterns and production requirements with exceptional accuracy, enabling proactive planning, resource allocation, and optimizing production lead times. It achieved a 15% reduction in changeover times across its manufacturing facilities globally. As a result, it has experienced substantial gains in production throughput and resource utilization, driving tangible improvements in operational efficiency and customer satisfaction.

A leading pharmaceutical company implemented advanced analytics and BI to enhance production scheduling in its manufacturing operations. It optimized its production schedules to ensure timely delivery of pharmaceutical products, starting with accurately forecasting demand. This approach allowed the company to streamline its production processes, minimize production lead times, and respond quickly to changes in market demand. The company implemented advanced scheduling models using multifactor features such as production capacity, equipment availability, and raw material availability, among others. By incorporating these models in its production scheduling system, it has generated optimized production schedules to maximize utilization and minimize idle time. As a result, it has achieved a more than 20% reduction in production cycle times, enhanced its ability to meet customer demand, and reduced excess inventory and associated carrying costs.

The migration toward implementing advanced analytics and BI solutions in manufacturing involves establishing a robust data infrastructure for collection, storage, and processing, alongside fostering a culture of data-driven decision making. This shift entails investing in data management platforms and cloud computing technologies to aggregate and harmonize data from various sources, followed by deploying advanced analytics techniques like predictive modeling. Training programs and cross-functional collaboration further empower employees to leverage data for process improvements. Leadership support is crucial, along with prioritizing data governance, security, and privacy to protect sensitive information. Establishing performance metrics ensures the effectiveness of analytics initiatives, driving sustainable value creation across the manufacturing value chain.

IMPLEMENTING ADVANCED ANALYTICS SOLUTIONS

Implementing advanced analytics and BI in manufacturing represents a transformative journey toward data-driven decision making and operational excellence. In today's competitive landscape, manufacturers increasingly leverage advanced analytics technologies to unlock valuable insights from their data, optimize processes, and drive innovation. However, embarking on this journey requires careful planning, strategic alignment, and effective execution to realize the full potential of analytics initiatives. Implementing advanced analytics and BI solutions mandates the recognition that data is a strategic asset. Vast amounts of data generated from various sources such as sensors, machines, production systems, supply chain operations, and external sources, inundates manufacturers. By harnessing this data effectively, manufacturers can gain actionable insights into their operations, identify trends, and make informed decisions to improve efficiency, quality, and productivity. However, the volume and complexity of data pose significant challenges and need robust data integration, processing, and analysis capabilities.

> *"Data is the soul of business transformation."*
>
> —Anand Mahindra, Chairman of Mahindra Group

The implementation process begins with defining clear objectives and aligning analytics initiatives with business goals. This involves engaging stakeholders across different departments to understand their requirements, challenges, and priorities. By establishing a shared vision for analytics-driven transformation, organizations can ensure that analytics solutions are tailored to address specific business needs—not just current but also future—and deliver tangible value. Setting realistic expectations and milestones is crucial to managing stakeholders' expectations and garnering support throughout the implementation journey. Another critical aspect of implementing advanced analytics and BI solutions is establishing a solid technological foundation. This includes selecting the right tools, platforms, and infrastructure to support integration, storage, processing, and analysis. Organizations must evaluate different analytics software, ML frameworks, and cloud-based solutions to find the most suitable options for their unique requirements. Additionally, investing in scalable and flexible infrastructure like the cloud is essential to accommodate future growth and evolving analytics needs. By laying the groundwork for advanced analytics capabilities, organizations can unlock the full potential of their data to drive innovation and competitive advantage in the manufacturing industry. Figure 3.6 shows a typical integration

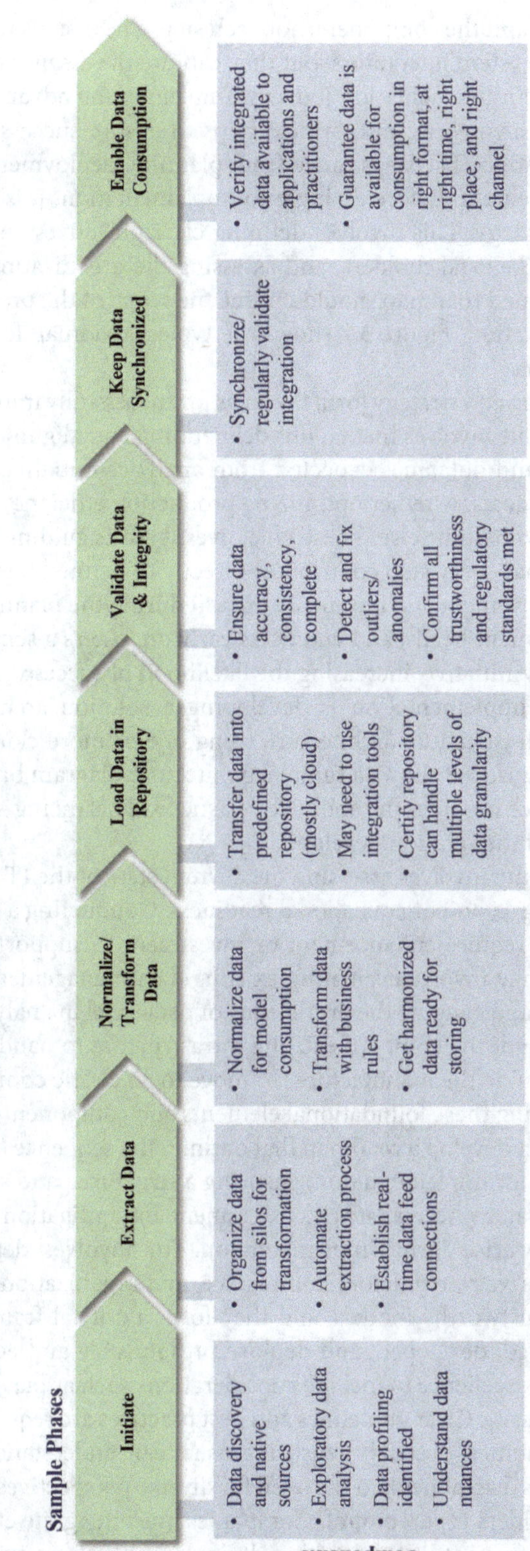

Figure 3.6 Typical data integration process.

process with seven key steps and the components for each step. These steps may differ from company to company, including system integrators, but they capture the essence of integration.

Let us dig deeper into the implementation journey. Implementing advanced analytics and BI solutions in manufacturing involves several critical steps to ensure successful deployment and adoption across the organization. The steps range from planning deployment and integration to OCM to continuous improvement. Before diving into implementation, it is crucial to develop a comprehensive plan and strategy. This involves defining clear objectives and goals for the analytics initiatives, identifying key stakeholders, and assessing the organization's readiness for advanced analytics. A well-defined roadmap should outline the scope of the project, key milestones, and timelines for implementation. Figure 3.7 shows the typical roadmap for the deployment of data transformation solutions.

In manufacturing, planning and strategy form the basis for successfully implementing advanced analytics and BI solutions. This involves meticulous deliberation and alignment of organizational objectives with capabilities and outcomes expected from analytics initiatives. It is imperative to define clear and measurable goals, whether optimizing production efficiency, reducing costs, improving product quality, or something else. These objectives serve as guiding beacons throughout implementation, ensuring that every step contributes directly to defined business priorities. Additionally, stakeholders' involvement and buy-in are critical during the planning phase. Engaging key decision makers, department heads, and end users early on fosters a sense of ownership and commitment to the analytics initiative, increasing the likelihood of success.

One key component in implementation is developing a solution architecture. Nowadays, companies need to build infrastructure architecture using one or more cloud providers—AWS, Google, Azure, and IBM. Figure 3.8 shows a sample architecture diagram based on AWS. As the figure shows, the cloud service provides the ability for elastic load balancing and other AWS components to meet the needs of the solution developed.

Moreover, strategic planning involves assessing the current state of the IT and data ecosystem, analytics capabilities, talent gap, and organizational readiness. Conducting a thorough gap analysis helps identify areas that require enhancement or investment to support advanced analytics initiatives effectively. This may involve evaluating existing data management systems, assessing data quality and availability, and gauging the proficiency of personnel in analytics tools and methodologies. It may also highlight the ability of existing infrastructure to handle future computing requirements and possibly guide the manufacturer to move to an elastic computing environment on the cloud. By understanding these foundational elements and components to be deployed (see Figure 3.9), organizations can develop a roadmap that outlines the sequence of activities, resource requirements, and timelines for implementation, ensuring a structured and systematic approach.

Effective planning and strategy necessitate risk assessment and mitigation strategies to address potential challenges that may arise during implementation. This involves identifying risks associated with technology, data privacy, regulatory compliance, and organizational change. There are real risks associated with trustworthy models, and they form a critical framework for ensuring that technologies are designed, developed, and deployed responsibly and ethically. Trustworthy models go beyond technical excellence to include considerations such as bias mitigation, data privacy, accountability, and security. Clear guidelines and best practices are required to address issues such as explainability and human oversight, ensuring users can understand and monitor these models. Collaboration across disciplines and the need for diverse perspectives to create these solutions that benefit all stakeholders is paramount. Companies must integrate ethical considerations throughout the model development life cycle, from design and development to deployment and

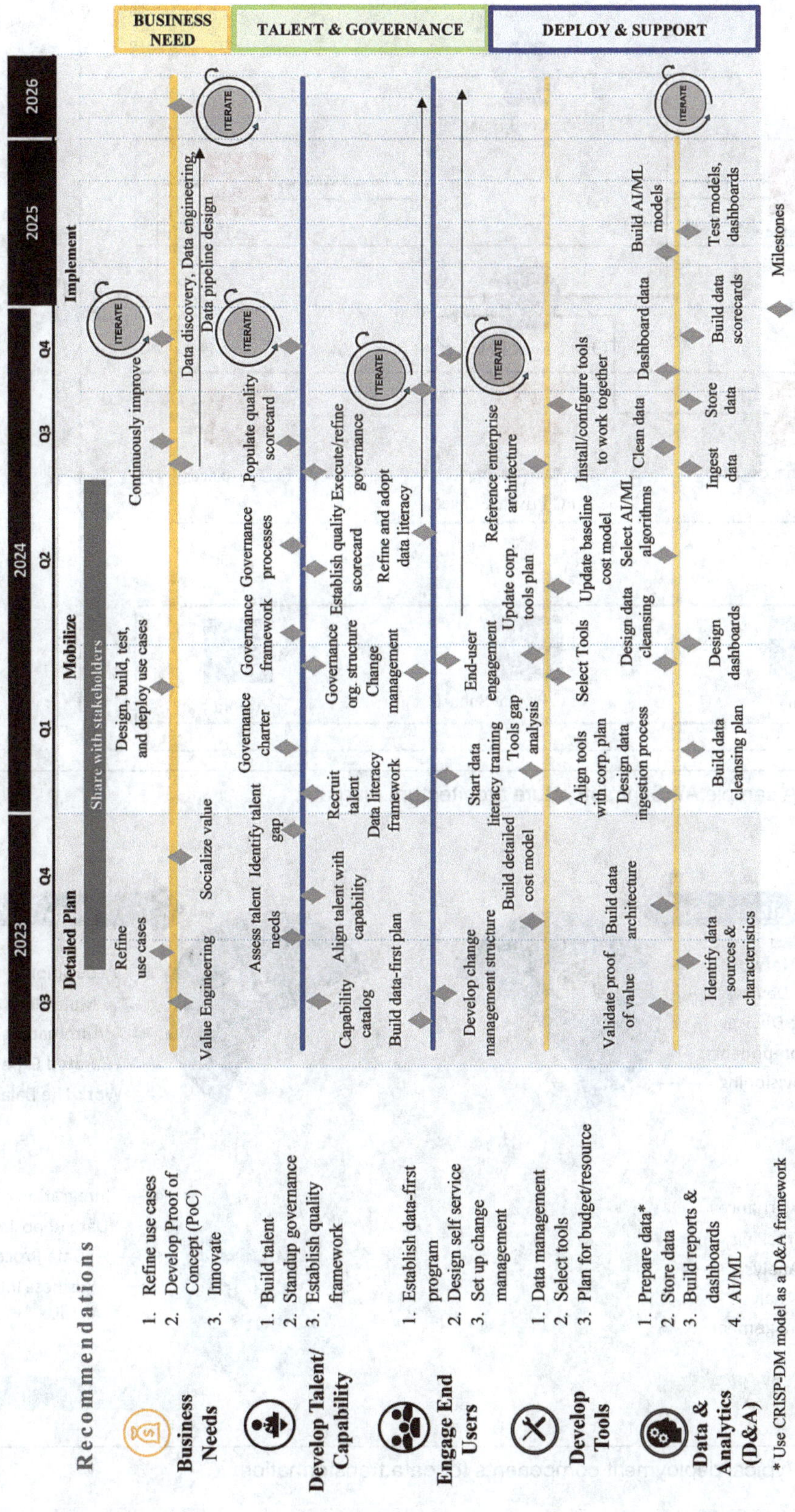

Figure 3.7 Sample high-level roadmap to deploy data transformation solutions.

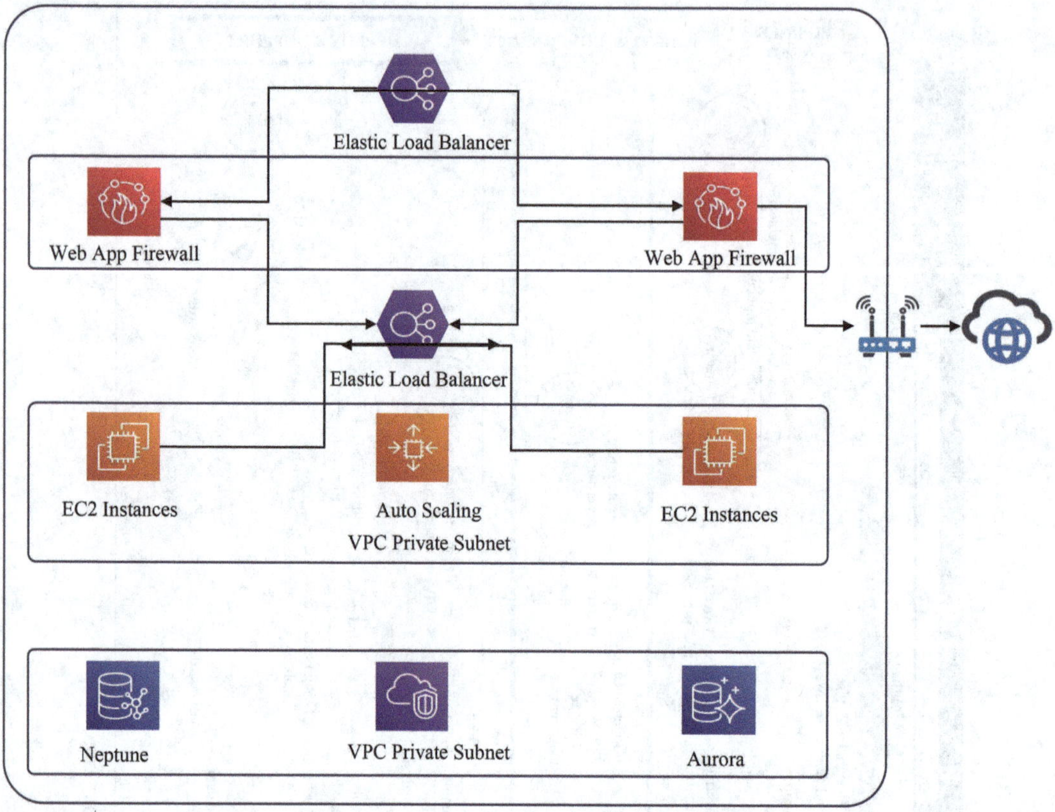

Figure 3.8 A sample AWS infrastructure architecture diagram.

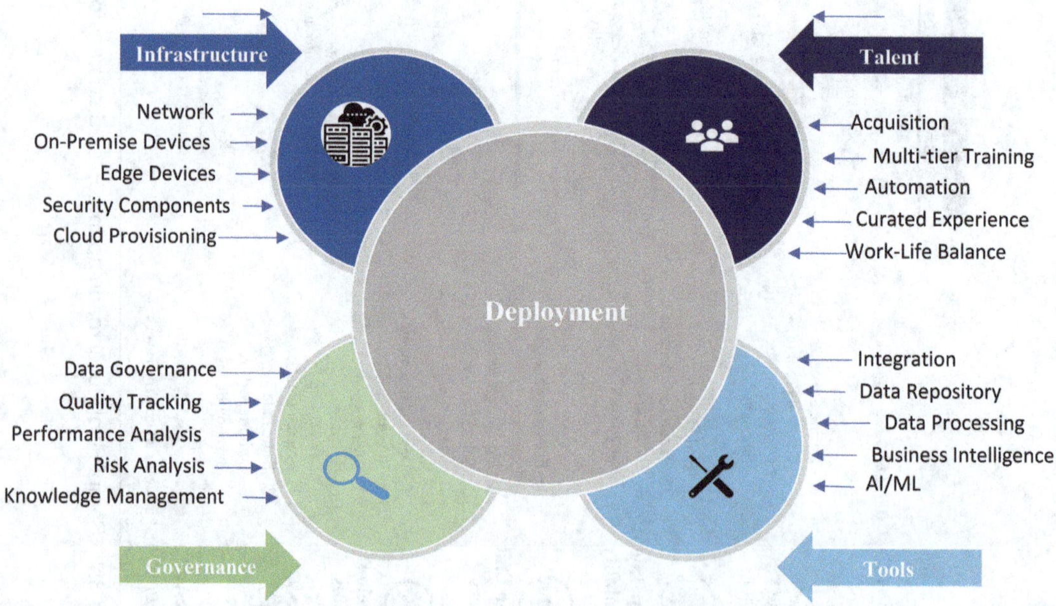

Figure 3.9 Typical deployment components for data transformation.

maintenance. Ultimately, trustworthy models foster public confidence and support sustainable solution adoption by balancing innovation with ethical and societal concerns.[17] By proactively identifying and mitigating risks, organizations can minimize disruptions and ensure the smooth execution of analytics initiatives. Establishing clear communication channels and governance structures helps maintain transparency, accountability, and alignment with strategic objectives throughout the implementation process.

Once the strategy and planning are complete, we move to data integration. Data integration is crucial to the success of advanced analytics programs. Implementing advanced analytics and BI solutions in manufacturing is a multidimensional process crucial for deriving actionable insights from disparate data sources. At its core, data integration involves collecting, aggregating, and harmonizing data from various sources like production systems, supply chain databases, sensor networks, and market research platforms. This harmonization of data ensures a comprehensive view of the manufacturing ecosystem, facilitating more informed decision making and predictive analysis. For instance, in a manufacturing facility, data integration may involve extracting real-time data from equipment sensors to monitor performance metrics such as throughput, downtime, and energy consumption. Correlating this with historical production data can be used to analyze trends and predictive maintenance. Further, data integration invariably extends beyond internal systems. It must encompass external data—such as market trends, weather forecasts, geopolitical events, and changing federal regulations—to provide additional context for analysis and forecasting. Integrating market demand data enables manufacturers to adjust production schedules using consumer preferences and seasonal trends, optimizing inventory levels and minimizing stockouts. The integration of supply chain data enables manufacturers to track raw material availability, shipping schedules, and supplier performance, enhancing visibility and resilience (refer to Figure 3.6 for an example of a typical integration pipeline).

Effective data integration is essential for ensuring data quality, consistency, and accessibility, laying the foundation for successful advanced analytics and BI initiatives. Manufacturers must invest in robust integration capabilities and leverage modern technologies. Some technologies include ETL or ELT tools, data lakes, and APIs. Manufacturers must overcome data silos and enable seamless data flow across the organization. Moreover, manufacturers must implement data governance policies and data management best practices. This safeguards data quality and integrity throughout integration. Overall, data integration is a critical enabler of data-driven decision making in manufacturing, empowering organizations to harness the full potential of their data assets for strategic insights and operational excellence.

Selecting the right advanced analytics and BI tools is critical to the success of these initiatives. The process of tool selection involves careful consideration of several factors. These include the specific needs and objectives of the organization, the complexity of manufacturing, the availability of data, and the expertise of the users. One crucial aspect is the scalability of tools, ensuring they can accommodate the growing user base, volume, and complexity of data as the organization expands. Additionally, compatibility with existing systems and infrastructure is essential to facilitate seamless integration and minimize disruptions during implementation. The versatility of tools plays a significant role in their selection. Manufacturing often requires a diverse range of analytics capabilities, from descriptive and diagnostic analytics to more advanced predictive and prescriptive analytics. As described earlier in this chapter, there are several advanced analytics techniques. Selecting the best among these for your needs is critical to your success. Further, tools with diverse functionalities like data visualization, statistical analysis, ML, and optimization enable organizations to tackle business challenges and derive valuable insights from their data.

In addition, user-friendliness, the learning curve, ease of customization, and vendor support are essential to ensure the tools meet the needs and preferences of users. This ultimately drives user adoption and maximization of the return on investment.

Several leading vendors offer advanced analytics and BI tools tailored specifically for the manufacturing industry. These provide powerful capabilities to analyze data, optimize processes, and make data-driven decisions. Examples include SAP Analytics Cloud, Tableau, Microsoft Power BI, IBM Cognos Analytics, and SAS Visual Analytics. Each of these tools offers unique features and functionalities, catering to different use cases and business requirements. There are a few programming languages and their libraries that allow manufacturers to build their own custom solutions and visualizations. Some of these include Python, R, and Jupyter Notebooks. Open-source tools like KNIME and Orange are platforms for building data analytics using visual design and programming. By carefully evaluating these tools based on criteria such as functionality, scalability, compatibility, and user experience, manufacturers can select the most suitable solution to empower their analytics initiatives and drive operational excellence. Figure 3.10 illustrates the tool selection matrix with a description of what each quadrant represents.

Setting up infrastructure is another crucial step in advanced analytics and BI initiatives. It involves establishing a robust technological foundation capable of handling large volumes of data and supporting sophisticated analytical processes. This infrastructure must be a scalable and reliable data storage system, often implemented using cloud-based solutions or sometimes on-premises data warehouses. These systems are designed to store structured and unstructured data from various sources. By centralizing data storage, manufacturers can ensure data accessibility, integrity, and security while facilitating seamless data integration for analytics purposes. As manufacturers are in the selection process, they should also conduct an impact assessment across key components that will be impacted by the transformation. Figure 3.11 is an illustrative impact assessment across benefits and cost criteria, each with multiple elements.

In addition to data storage, infrastructure setup entails implementing powerful computing resources for data processing and analysis. High-performance—typically GPU-based—computing clusters equipped with advanced processing units and memory handle complex analytical workloads efficiently. These computing resources enable manufacturers to perform tasks such as predictive modeling, simulation, and optimization on large datasets in a timely manner. Leveraging the elastic nature of cloud computing services offers the flexibility to scale computing resources based on demand. This guarantees optimal performance and cost-effectiveness for analytics initiatives. The infrastructure setup may involve deploying specialized software tools and platforms tailored to the industry's unique requirements. These tools encompass a wide range of functionalities like data visualization, real-time monitoring, and integration with esoteric equipment and systems. These tools allow manufacturers to extract actionable insights from their data for informed decision making. A typical data management and processing solution is shown in Figure 3.12, which shows the sequential steps of identifying data sources, data acquisition, integration, analysis, BI, and decision making.

Now that tools are selected and an infrastructure is in place with your data, we move to model development and validation. These play a pivotal role in the success of advanced analytics and BI initiatives. In this phase, data scientists and analysts leverage various statistical and advanced analytics techniques to develop predictive models for specific business objectives. Initially, exploratory data analysis (EDA) and data profiling are conducted to gain insights into the underlying data patterns and relationships. Subsequently, appropriate algorithms are selected based on factors such as data type, volume, and complexity. For instance, in predictive maintenance applications, algorithms—such as random forest or gradient boosting—may be selected to forecast equipment

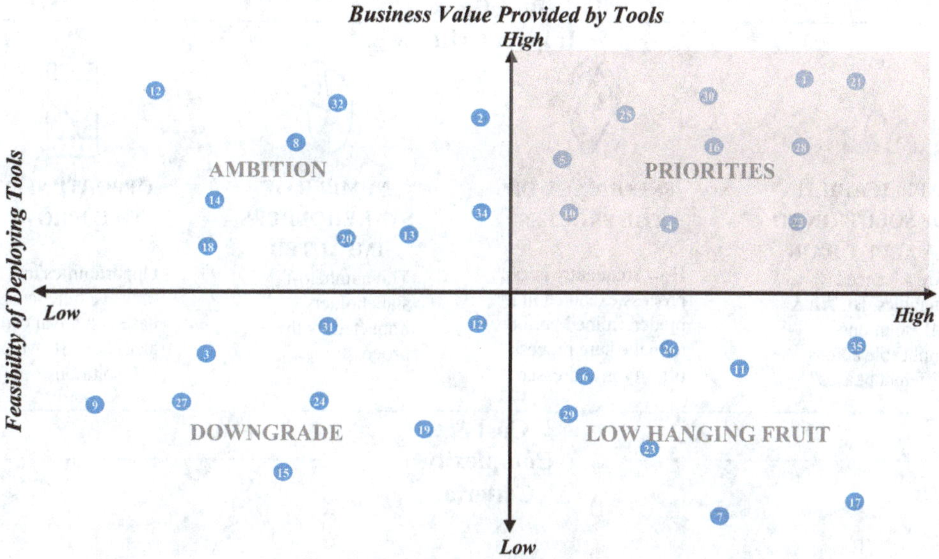

Quadrant Description

Priorities	Low Hanging Fruit
Tools providing highest value and risk-to-reward profile. Focus on further design-specific deep-dives & to support roadmap development	Tools that can quickly deliver value and set framework to move from PoC to production

Ambition	Downgrade
Tools that the organization should think about in the longer term but do not need the capability now	Tools that should not be pursued due to high implementation complexity, low return, and not on architectural roadmap

Figure 3.10 Illustrative tool selection matrix.

failures based on historical sensor data and maintenance logs. Once the models are developed, they undergo rigorous validation to assess their accuracy, robustness, and generalizability.

Validation of predictive models involves several steps to ensure their reliability and effectiveness in real-world scenarios. One common approach is to split the dataset into training and testing (typically 70% to 30%) subsets. The training set is used to train the model, and the testing set is used to evaluate its performance. Additionally, techniques such as k-fold cross-validation are employed to assess the model's stability and prevent overfitting. Figure 3.13 shows the train-test split mechanism along with the k-fold cross-validation (with $k = 5$). In manufacturing, validation may include simulating different production scenarios or pilot tests in controlled environments to validate the model's predictions against actual outcomes.

Stakeholders and domain experts are heavily involved with validation and provide feedback and domain-specific insights. This ensures that models align with business requirements and operational constraints. To ensure these algorithms are trustworthy, manufacturers need to implement robust data governance practices, including data quality control, bias mitigation, privacy protection, and transparency in how decisions are made. Collaboration with internal or external experts

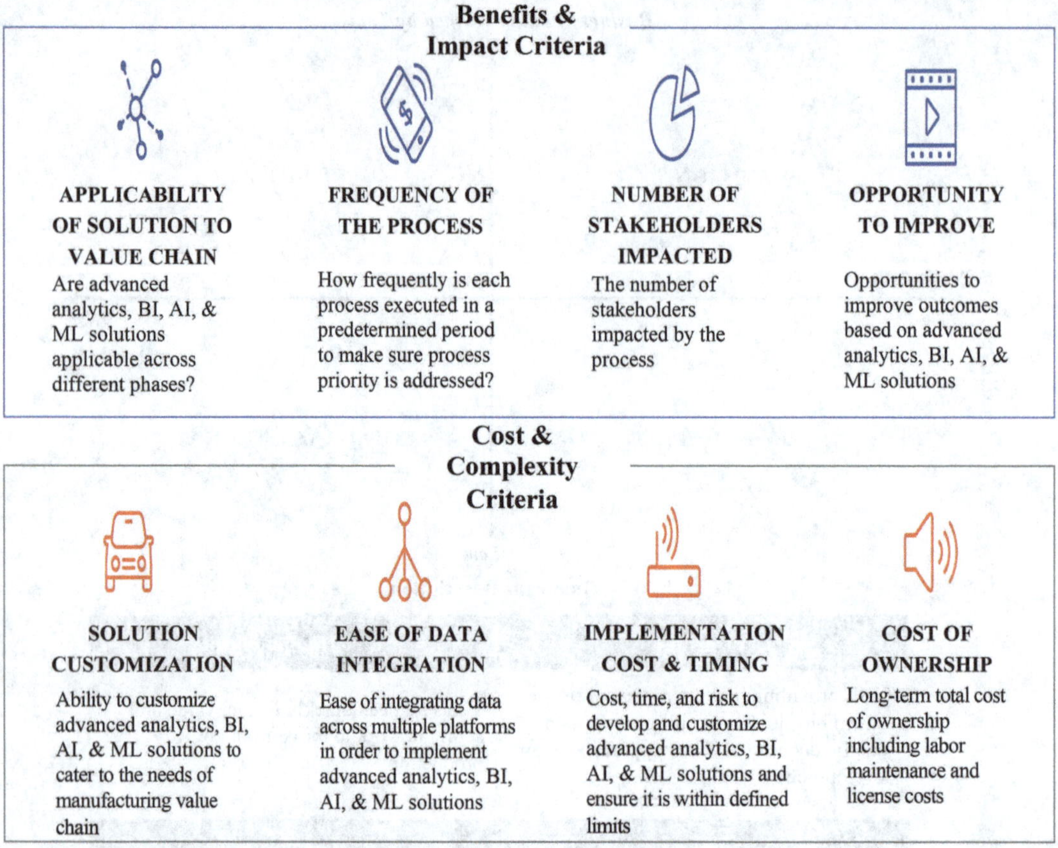

Figure 3.11 Illustrative organization impact assessment matrix dimensions.

in AI ethics and governance can also help manufacturers align their advanced analytics efforts with ethical standards and regulatory requirements. Figure 3.14 depicts the interaction of stakeholders in the validation process. As shown in this diagram, the first five steps establish the process, while the last two show the actual interaction between stakeholders and the solution team.

Further, ongoing validation and monitoring are essential to maintain the effectiveness of predictive models over time. As new data becomes available and business conditions change, models tend to drift and may need to be rebuilt, retrained, or recalibrated to ensure their relevance and accuracy. Continuous monitoring of KPIs, such as prediction error rates or model drift, helps identify potential issues and opportunities for improvement. Feedback loops are established to gather end-user insights to incorporate within the iterative model refinement processes. By following a systematic approach to model development (see Figure 3.15) and validation, manufacturers leverage advanced analytics and BI solutions to drive actionable insights, optimize operations, and achieve sustainable competitive advantage. The CRISP-DM framework (a common framework used in the industry) refers to the cross-industry standard process for data mining (CRISP-DM) made popular by IBM's SPSS Modeler.[18]

Deployment and integration are critical phases in implementing advanced analytics and BI initiatives in manufacturing. During deployment, validated models and algorithms are rolled out into production environments. This process involves configuring the analytics solutions to interface

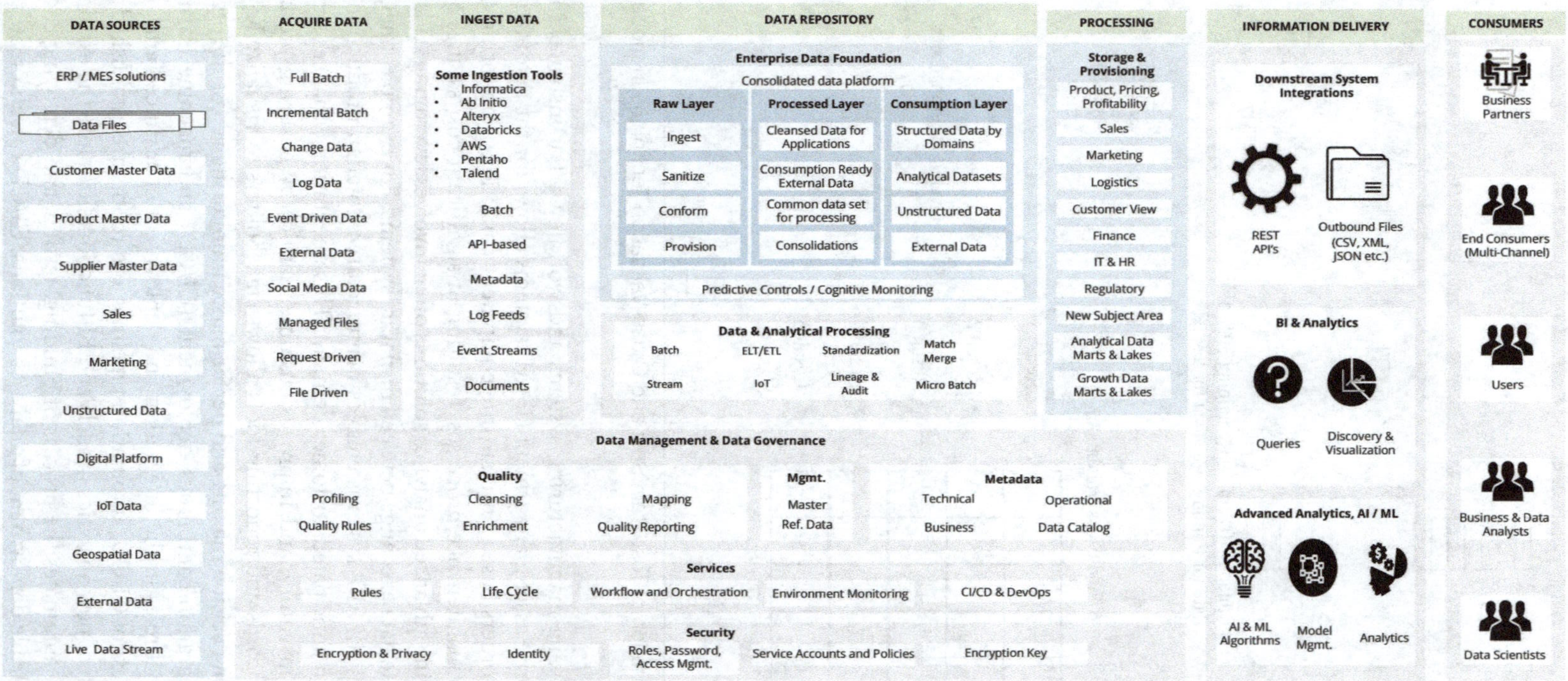

Figure 3.12 Data management and processing solution.

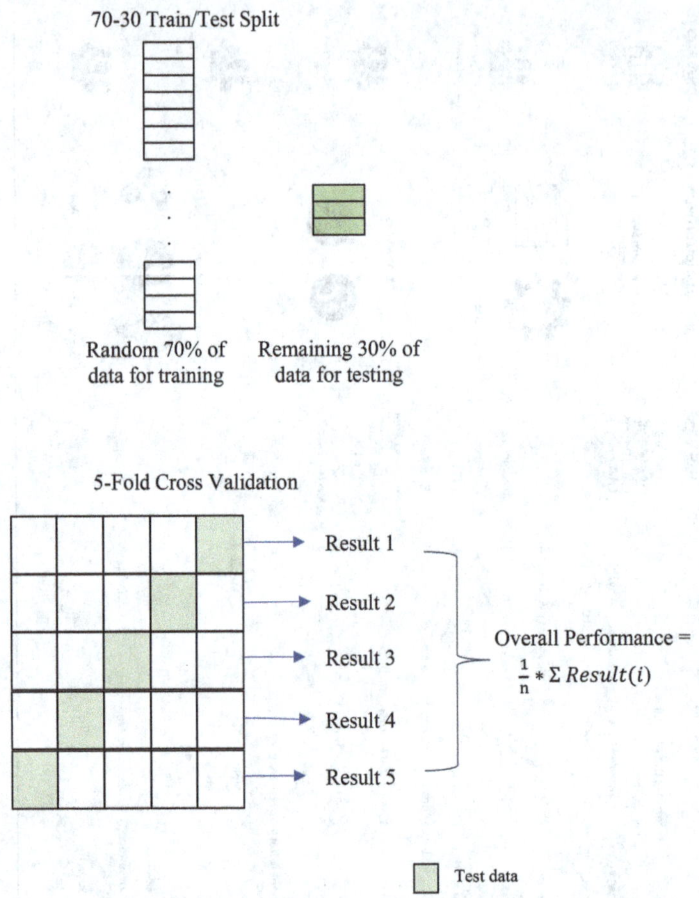

Figure 3.13 Visualization of Test/Train Split and Cross-Validation Technique in model development.

seamlessly with existing manufacturing systems and workflows. For instance, integrating PdM models with enterprise asset management (EAM) systems allows for the automatic generation of maintenance work orders based on real-time equipment condition-monitoring data. This integration streamlines maintenance processes, resulting in significant cost savings and operational efficiencies. Deploying advanced analytics solutions often involves leveraging APIs or middleware platforms to facilitate data exchange between disparate systems. For example, integrating demand forecasting models with ERP systems enables manufacturers to align production schedules and inventory levels with anticipated customer demand. This integration enhances supply chain agility, reduces excess inventory, and improves customer service by ensuring product availability. Deploying BI dashboards and reporting tools allows stakeholders to visualize KPIs in real time, enabling data-driven decision making and performance monitoring across the organization.

Continuous integration/continuous deployment (CI/CD) and machine learning operations (MLOps) play crucial roles in the integration and deployment of advanced analytics and BI initiatives. In the context of CI/CD, automation pipelines and robots are established to facilitate the seamless integration of new analytics models and BI functionalities into the existing manufacturing systems. CI/CD automates the process of building, testing, and deploying code changes, ensuring rapid and reliable deployment of analytics solutions. CI/CD can automate the deployment

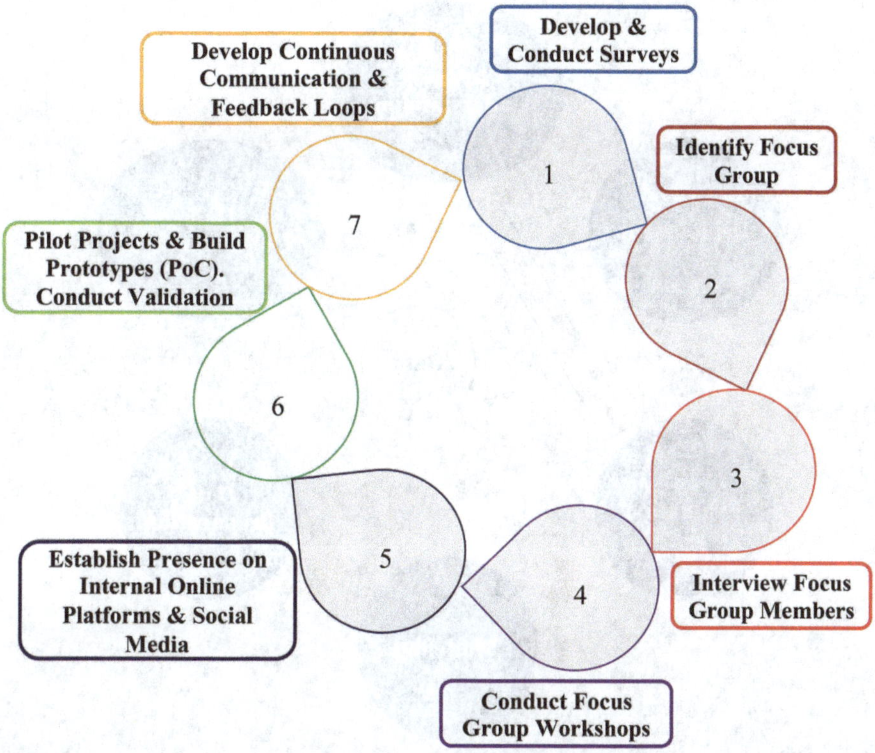

Figure 3.14 Typical interaction between stakeholders during validation.

of updated models or BI dashboards, enabling manufacturers to roll out improvements efficiently with minimal disruption to production operations.

MLOps, on the other hand, focus specifically on the operationalization and management of models within the production environments. MLOps practices involve versioning, monitoring, and managing the life cycle of models to ensure their reliability, scalability, and performance in production settings. This includes establishing model governance processes, implementing model versioning mechanisms, and deploying model monitoring tools to track model drift and performance degradation over time. By incorporating MLOps practices into the deployment and integration phases, manufacturers can ensure that models operate effectively in real-world production environments, delivering tangible business value and insights. Remember that CI/CD and MLOps are integral components of deploying and integrating advanced analytics and BI initiatives. They enable manufacturers to automate and streamline deployment, accelerate time-to-market for analytics solutions, and ensure the reliability and scalability of models in production. By adopting CI/CD and MLOps practices, manufacturers achieve greater agility, efficiency, and innovation to leverage advanced analytics and BI to drive business outcomes.

Just because a company has CI/CD and MLOps does not mean it can ignore the continuous improvement that encompasses its processes. A traditional continuous cycle is required to ensure improvements are well-planned, validated, and rolled out. One version of continuous improvement follows the plan-do-check-act (PDCA) cycle as shown in Figure 3.16.

Prior to the development phase, at kick-off, manufacturers must conduct organizational change management as it plays a pivotal role in the success of the implementation of advanced analytics

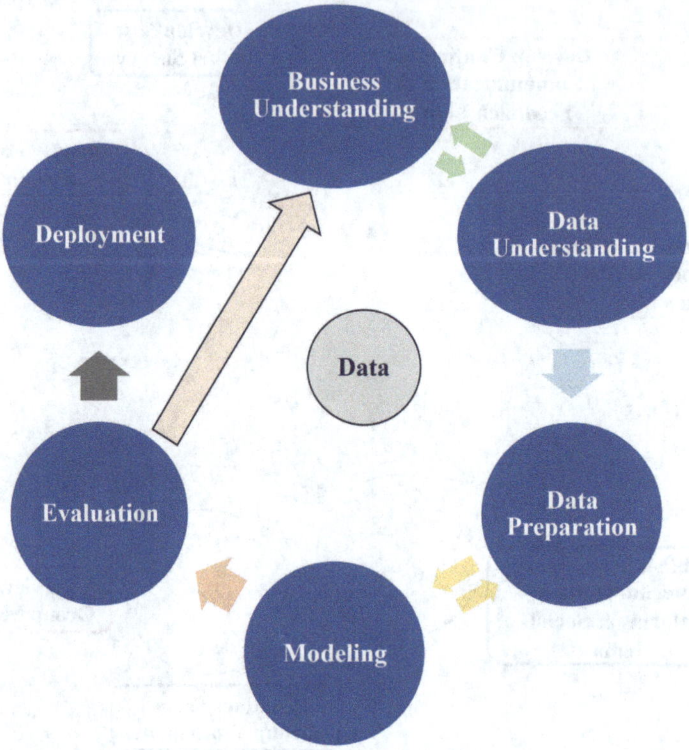

Figure 3.15 Typical model development life-cycle framework called CRISP-DM.

and BI initiatives. First, effective communication is essential to foster understanding and buy-in across all levels of the organization. This involves clearly articulating the benefits of analytics adoption, addressing concerns, and highlighting how it aligns with strategic objectives. Regular communication via multiple channels—including intranet, town hall meetings, training sessions, and interactive workshops—facilitates dialogue and engagement, ensuring employees feel empowered and motivated to embrace the changes. Additionally, establishing a dedicated change management team or steering committee provides leadership and guidance throughout the transition, driving accountability and alignment with business goals.

Next, comprehensive training and upskill programs are critical to equip employees with the necessary knowledge and skills to leverage advanced analytics tools effectively. This includes technical training on analytics platforms and software, as well as education on data interpretation, visualization techniques, and decision-making frameworks. By investing in training programs tailored to different roles and proficiency levels, organizations can build a culture of data literacy and empower employees to contribute meaningfully to analytics-driven initiatives. Further, this creates a collaborative environment and fosters a culture where cross-functional teams that work together on analytics projects can facilitate knowledge sharing and promote a culture of innovation and continuous learning.

OCM involves nurturing a supportive and inclusive culture that values experimentation (whether successful or not), agility, and adaptability. Encouraging feedback and soliciting input from employees throughout the implementation process helps identify pain points, address resistance, and drive continuous improvement. Recognizing and rewarding individuals and teams for their

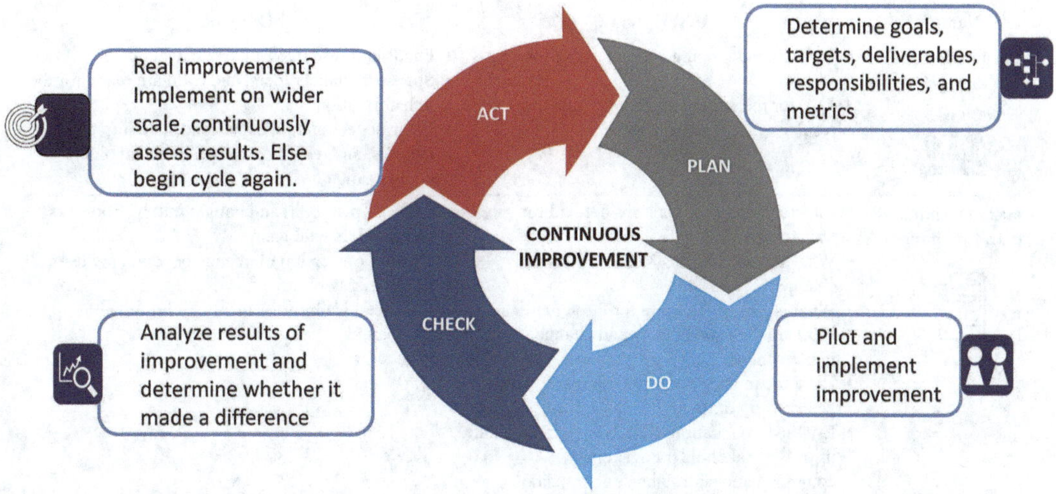

Figure 3.16 Continuous improvement cycle.

contributions to analytics projects reinforces desired behaviors and incentivizes engagement. By embedding a culture of data-driven decision making and innovation, organizations realize the full potential of advanced analytics and BI, which will cultivate a competitive advantage in today's rapidly evolving manufacturing landscape.[19]

Figure 3.17 captures the key transformation change components. Preparing for change is critical to change management; if this step is missed, there will be issues downstream. Following this chain of events is the key step in developing a vision for the change and the corresponding plan that will be executed. Once the plan is executed, it has to be embedded into the manufacturer's standard operating procedure, and steps need to be put in place to ensure practitioners do not revert to the old processes. Evaluating the outcome of the change is necessary to demonstrate the performance of the change to the stakeholders.

At this point, we have arrived at the critical topic of continuous improvement. This ensures that the advanced analytics solutions deliver ongoing value and impact to the organization. Continuous improvement is vital in sustaining the effectiveness and relevance of advanced analytics and BI initiatives in manufacturing. It involves ongoing monitoring of KPIs to assess the impact of implemented solutions on business objectives. Metrics such as forecast accuracy, production efficiency, and inventory turnover rates serve as benchmarks for evaluating the performance of analytics models and identifying areas for enhancement. Regular review and analysis of KPIs enable organizations to detect inefficiencies, bottlenecks, or emerging trends, prompting timely adjustments to optimization strategies. Continuous improvement further entails leveraging feedback loops and iteration cycles to continually refine predictive models and algorithms. By gathering feedback from end users, stakeholders, and operational teams (similar to what is shown in Figure 3.16), organizations can gain valuable insights into the effectiveness and usability of analytics solutions. This feedback helps with iterative enhancements to algorithms, data models, and visualization techniques, aiming to address evolving business needs to improve overall decision-making capabilities. Also, advanced analytics platforms often incorporate features to automatically retrain and refine models, allowing organizations to adapt to changing data patterns and maintain model relevance over time.

Name	Description	Outcomes
Prepare for the change	• Define adoption and outcome-based business metrics • Plan for messaging and communicating results	• Funds and resources are made available • Sponsors, stakeholders, and team are ready for the transformation • Clear understanding of what is changing, who it impacts, and what is expected of stakeholders to adopt change
Develop vision and plan for change	• Ensure vision and goals are defined for transformation • What success metrics/KPIs will be new, changed • What is current baseline for these metrics • Who are the stakeholders and change implementors • Have the in-scope/out-of-scope, cost, risk been clearly defined • Establish foundation to be used to inform and prioritize stakeholder engagement, communications, training, and transition activities	• SMART plan built and bought into by Sponsors, stakeholders, and team • Broader ownership of change beyond core team • Reduced risk of adoption • Enhanced visibility & increased buy-in and support for change
Execute change	• Execute against plan to implement changes • Celebrate any and all wins • Conduct post-mortem on any failures • Ensure roadblocks are anticipated and prevented, removed, and mitigated • Periodic communication with the sponsors, stakeholders, and the team	• Ensure deliverables are clearly defined and accepted • Develop solution and conduct successful unit test, integration test, and user acceptance test • Get buy-in from sponsors, stakeholders, and team on completion • Complete documentation as required
Embed new processes in SOP	• Ensure practitioners do not revert back to old ways for workflows, culture, strategy, etc. • Pay extra attention to transient period • Establish SOPs and remove as much of older processes as possible	• Rapid and increased knowledge transfer • Increased adoption rates • Change metrics being tracked • Sunset old processes
Impact of change	• Assess the degree of change as a result of transition in people, process/structure, or technology by stakeholder group • Measure, document, and communicate results against the baseline • Define periodic metrics evaluation	• Change metrics being tracked • Metrics trending in the right direction • Root cause analysis conducted for metrics trending incorrectly • Adoption rates continue to increase

Figure 3.17 Sample key transformational change components.

Fostering a culture of continuous innovation and learning is essential to drive improvements in advanced analytics and BI initiatives. Encouraging cross-functional collaboration, knowledge sharing, and experimentation empowers teams to explore new analytical techniques, tools, and methodologies. Organizations can establish dedicated internal forums, such as data-science communities or innovation labs, to facilitate idea exchange and exploration of emerging technologies. By fostering a culture of continuous learning and innovation, manufacturers can stay ahead of industry trends, drive meaningful advancements in analytics capabilities, and ultimately achieve sustainable competitive advantage.[20]

This section covered the implementation of advanced analytics solutions in manufacturing, which is a multifaceted process that demands careful planning, strategic alignment with business objectives, and robust execution. By following the previously outlined steps—including data integration, tool selection, infrastructure setup, OCM, and continuous improvement—organizations can unlock the full potential of their data assets to drive operational excellence, enhance decision-making capabilities, and gain a competitive edge in the market. However, as organizations harness the power of advanced analytics and BI, it is essential to also prioritize data security and governance to safeguard sensitive information and ensure regulatory compliance. Therefore, the next section will delve into this critical aspect in the context of advanced analytics initiatives, exploring strategies for mitigating risks, ensuring data integrity, and fostering trust in data-driven decision-making processes.

DATA SECURITY AND GOVERNANCE

In today's digital era, data has become one of the most valuable assets for organizations across all industries, including manufacturing. As companies increasingly rely on data-driven insights to drive business decisions and enhance operational efficiency, the importance of safeguarding this data against cyber threats and IP theft and ensuring its integrity and compliance with regulatory requirements cannot be overstated. This is where data security and governance come into play. Data security refers to the practices, technologies, and measures implemented to protect sensitive information from unauthorized access, disclosure, alteration, or destruction. In manufacturing, vast amounts of proprietary data, IP, and trading partner information are generated and stored. Maintaining robust data security measures is essential to prevent data breaches, mitigate risks, and safeguard the organization's reputation.

Data governance focuses on establishing policies, procedures, and frameworks to ensure that data is managed, utilized, and protected effectively throughout its life cycle. It encompasses aspects such as data quality, integrity, and access controls, along with compliance with regulatory mandates. By implementing sound data governance practices, manufacturing companies can enhance data visibility, ensure data accuracy, and align data management activities with business objectives. Effective data security and governance are imperative for protecting sensitive information and critical for maintaining regulatory compliance, building trust with customers and stakeholders, and driving business success. As manufacturers continue to embrace digital transformation and leverage data-driven technologies, the need for robust data security and governance frameworks become increasingly apparent. In the following sections, we will dig deeper into the key principles of data security and governance, exploring best practices, technologies, and strategies that manufacturing companies can adopt to mitigate risks, ensure compliance, and safeguard their data assets effectively.

Let us start with why compliance is important. Noncompliance with regulatory mandates carries significant repercussions for manufacturers, ranging from hefty financial penalties to legal liabilities and reputational damage. Violations of GDPR can result in fines of up to €20 million or 4% of global annual turnover, whichever is higher, posing substantial financial risks for noncompliant organizations. CCPA violations can lead to statutory damages of up to $7,500 per intentional violation, exposing businesses to costly legal battles and class-action lawsuits. Moreover, regulatory breaches can erode customer trust and tarnish brand reputation, impacting market credibility and long-term viability.

Regulations such as GDPR, CCPA, the United States Health Insurance Portability and Accountability Act (HIPAA), and other industry-specific mandates impose stringent requirements on how businesses handle and protect sensitive data. GDPR, applicable to organizations processing personal data of EU residents, mandates strict data protection measures, including obtaining explicit consent for data processing, implementing robust security controls, and notifying authorities of data breaches within 72 hours. Similarly, CCPA grants California residents greater control over their personal information, requiring businesses to disclose data collection practices, honor consumer rights to access and delete personal data, and refrain from selling personal information without explicit consent.

Manufacturers must adopt a proactive approach to compliance management. Conducting regular audits and assessments of data processing activities, security controls, and privacy practices help identify compliance gaps and areas for improvement. Implementing data protection impact assessments enables organizations to evaluate and mitigate risks associated with data processing activities, ensuring alignment with regulatory requirements. Staying abreast of evolving regulations and industry standards through continuous monitoring and participation in industry forums and regulatory updates is essential to maintain compliance in the regulatory landscape. By prioritizing regulatory compliance and embracing a culture of data stewardship, manufacturing organizations can navigate the complex regulatory environment, mitigate compliance risks, and uphold trust and confidence among stakeholders.

A key aspect of compliance is safeguarding sensitive information from cyber threats and unauthorized access. With the proliferation of digital technologies and the increasing reliance on data-driven insights, manufacturers drive operational efficiency and innovation. Diving deeper into data security principles, confidentiality, integrity, and availability (CIA) form the cornerstone of effective data protection strategies in manufacturing. *Confidentiality* ensures that sensitive manufacturing data is accessible only to authorized users, preventing unauthorized access or disclosure. *Integrity* guarantees the accuracy and consistency of manufacturing data throughout its life cycle, safeguarding against unauthorized modifications or tampering. *Availability* ensures that manufacturing data is accessible to authorized users when and where needed, minimizing disruptions to production and ensuring continuity of operations. Figure 3.18 summarizes the data security principles of CIA.

Data security encompasses a comprehensive set of measures, practices, and technologies designed to protect manufacturing data assets throughout their life cycle. Data breaches pose a significant risk, potentially exposing proprietary designs, IP, or trading partner information to unauthorized parties. Insider threats, whether intentional or unintentional, can also jeopardize data security by exploiting internal privileges or mishandling sensitive information. Malware, including viruses, worms, trojan horses, and ransomware, can infiltrate manufacturing systems, compromising data integrity and disrupting or even shutting down operations. Ransomware attacks are on the rise. They encrypt manufacturing data, demanding ransom payments for decryption keys, leading to operational downtime and financial losses. In February 2024, UnitedHealth Group and other healthcare and pharmacy providers were disrupted due to ransomware attacks; some are still feeling the aftereffects of this attack.[21] The Colonial Pipeline attack in May 2021 is a recent example of a ransomware attack that shut down the oil flow to the entire Eastern seaboard.[22]

One key aspect of data security in advanced analytics and BI for manufacturing is encryption. Encrypting data ensures that it remains protected, even if it is intercepted during transmission or storage. Advanced encryption algorithms—such as SHA-256 (Secure Hash Algorithm), AES (Advanced Encryption Standard), and RSA (Rivest-Shamir-Adleman)—are commonly employed to secure data at rest and in transit. Implementing robust access controls and authentication mechanisms helps restrict access to sensitive manufacturing data, ensuring only authorized personnel can view or manipulate it. Another critical consideration in data security is data masking and anonymization. By masking sensitive information such as personally identifiable information or proprietary data product designs, manufacturing companies can mitigate the risk of data breaches while allowing meaningful analysis and insights generation. Techniques like tokenization and

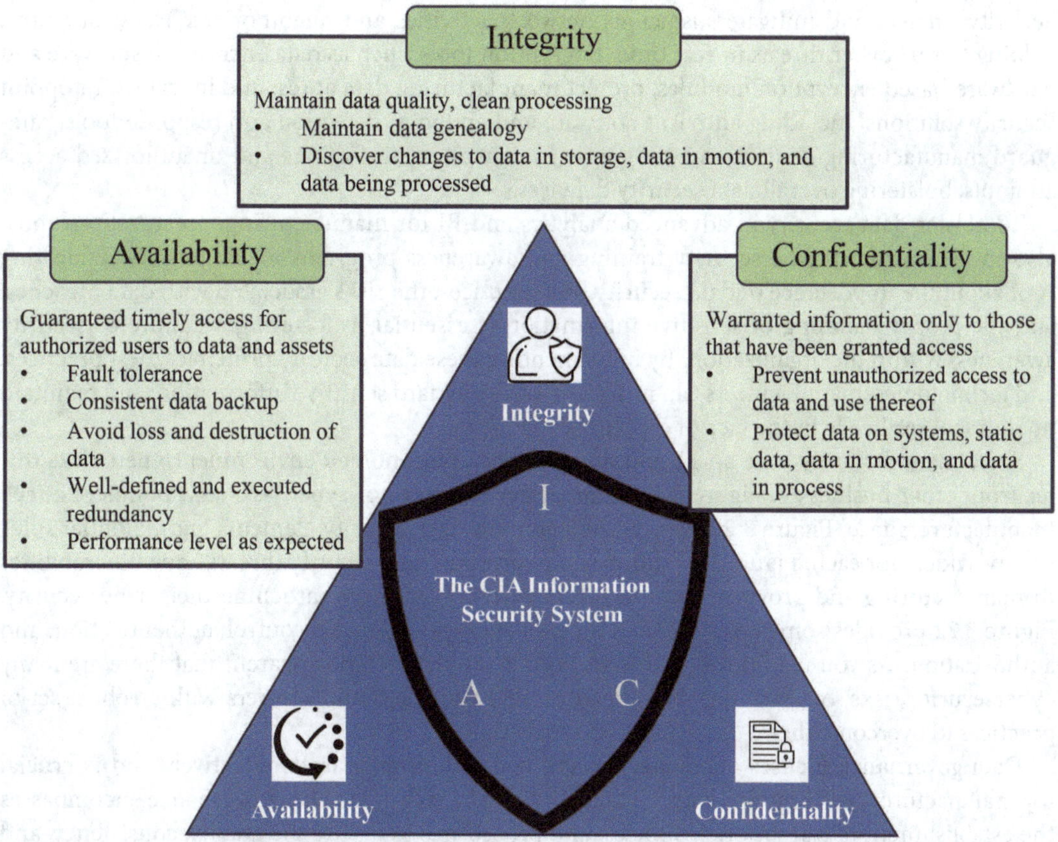

Figure 3.18 Data security principles: confidentiality, integrity, and availability (CIA).

pseudonymization are utilized to anonymize data, preserving its utility for analytics while protecting individual privacy. Access control mechanisms, including role-based access control and attribute-based access control, limit user privileges and restrict access to sensitive data based on predefined policies. Authentication mechanisms, such as multifactor and biometric authentication, verify the identity of users accessing manufacturing systems, enhancing security against unauthorized access. Authorization mechanisms, including access policies and permissions, determine the actions users can perform on manufacturing data, ensuring compliance with security policies and regulations.

Furthermore, continuous monitoring and auditing of data access and usage are pivotal in ensuring data security in advanced analytics and BI for manufacturing. By employing tools and technologies that track user activities, detect anomalies, and generate audit trails, organizations can identify and respond to potential security incidents in a timely manner. This proactive approach enables manufacturing companies to maintain compliance with regulatory requirements like GDPR and industry standards like ISO 27001. Numerous data security technologies and solutions are available to manufacturing organizations to enhance their cybersecurity posture. Firewalls, both network-based and host-based, monitor and filter incoming and outgoing network traffic, preventing unauthorized access and malicious activities. Intrusion detection systems and intrusion prevention systems like Acalvio's Active Defense[23] and AccuKnox's Zero Trust Cloud

Security[24] detect and mitigate suspicious network activities and unauthorized access attempts, helping thwart cyber threats in real time. Encryption tools, such as data encryption software and hardware-based encryption modules, protect manufacturing data at rest and in transit. Endpoint security solutions, including antivirus software and endpoint detection and response tools, safeguard manufacturing devices and endpoints from malware infections and unauthorized access attempts, bolstering overall data security defenses.

To bolster data security in advanced analytics and BI for manufacturing, organizations must also invest in robust cybersecurity training and awareness programs for employees. Educating staff about the importance of data security best practices, the risks associated with data breaches, and the proper handling of sensitive information is essential in fostering a culture of security awareness within the organization. By implementing these data security principles, best practices, and technologies, manufacturers can mitigate risks, safeguard sensitive information, and maintain operational resilience in the face of evolving cyber threats.

Since manufacturers have an IT and an operational technology environment, their risks differ from other businesses. Figure 3.19 summarizes the types of cyber risks and points of attack manufacturers face. Figure 3.20 depicts the common data security elements and potential solution providers for each. Figure 3.21 shows some common data security threats and vulnerabilities in manufacturing and provides manufacturers with guidance on hardening their cybersecurity. Figure 3.22 provides some best practices on data encryption, access control, authentication, and authorization. As you read through this section, it should become apparent that there are many cybersecurity risks, but tackling them head-on will provide manufacturers with a robust set of practices to overcome them.

Data governance focuses on managing and controlling data assets effectively and is crucial for manufacturers leveraging advanced analytics and BI solutions. Data governance encompasses the establishment of frameworks, policies, and procedures to ensure the quality, consistency, and compliance of data used in analytics and decision-making processes. In manufacturing, where vast amounts of data are generated across the value chain, data governance is pivotal in ensuring that data assets are managed strategically and responsibly. A robust data governance program has frameworks that provide guidelines and standards for managing data throughout its life cycle. By implementing governance elements and establishing robust frameworks, policies, and procedures, manufacturers can effectively manage and control their data assets to drive informed decision making and maintain a competitive edge. These frameworks define the roles and responsibilities of the data governance board, sponsors, steering committee, and stakeholders to establish data quality standards and outline the processes for data acquisition, storage, processing, and dissemination. By adhering to data governance frameworks, manufacturers can maintain data integrity, enhance decision-making capabilities, and mitigate risks associated with data misuse or mismanagement.

Key components of a data governance program in manufacturing include data ownership and stewardship, data classification, metadata management, and data life-cycle management. Data owners are accountable for the integrity and security of specific data assets, while data stewards oversee governance processes and enforce policies. Data stewards also ensure compliance with regulatory requirements. Establishing clear roles and responsibilities, defining data ownership, and implementing accountability mechanisms are essential aspects of effective data governance in manufacturing. By defining data ownership and accountability, manufacturing organizations can establish a culture of data responsibility and promote transparency and trust in data-driven decision making. Data classification involves categorizing data based on sensitivity, criticality, and

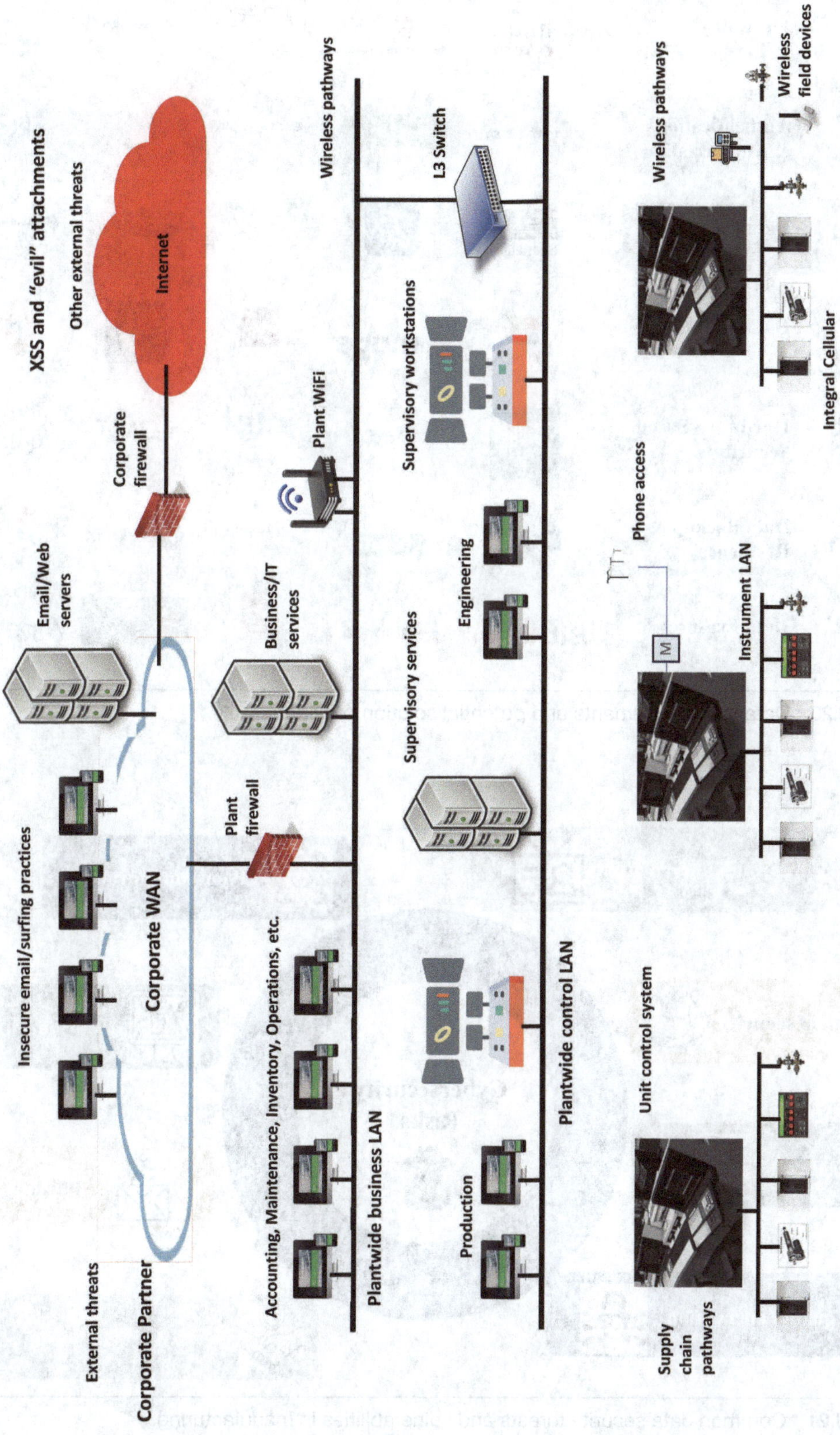

Figure 3.19 Cybersecurity risks and points of attack in manufacturing.[25]

Figure 3.20 Data security elements and potential solution providers.

Figure 3.21 Common data security threats and vulnerabilities in manufacturing.

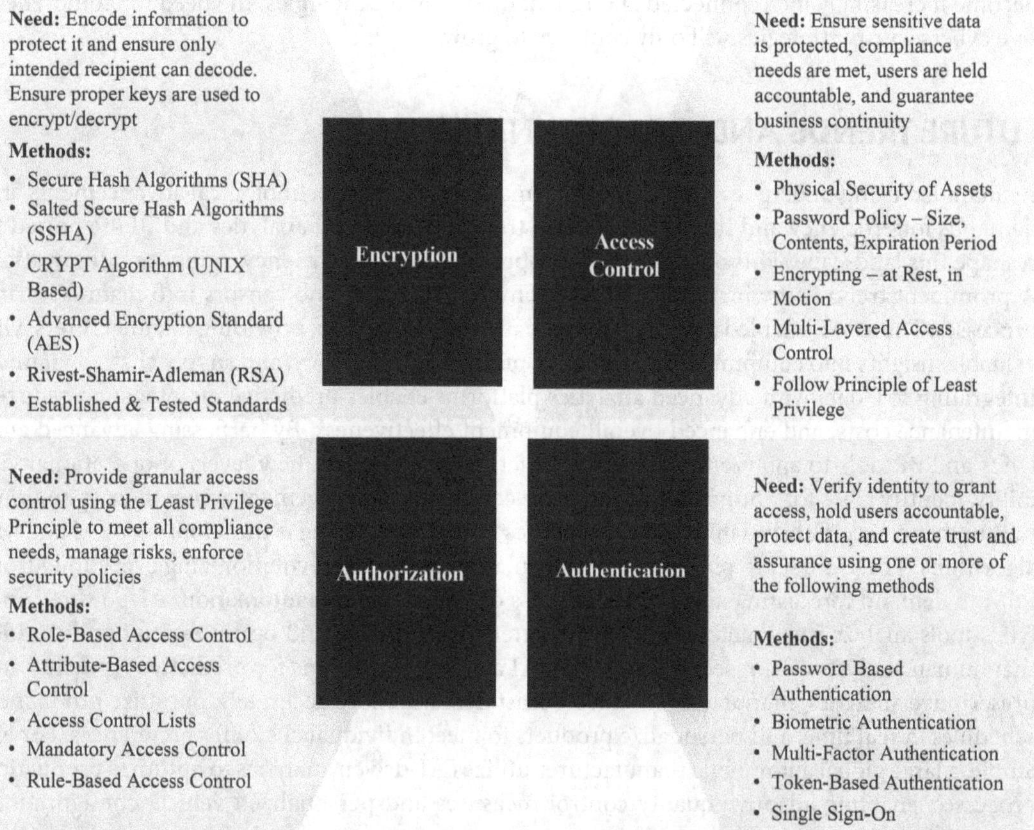

Figure 3.22 Best practices for data encryption, access control, authentication, and authorization.

regulatory implications to determine appropriate handling and protection measures. Included in data classification is the need for a well-defined and well-executed data retention policy. Metadata management focuses on capturing and maintaining metadata, such as data lineage, definitions, and usage, to facilitate data discovery, understanding, and governance. Data life-cycle management encompasses the processes and controls for managing data from creation to disposal, ensuring that data remains relevant, accurate, and secure throughout its life cycle. The data governance program is managed by a data governance board comprised of leaders, architects, and data stewards. It is crucial to ensure clear communication of data governance exists across all stakeholders.

The successful implementation of advanced analytics and BI in manufacturing hinges upon a holistic approach that encompasses technological innovation and robust data security and governance frameworks. By prioritizing data security and governance, manufacturers can unlock the full potential of their data assets while safeguarding against potential risks and threats, ultimately driving sustainable growth and competitive advantage in the digital era.

The landscape of advanced analytics and BI in manufacturing is poised for further evolution and innovation. Emerging technologies such as AI, ML, and blockchain hold immense potential to revolutionize data security practices, enabling more proactive threat detection, automated response mechanisms, and immutable data integrity. Moreover, as manufacturing operations become increasingly interconnected and reliant on digital technologies, the need for comprehensive cybersecurity strategies will only continue to grow.

FUTURE TRENDS AND OPPORTUNITIES

As manufacturing undergoes a digital transformation driven by technological advancements and demands for efficiency and innovation, future trends in advanced analytics and BI are poised to reshape this landscape. This offers numerous opportunities for efficiency, agility, and innovation. A prominent trend is the increasing integration of IoT devices and sensors into manufacturing processes. These IoT-enabled devices generate vast amounts of data, providing manufacturers with valuable insights into equipment performance, production efficiency, and supply chain dynamics. Integrating IoT data with advanced analytics platforms enables predictive maintenance, reduced maintenance costs, and enhanced overall equipment effectiveness. By harnessing advanced analytics and BI tools to analyze this data, manufacturers can unlock new levels of operational visibility, identify areas for optimization, and drive continuous improvement across their operations.

Another emerging trend in advanced analytics for manufacturing is the adoption of AI and ML algorithms. These are set to play a pivotal role in manufacturing, revolutionizing everything from PdM to demand forecasting and driving process optimization and automation. AI-powered analytics tools analyze vast datasets to identify patterns, anomalies, and optimization opportunities that human analysts do not see. By deploying AI and ML algorithms for predictive modeling and prescriptive analytics, manufacturers can forecast demand more accurately, optimize production schedules in real time, and personalize products to meet individual customer preferences. For example, a large global automobile manufacturer utilizes AI-driven analytics to optimize production processes, enabling adaptive quality control measures and personalized vehicle configurations based on customer specifications. This approach improves operational efficiency and enhances customer satisfaction by delivering tailor-made products with minimal lead times. Other AI technologies, such as computer vision, natural language processing, conversational AI, and generative AI, are rapidly gaining momentum in manufacturing. As AI and ML technologies continue to mature, their potential to drive innovation and transformation within manufacturing is virtually limitless. There will be more on this topic in Chapter 4.

The convergence of advanced analytics with digital twin technology presents compelling opportunities for manufacturers to simulate and optimize their production environments in virtual environments. Digital twins are virtual representations of physical assets or processes with real-time data feeds and analytics capabilities that enable predictive modeling and scenario analysis. By creating digital twins of manufacturing facilities, companies can simulate various production scenarios, test operational changes, and identify potential bottlenecks or inefficiencies before implementing them in the real world. This proactive approach to production optimization enhances safety, minimizes risks, reduces costs, and accelerates time-to-market for new products. Siemens, a global leader in digital twin technology, offers solutions like Siemens Digital Enterprise Services, which enables manufacturers to create comprehensive digital twins of their entire production ecosystem from product design to manufacturing and supply chain management.

The advent of edge computing infrastructure is changing data processing and analytics in manufacturing. It enables real-time insights and decision making at the edge of the network, closer to where data is generated. Edge analytics platforms leverage distributed computing resources and AI algorithms to analyze data locally on the edge. This reduces latency and bandwidth requirements while enhancing data privacy and security. The decentralized approach to analytics allows the extraction of valuable insights from IoT sensors and devices in real time and enables proactive maintenance, quality control, and process optimization on the factory floor. For instance, Intel, through its Smart Edge, offers edge computing solutions that can be tailored for manufacturing.[26] This allows manufacturers to deploy AI-powered analytics at the edge for PdM, defect detection, and energy management. By harnessing the power of edge analytics, manufacturers can unlock new opportunities for efficiency gains, cost savings, and innovation in their operations.

The adoption of cloud-based analytics platforms and solutions has accelerated in the manufacturing sector, offering infrastructure elasticity through scalability, flexibility, and accessibility to data analytics capabilities. Cloud-based BI platforms enable manufacturers to store, process, and analyze vast amounts of data securely while providing seamless integration with existing systems and applications. This allows manufacturers to democratize data access across the organization, empowering employees at all levels to make data-driven decisions and drive continuous improvement initiatives. Moreover, cloud-based analytics solutions offer predictive modeling capabilities and advanced visualization tools to derive actionable insights from their data.

The future of manufacturing is bright and is shaped by the convergence of multiple technologies—including IoT, AI, digital twin technology, cloud computing, and edge computing. These technologies unlock exceptional opportunities for efficiency, agility, and innovation. By embracing these solutions, manufacturers can optimize their production processes, enhance product quality, and deliver superior customer experiences in an increasingly competitive market landscape. However, realizing the full potential of these technologies requires strategic investments in cybersecurity risk management, talent, infrastructure, and organizational culture to foster a data-driven mindset and facilitate seamless integration into existing workflows. As manufacturing continues to grow, companies that leverage advanced analytics and BI solutions effectively will gain a competitive edge and drive sustainable growth in the Industry 4.0 era.[27]

SUMMARY AND KEY INSIGHTS

As the manufacturing landscape undergoes a digital transformation, the integration of advanced analytics and BI has emerged as a potent catalyst for innovation and efficiency. Throughout this chapter, we probed into the multidimensional applications of data-driven insights in optimizing various facets of the manufacturing value chain. From demand forecasting and production scheduling to quality control and supply chain management, manufacturers leverage advanced analytics to glean actionable insights, streamline processes, and enhance decision making. By harnessing the power of BI solutions, companies gain unprecedented visibility into their operations, enabling them to adapt swiftly to market changes and evolving customer demands.

A central theme that permeates discussions around leveraging advanced analytics and BI in manufacturing is the paramount importance of data security and governance. In an era where data is increasingly recognized as a strategic asset, manufacturers must prioritize robust data security measures to safeguard sensitive information, ensure regulatory compliance, and maintain stakeholder trust. Organizations can mitigate cybersecurity risks, protect IP, adhere to required compliance, and foster a data-driven culture of innovation and transparency by implementing comprehensive data security protocols and establishing clear governance frameworks.

Looking ahead, the future of manufacturing holds immense promise, driven by emerging technologies such as AI, ML, cloud computing, edge computing, digital twins, and the proliferation of IoT. These transformative technologies are poised to revolutionize production processes, improve

PdM, and unlock new opportunities for personalized customer experiences. Further, the convergence of AI with digital twin technology and edge computing will empower manufacturers to create virtual replicas of their production environments, optimize operations in real time, and drive continuous improvement.

Some of the key takeaways from this chapter include:

1. Advanced analytics and BI are essential tools for driving operational excellence, enhancing decision-making processes, and gaining a competitive edge in the market.
2. Advanced analytics encompasses a variety of sophisticated techniques and algorithms that allow manufacturers to extract valuable insights from large and complex data sets, enabling predictive capabilities and proactive decision making.
3. BI encompasses a broader set of methodologies, technologies, and processes aimed at transforming raw data into actionable information that will facilitate quick and informed decision making at all levels of the organization.
4. There are many potential use cases for advanced analytics and BI in manufacturing, including predictive maintenance, quality control, and supply chain optimization.
5. Manufacturers, however, need to proceed with caution when implementing advanced analytics and BI solutions because they can become complex and expensive.
6. Whichever advanced analytics manufacturers choose to pursue, effective data security and governance will be a mandatory implementation requirement.

4

COMMON ALGORITHMS, TOOLS, AND FRAMEWORKS IN AI/ML FOR MANUFACTURING

"Going forward, AI will play a pivotal role in defining what companies will come out advantageous."

—Hadi Tabani, CEO of Liquid Technologies

". . . the center of all this is, there should be a core group of leaders who are kind of constantly connecting the dots and taking help from the extended team of experts."

—Ranjeet Banerjee, CEO of Cold Chain Technologies

At the heart of modern manufacturing, a digital revolution is unfolding—one where the marriage of artificial intelligence (AI) and machine learning (ML) is reshaping the very foundations of the industry. As we stand at the crossroads of tradition and innovation, this chapter embarks on a journey into the dynamic realm where algorithms and tools become the architects of tomorrow's factories. Imagine a factory floor pulsating with life, where machines not only perform tasks but anticipate needs, optimize processes, and learn from every iteration. It is not science fiction; it is the reality crafted by the ingenious fusion of AI and ML technologies working in close concert with the large amounts of data within your enterprise.

In our exploration, we delve into algorithms and tools that are the unsung heroes of this manufacturing renaissance. In the previous chapter, you read about common use cases for advanced analytics. In this chapter, we get into specific algorithms that cover the same manufacturing processes but perform even better. These digital artisans work silently, predicting equipment failures before a wrench touches a bolt, meticulously scrutinizing products for imperfections, and orchestrating production schedules with a symphony of efficiency. But this chapter is not just about codes and models; it is about the profound impact these algorithms can have on the very essence of manufacturing—efficiency, quality, sustainability, and innovation. This chapter is not intended to be a technical manual but a captivating story of machines that not only compute but comprehend and not just analyze but adapt.

Join us on this captivating journey through the binary realms where predictive maintenance (PdM) becomes clairvoyance, quality control (QC) becomes a visual art form, and supply chains evolve into dynamic, self-optimizing networks. It is a narrative where robots are not just tools; they are learning companions, and energy management is a dance with data to the rhythm of sustainability. The future is not on the horizon; it is already here, embedded in lines of code, pulsating through circuits, and transforming manufacturing as we know it. Fasten your seatbelts; your journey into the AI/ML-powered manufacturing future begins now.

MACHINE LEARNING ALGORITHMS IN MANUFACTURING

As the manufacturing sector embraces the era of Industry 4.0, the integration of ML algorithms has become a linchpin for driving efficiency, ensuring quality, and optimizing processes. This section meticulously explores the diverse array of ML algorithms employed in manufacturing, categorizing them into supervised, unsupervised, and reinforcement learning. The dynamic nature of manufacturing processes necessitates a nuanced understanding of these algorithms to harness their full potential.

Supervised Learning Algorithms

Supervised learning algorithms, a cornerstone of modern manufacturing, play a pivotal role in making predictions and classifying data. In addition, classification algorithms such as support vector machines (SVMs) and decision trees contribute extensively to QC, precisely categorizing products and identifying defects in real time. These algorithms operate under the guidance of labeled datasets (hence supervised), learning patterns, and relationships to make informed decisions. In the context of manufacturing, their application spans a wide range of processes, from PdM to QC, contributing significantly to the industry's advancement.

> "*. . . the key to creating these valuable AI systems is that we need teams that can program with data rather than program with code.*"
>
> —Andrew Ng, Founder and CEO of Landing AI

In the domain of PdM, regression analysis stands as a robust tool. By analyzing historical data and identifying patterns, manufacturers can foresee potential equipment failures, thus enabling proactive maintenance measures. This supervised learning technique analyzes historical data to identify patterns and relationships, allowing manufacturers to predict when equipment is likely to fail. By understanding the factors contributing to equipment failures, manufacturers can schedule maintenance activities proactively, preventing unexpected downtimes and ensuring the continuous operation of critical machinery.

Several classification algorithms exist, but SVMs and decision trees prove invaluable in QC processes. SVMs, with their ability to handle linear and nonlinear data, are adept at categorizing products and identifying defects, contributing to the production of high-quality goods.

Regression Analysis for Predictive Maintenance

Regression analysis stands as a pivotal technique in the realm of manufacturing, offering a powerful tool for predictive modeling and decision making based on labeled historical data. In the context of manufacturing processes, where understanding and predicting outcomes is crucial, regression analysis becomes instrumental in identifying relationships between variables. This statistical method aids in quantifying the impact of independent variables on dependent variables, providing valuable insights for optimizing processes, ensuring product quality, and enabling PdM. A sample of a single variable regression model is shown in Figure 4.1, depicting maintenance costs ($K) across the second quarter of 2023. The regression line (the dotted line) and the regression

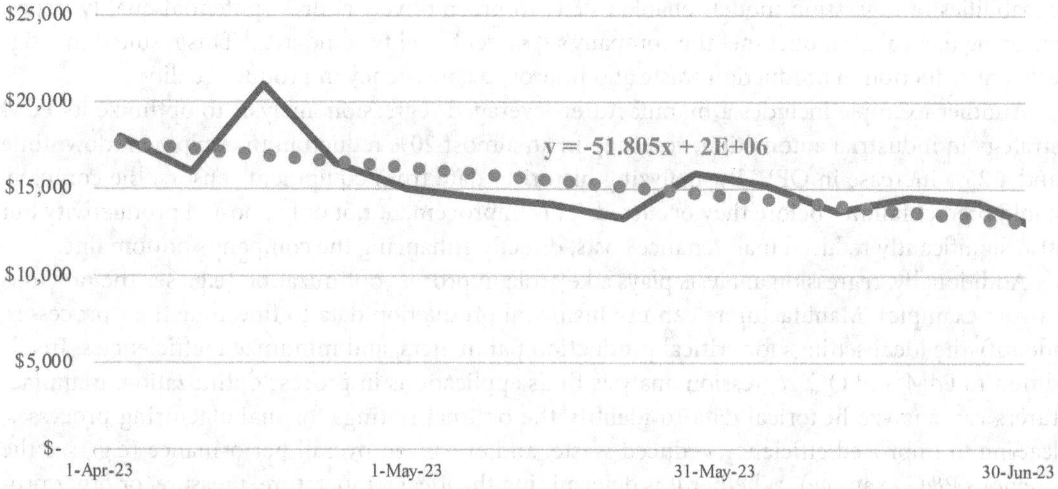

Figure 4.1 Representation of regression analysis with trendline and an equation.

equation can be used to interpolate the cost value within the time window shown in the graph. Extrapolation is not recommended since it can result in misleading results, but it is hard to stop users from predicting values of a target variable outside the data range.

One prominent application of regression analysis in manufacturing is in PdM. By analyzing historical data related to equipment performance, regression models can identify patterns and trends indicative of potential issues. This allows manufacturers to anticipate when equipment is likely to fail, enabling proactive maintenance measures. The ability to predict and address potential failures before they occur contributes significantly to minimizing unplanned downtime, optimizing overall equipment efficiency (OEE), and enhancing operational efficiency.

> Toyota employs regression analysis to optimize the welding process in its assembly lines. Through real-time monitoring of various parameters such as temperature, welding speed, and welding tool metrics, the system can predict optimal welding conditions for different car models. As a result, Toyota has significantly improved welding precision, reducing defects and ensuring consistent quality across its vehicle lineup. For Toyota—and virtually all other automotive companies—metrics like the weld quality index (WQI), defect rate per thousand units (DPU), and production cycle time reduction are critical. By leveraging regression analysis, Toyota has achieved a significant reduction in DPU, along with a decrease in production cycle time and an increase in the WQI.

Regression analysis is also extensively utilized in QC within manufacturing processes. By examining the relationship between various input factors and the quality attributes of a product, regression models can help identify critical parameters that influence product quality. Manufacturers like Procter & Gamble (P&G) have used vast amounts of production data to identify patterns and correlations that were previously undetectable. P&G identified inefficiencies in its production processes using this data and regression analysis models. This allowed them to streamline operations and minimize waste at various stages of manufacturing. Additionally, the predictive capabilities of regression models enabled P&G to preemptively address potential quality issues, ensuring that each product met the company's stringent quality standards.[1] This resulted in a significant reduction in production waste and improved consistency in product quality.

Another example includes a manufacturer-leveraged regression analysis to optimize its PdM strategy in industrial automation, resulting in an almost 20% reduction in equipment downtime and a 25% increase in OEE. By analyzing historical data from equipment sensors, the company could predict failures before they occurred. This improvement not only boosted productivity but also significantly reduced maintenance costs, directly enhancing the company's bottom line.

Additionally, regression analysis plays a key role in process optimization (e.g., see the previous Toyota example). Manufacturers can use historical production data to fine-tune their processes, identify the ideal settings for critical production parameters, and minimize inefficiencies. In addition to PdM and QC, regression analysis finds applications in process optimization. Manufacturers can analyze historical data to identify the optimal settings for manufacturing processes, leading to improved efficiency, reduced waste, and enhanced overall performance (e.g., see the previous P&G example). Whether it is determining the ideal temperature, pressure, or other production variables, regression analysis provides a quantitative method for optimizing each aspect of the manufacturing process.

While regression analysis is valuable, several challenges need careful consideration for successful implementation in PdM strategies. One of these challenges includes nonlinear relationships where the degradation of equipment and components often follows nonlinear patterns. Another cause of nonlinearity is the diversity of sensors used, each with its own characteristics and sources of potentially nonlinear variability. Traditional regression models assume linearity, but capturing complex nonlinear relationships is challenging. Addressing this requires advanced regression techniques, like polynomial regression or other algorithms capable of handling nonlinearities. Further, continuous model retraining and adaptation strategies become crucial to ensure the model's relevance and accuracy over the equipment's life cycle. Another challenge is multicollinearity since many PdM datasets often involve multiple correlated features, leading to multicollinearity issues. High correlation among independent variables can result in unstable coefficient estimates and reduced interpretability of the model. Techniques like variable selection or dimensionality reduction must be applied to mitigate multicollinearity effects.

Yet another challenge lies in data quality and preprocessing. This is not a unique challenge but plays an important role in the success of regression models. These models rely heavily on the quality of input data. Missing values, outliers (see Figure 4.1 for an example of an outlier), or inaccuracies can adversely impact model performance. Applying multiple data preprocessing mechanisms is essential to ensure the reliability of regression analysis. A limited amount of historical data, including censored data where the exact failure time is unknown or unobserved, negatively impacts regression in PdM. Acquiring a sufficient amount of labeled data for regression analysis can be a challenge, especially for rare failure events. However, techniques like transfer learning and domain knowledge can help address relevant data scarcity. Techniques like survival analysis or time-to-event modeling should be considered for censored data.

A common challenge within any model is the interpretability versus complexity trade-off. While complex models can capture intricate relationships, they often sacrifice interpretability. Such complex models also often result in overfitting, making model performance poor when presented with unseen data. In PdM, understanding the factors contributing to equipment degradation is crucial for informed decision making. Striking a balance between model complexity and interpretability is a persistent challenge in regression analysis and other AI/ML models.

Support Vector Machines

SVMs represent a formidable class of supervised learning algorithms with diverse applications in manufacturing. They excel in classification tasks by identifying an optimal hyperplane to divide data points into distinct classes (see Figure 4.2). In manufacturing, SVMs find significant utility in QC, where precision in categorizing products is paramount. The ability of SVMs to handle both linear and nonlinear data makes them particularly adaptable to the intricate and dynamic nature of manufacturing processes.

One key strength of SVMs lies in their capacity to transform data into a higher dimensional space, effectively capturing complex relationships between variables. This transformation is achieved through the use of kernel functions, which enable SVMs to discern patterns that might challenge other algorithms. SVMs are ideal in manufacturing where products may exhibit nonlinear characteristics or where the relationship between input features is intricate. The concept of support vectors, which are key in SVMs, involves identifying data points that play a crucial role in determining the optimal hyperplane. These support vectors essentially guide the algorithm in establishing a boundary that maximally separates different classes. In a manufacturing context, this translates to precise categorization, facilitating quality assurance by distinguishing between acceptable and defective products in real time.

Further, an SVM's robustness extends to scenarios with limited data where it can generalize well, based on a relatively small set of training examples. This attribute is particularly advantageous

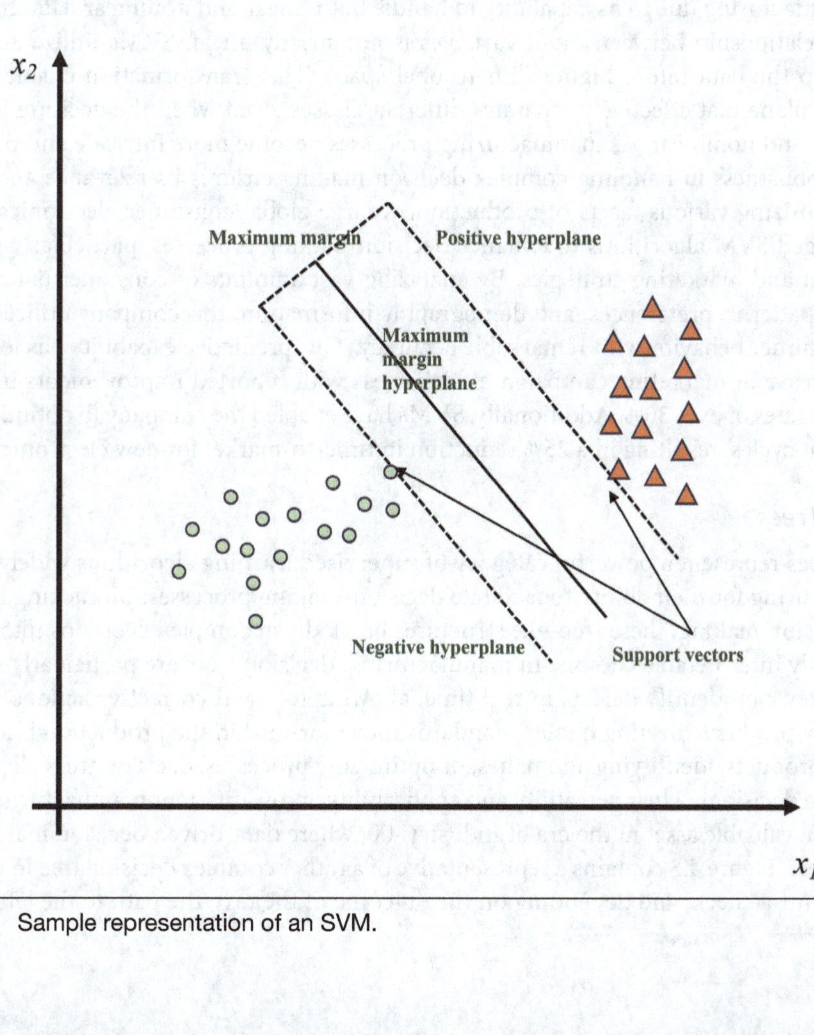

Figure 4.2 Sample representation of an SVM.

in manufacturing, where acquiring extensive labeled datasets may be challenging. An SVM's ability to make accurate predictions with fewer instances ensures its practical applicability across a spectrum of manufacturing processes. In essence, SVMs stand as the basis for supervised learning, offering manufacturers a powerful tool for classification tasks critical to ensuring product quality and operational efficiency.

> An example of where SVMs have played a key role is in QC. Their capability to accurately distinguish between different classes ensures that only products meeting stringent quality standards proceed through the production line. This application is particularly crucial in industries such as aerospace or automotive manufacturing, where defects can have significant consequences. Automotive companies have implemented SVM algorithms to enhance their defect detection processes in manufacturing. By leveraging SVMs, they have achieved remarkable improvement in defect identification accuracy, with reported reductions in false positives of almost 25%. This technology has not only optimized the QC process but has also led to a significant decrease in production costs, estimated at around 15%.

Another example is an SVM's exceptional ability to handle complex decision-making processes within manufacturing due to its capability to handle both linear and nonlinear data. In scenarios where the relationship between input variables is not strictly linear, SVMs utilize kernel functions to map the data into a higher dimensional space. This transformation enables SVMs to find a hyperplane that effectively separates different classes, even when the decision boundaries are complex and nonlinear. As manufacturing processes become more intricate and data-driven, an SVM's robustness in handling complex decision making ensures its relevance and effectiveness in optimizing various facets of production. A large global consumer electronics manufacturer leveraged SVM algorithms to enhance decision-making processes, particularly in product development and marketing strategies. By analyzing vast amounts of consumer data, including purchasing patterns, preferences, and demographic information, this company utilized SVMs to predict consumer behavior with remarkable accuracy. This predictive capability has led to a substantial increase in marketing campaign effectiveness with reported improvements in customer engagement rates of over 30%. Additionally, SVMs have enabled the company to optimize product development cycles, resulting in a 25% reduction in time-to-market for new electronic devices.

Decision Trees

Decision trees represent a powerful category of supervised learning algorithms widely employed in manufacturing for their ability to facilitate decision-making processes. Mimicking the logic of human decision making, these tree-like structures break down complex decisions into a series of simpler, easily interpretable choices. In manufacturing, decision trees are particularly valuable in QC since they can identify defects in real time, allowing for swift corrective actions and ensuring that only products meeting quality standards move forward in the production line. Whether classifying products, identifying anomalies, or optimizing processes, decision trees allow for swift and accurate decisions. Their versatility and applicability across different manufacturing domains make them a valuable asset in the era of Industry 4.0, where data-driven decision making is integral to success. Figure 4.3 contains a representative of a rather complex decision tree in which each node performs a check, and depending on the outcome of the text, the path to the left or right is followed.

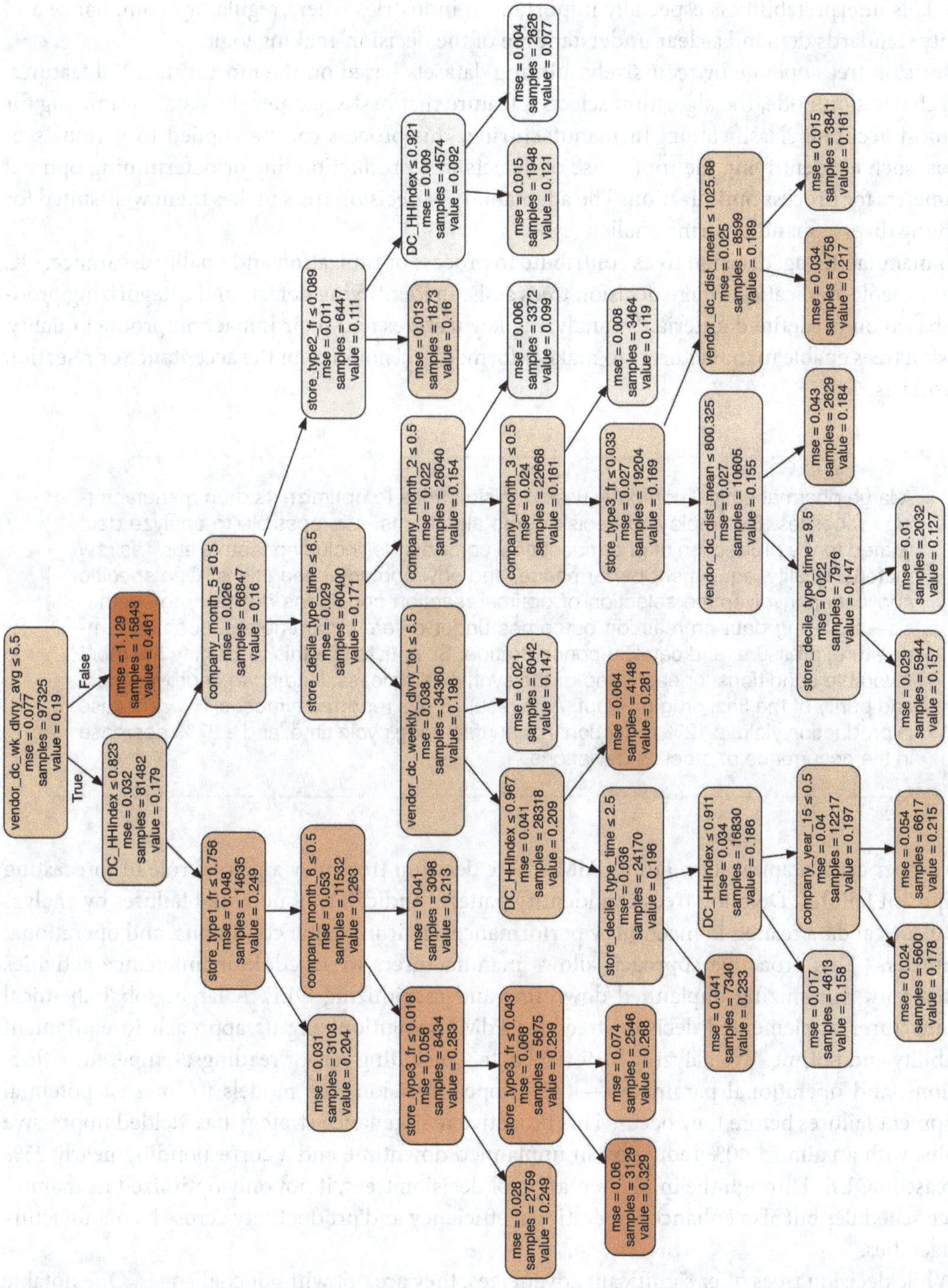

Figure 4.3 A representative decision tree.

One of the primary strengths of decision trees lies in their transparency and interpretability. The intuitive nature of these structures allows manufacturers to comprehend the decision-making process at each node, making it easier to identify and rectify issues within the manufacturing process. This interpretability is especially important in industries where regulatory compliance and quality standards demand a clear understanding of the decision-making logic.

Decision trees operate by recursively splitting datasets based on the most influential features. At each decision node, the algorithm selects a feature that best separates the data, optimizing for the most accurate classification. In manufacturing, this process can be applied to various scenarios, such as identifying the root cause of defects in a production line or determining optimal parameters for process optimization. The adaptability of decision trees makes them well-suited for handling diverse manufacturing challenges.

In manufacturing, decision trees contribute to process optimization and quality assurance. QC is one notable application where decision trees assist in identifying defects and categorizing products based on predefined criteria. By analyzing key features and their impact on product quality, decision trees enable manufacturers to make informed decisions about the acceptance or rejection of products.

> A large pharmaceutical company uses decision trees to optimize its drug manufacturing processes. By employing decision tree algorithms, it is possible to analyze data related to the production of pharmaceutical compounds, including factors such as raw material quality, equipment performance, and environmental conditions. One specific application involves the selection of optimal reaction conditions during drug synthesis—analyzing data on reaction outcomes under different parameters, such as temperature, pressure, and catalyst concentration. Scientists use this to identify the most favorable conditions for each step of the synthesis process, leading to improved yield and purity of the final drug product. As a result, it has reported almost a 15% increase in production yield, a 12% reduction in manufacturing cycle time, and a 27% decrease in the occurrence of process deviations.

Another crucial application lies in PdM where decision trees play a pivotal role in forecasting equipment failures. Decision trees can identify patterns indicative of potential failures by analyzing historical data related to machinery performance, environmental conditions, and operational parameters. This proactive approach allows manufacturers to schedule maintenance activities strategically, minimizing unplanned downtime and maximizing OEE. A large global chemical manufacturer implemented decision trees for PdM, revolutionizing its approach to equipment reliability and uptime. By analyzing historical data—including sensor readings, temperature fluctuations, and operational parameters—it developed decision tree models to forecast potential equipment failures before they occur. This proactive maintenance strategy has yielded impressive results, with an almost 40% reduction in unplanned downtime and a corresponding nearly 25% increase in OEE. Through the implementation of decision trees, it not only optimized its maintenance schedules but also enhanced operational efficiency and productivity across its manufacturing facilities.

While decision trees offer significant advantages, they are not without challenges. One notable consideration is the potential for overfitting, where the algorithm captures noise in the training data as if it were a meaningful pattern. This can lead to suboptimal performance on new, unseen data. To mitigate this, techniques such as tree pruning and setting appropriate stopping criteria such as tree depth are employed. Additionally, decision trees may struggle with capturing complex relationships in data that involve nonlinearities. Alternative algorithms or ensemble decision tree methods (such as *random forests*) may be considered in scenarios where the relationships are intricate and not easily separable by simple decision boundaries. Striking a balance between interpretability and complexity is essential when deploying decision trees in manufacturing contexts.

Random Forest

Random forest is an ensemble learning technique that enhances the capabilities of decision trees and supervised learning in manufacturing. In essence, a random forest is an aggregation of multiple decision trees, each trained on a different subset of the dataset and contributing to a collective, robust prediction. Random forests provide strong and accurate predictions, especially in scenarios where manufacturing processes involve numerous variables and intricate relationships. Their ability to handle large datasets and maintain predictive accuracy makes them invaluable for optimizing complex manufacturing operations. When handling large datasets with diverse features in manufacturing, where data streams from various sensors and sources are abundant, the algorithm excels in maintaining predictive accuracy. The ensemble nature of random forest helps mitigate overfitting, which is a common challenge in ML because the model aggregates insights from multiple trees, leading to more generalizable and reliable predictions.

Within manufacturing, random forests find applications across diverse scenarios. In PdM, where identifying the optimal time for equipment servicing is critical, random forests analyze historical data to predict potential failures. By understanding the contributing factors and patterns leading to equipment malfunctions, manufacturers can proactively schedule maintenance activities, thereby reducing unplanned downtimes and optimizing OEE.

> A global leader in the healthcare industry employed random forest algorithms to enhance predictive analytics in drug discovery and development processes. By analyzing diverse datasets encompassing molecular structures, biological interactions, and clinical trial outcomes, it utilized random forest models to predict the efficacy and safety profiles of novel drug candidates. This predictive capability has led to a significant increase in the success rate of drug candidates entering clinical trials, with reported improvements in identifying promising compounds by up to 40%. Random forest algorithms have enabled streamlining research and development efforts, resulting in a nearly 20% reduction in time and resources required for bringing new drugs to market.

In QC processes, random forests contribute to the identification of defects and anomalies in real time. The algorithm's ability to analyze multidimensional datasets aids in precisely categorizing products based on predefined criteria. This ensures adherence to quality standards and also facilitates swift corrective actions, maintaining the integrity of the manufacturing process. As another example, a global manufacturing conglomerate implemented random forest algorithms to enhance QC across its diverse product portfolio. Leveraging data from various sensors and production parameters, the company utilized random forests to predict equipment failures and identify potential quality issues before they could escalate. This proactive approach resulted in a significant reduction in unplanned downtime, with an associated improvement in OEE of almost 15%. By embracing advanced analytics, it achieved greater operational reliability and product consistency, remaining a leader in industrial manufacturing.

Random forests operate on the principles of ensemble learning, combining the strength of multiple decision trees to enhance predictive accuracy. The algorithm's name is derived from the randomness injected during the construction of individual trees and the aggregation of their predictions. The key components of random forest include:

- *Decision trees*:
  - Random forests consist of multiple decision trees, each trained on a different subset of the data. Decision trees serve as the fundamental building blocks, making predictions based on a series of binary decisions at each node.
- *Bootstrap aggregating (bagging)*:
  - The random sampling of data points with replacement, known as bootstrapping or bagging, creates diverse subsets for training each decision tree. Bagging ensures that each tree learns from a slightly different perspective of the dataset, reducing overfitting.
- *Feature randomness*:
  - In addition to sampling data, random forests introduce randomness in selecting features for each decision node. This feature randomness further diversifies the trees, preventing them from relying too heavily on a specific subset of features.
- *Voting mechanism*:
  - During prediction, each tree in the forest provides an output, and the final prediction is determined through a majority voting mechanism. This ensures a collective decision based on the diverse insights contributed by individual trees.

Understanding these technical underpinnings is crucial for grasping how random forests harness randomness to create robust and adaptable models for complex manufacturing scenarios.

While random forests offer numerous advantages, they are not without challenges. The algorithm's interpretability can be limited compared to individual decision trees since its nature makes it a *black-box* model (a system using inputs and outputs to create useful information without any knowledge of its internal workings). Understanding the decision-making process of a random forest may be challenging, especially when explaining predictions to stakeholders or complying with regulatory requirements. Further, the black-box nature makes it difficult to extract actionable insights from the model. Additionally, as with any ML algorithm, the quality of predictions is contingent on the quality of the input data. Careful consideration of data quality and relevance is paramount to the success of random forests in manufacturing applications.

Despite their ensemble nature, random forests can still be prone to overfitting, especially when the model is overly complex or when the training data is limited. Random forests may struggle to handle imbalanced data, leading to biased predictions. Implementing techniques such as pruning, adjusting the number of features considered for splitting, or increasing the minimum samples per leaf can mitigate overfitting. Further, techniques like resampling (oversampling or undersampling) or using different evaluation metrics can address the unbalanced data.

A real-life example of overfitting is from a steel production facility. The company sought to optimize its manufacturing processes by implementing predictive analytics to forecast equipment failures and minimize downtime. Upon deploying random forest models to predict machine failures based on historical data, the algorithms initially showed promising results during testing. Yet, upon integration into the production environment, the models began to exhibit signs of overfitting. The random forest algorithms had *memorized* the specific patterns in the training data and struggled to adapt to new data patterns and variations encountered in real-time production scenarios. This led to inaccurate predictions, with the models flagging false positives and failing to identify genuine equipment failures promptly. This resulted in disruptions to production schedules, increased maintenance costs, and reduced overall efficiency in the manufacturing process. To address the overfitting challenge, the company undertook measures to improve data quality, increased the diversity of the training dataset, and fine-tuned the hyperparameters of the random forest models. By incorporating these corrective actions, the company successfully mitigated the overfitting problem and enhanced the performance of its PdM system.

Training a sizeable random forest with numerous trees and deep branches can be computationally intensive. Adding noisy or irrelevant features to the mix increases the computational load. In manufacturing environments where real-time predictions are crucial, optimizing the model for efficient inference becomes essential. Balancing model complexity with computational efficiency is a key consideration. Feature engineering and careful selection of relevant features become crucial processes to enable the model to focus on the most informative attributes for PdM.

Random forests may struggle to extrapolate beyond the range of training data or handle temporal dependencies. In manufacturing, where equipment conditions can evolve or drift, ensuring the model's ability to generalize unforeseen scenarios is vital. Further, random forests may not inherently capture temporal dependencies seen within PdM data. Regular model retraining with updated data is one approach to address this limitation. For temporal dependencies, adding another level of ensemble learning by combining random forests with time-series techniques enhances the model's ability to handle temporal aspects in the data.

Random forests are versatile and robust tools within manufacturing, addressing the intricacies of PdM, QC, and process optimization. Their ensemble nature, handling of large datasets, and ability to provide reliable predictions make them an invaluable asset in the pursuit of efficiency and quality in modern manufacturing processes as long as challenges are overcome.[2,3] Table 4.1 depicts a high-level comparison of the four algorithms across different dimensions.

Table 4.1 A high-level comparison of the four algorithms across different dimensions.

Feature	Regression	SVM	Decision Trees	Random Forests
Learning Curve	Easy to learn	Moderate learning curve	Low-to-moderate learning curve	Low-to-moderate learning curve
Popularity	Widely used in ML	Dominant in deep learning and neural nets	Moderately used	Widely used
Flexibility	High flexibility	Moderate flexibility	High flexibility	High flexibility
Community Support	Strong community support	Strong community support	Strong community support	Strong community support
Use Cases	Predictive modeling; forecasting	Classification; regression	Classification; regression; anomaly detection	Classification; regression; ensemble learning
Pros	Easy-to-use, interpretable results; suitable for linear relationships	Effective in high dimension spaces; versatile	Suitable for nonlinear relationships; minimal data processing and data assumptions	High accuracy; real-time results
Cons	Limited when used with nonlinear relationships	Not suitable for large datasets	Prone to overfitting	Complex; longer training time

Unsupervised Learning Algorithms

Unsupervised learning algorithms play a pivotal role in unraveling intricate patterns within vast datasets in the manufacturing domain. The dynamic nature of manufacturing processes demands a nuanced approach to data analysis, especially with unlabeled data—unsupervised learning provides a powerful arsenal of tools for this purpose. These algorithms are adept at discovering hidden structures and relationships and offering valuable insights for process optimization and decision making.

Unsupervised learning aims to discover the inherent structure of unlabeled data, grouping similar elements together or reducing its dimensionality. One of the primary tasks it undertakes is clustering, where the algorithm categorizes data points into clusters based on similarities. This process is akin to finding natural groupings within the data without any prior knowledge of the expected outcomes. In manufacturing, unsupervised learning techniques like clustering can be instrumental in QC, identifying anomalies, or optimizing production processes. By allowing algorithms to autonomously discern patterns, manufacturers gain valuable insights into their operations, fostering continuous improvement and innovation. Another facet of unsupervised learning is dimensionality reduction, a technique employed to streamline and simplify complex datasets. By identifying and retaining the most relevant features, unsupervised learning algorithms help distill the essence of the data, making it more manageable for subsequent analysis or visualization. One of the classic applications of unsupervised learning lies in exploratory data analysis. For instance, when confronted with large datasets in manufacturing, unsupervised learning can uncover patterns that might elude human observation. This capability is particularly valuable in scenarios where the relationships between variables are intricate and not readily apparent.

Unsupervised learning encompasses a spectrum of techniques designed to extract meaningful insights from data without predefined labels. One of these key techniques is *k-means clustering*. This algorithm excels in grouping similar data points, enabling manufacturers to identify patterns or anomalies in their processes. For instance, k-means can categorize machinery behavior in production lines, allowing for PdM strategies or highlighting and identifying inefficiencies, optimizing resource utilization, and enhancing overall operational efficiency. The iterative nature of k-means clustering ensures that clusters evolve over time, accommodating changes in manufacturing processes. *Hierarchical clustering* is another unsupervised learning gem. Unlike k-means, it organizes data hierarchically, showcasing relationships between different elements. In manufacturing, this can be applied to understand the hierarchical structure of processes, optimizing not just individual steps but the entire production chain. By identifying dependencies and interconnections, manufacturers gain a holistic view, enhancing decision making and process optimization.

Innovative applications of unsupervised learning extend to anomaly detection, where algorithms like *isolation forests* or *one-class SVM* can identify deviations from normal behavior. In a manufacturing setting, this translates to early detection of defects or malfunctions, minimizing waste and downtime.

k-Means Clustering

In the landscape of data analytics and ML, the k-means clustering algorithm is one of the fundamental and widely employed learning techniques. This makes it an indispensable tool in manufacturing. Its versatility lies in its ability to categorize data points into distinct groups based on similarities, providing manufacturers with valuable insights into their processes. Its simplicity and effectiveness have made it a key tool in various domains, including manufacturing, where the quest for insights from complex datasets is predominant.

The core principle of k-means revolves around grouping data points into "k" clusters based on their similarities. This clustering is achieved through an iterative process that refines the position of centroids until optimal cluster formations are achieved. The algorithm seeks to minimize the sum of squared distances between data points and their respective centroids, making it an excellent tool for identifying patterns and structures within datasets. Figure 4.4 shows a comparison of the multiple clustering algorithms available along with their result, with k-means showing up in the first column for *non-flat* geometry clustering. As you can see in Figure 4.4, the outcome of the clustering depends on the goal. For example, for the first row, if the goal is to cluster by circular data, only three algorithms achieve that goal—DBSCAN, HDBSCAN, and OPTICS.

Being a data-rich domain, manufacturing benefits significantly from the application of k-means. From production planning to supply chain optimization and PdM, the algorithm proves its value in diverse scenarios. Consider a scenario where a production line generates a plethora of data points, each representing a different aspect of the manufacturing process. k-means can group similar data points, allowing manufacturers to identify patterns, trends, or potential inefficiencies. A specific example would be one where a mobile phone manufacturer aims to categorize devices based on production characteristics. k-means has been instrumental in creating clusters that enable tailored production strategies for different device categories. In semiconductor manufacturing, k-means has been employed to cluster data related to the production of microchips. It becomes possible to optimize the fabrication process for improved yields and reduced defects by grouping similar manufacturing conditions or characteristics.

A real-life example is that of a large oil company with refineries across the globe. This company revolutionized its oil refinery operations by harnessing the power of k-means clustering to refine its production scheduling strategies. It meticulously collected, prepared, and pre-processed data using demand forecasts, resource availability, market trends, and historical production schedules. Data accuracy was crucial to set the stage for insightful analysis. It applied k-means clustering to segment historical production data into distinct clusters based on shared production patterns, resource utilization, and market conditions. Each cluster represented a unique operational scenario within the refinery, offering invaluable insights into optimal scheduling strategies. Leveraging these insights, the company developed and implemented optimization strategies to enhance production efficiency, streamline resource allocation, and synchronize production schedules with market demand fluctuations. The results were remarkable, with a nearly 20% increase in production efficiency and a reduction in turnaround time of just under 10%.

Another notable application lies in supply chain management, where k-means aids in categorizing suppliers based on their performance metrics. This strategic classification empowers manufacturers to make informed decisions about procurement, ensuring reliable and efficient supply chains. As a result, the algorithm contributes to the operational aspects and the overall resilience and adaptability of the manufacturing ecosystem.

Intel embarked on a journey to optimize its semiconductor manufacturing processes by integrating fault detection and PdM powered by k-means clustering and other algorithms. By leveraging ML techniques, Intel sought to proactively identify defects and equipment failures. The initiative began by collecting vast amounts of sensor data from manufacturing equipment—capturing parameters such as temperature, pressure,

continued

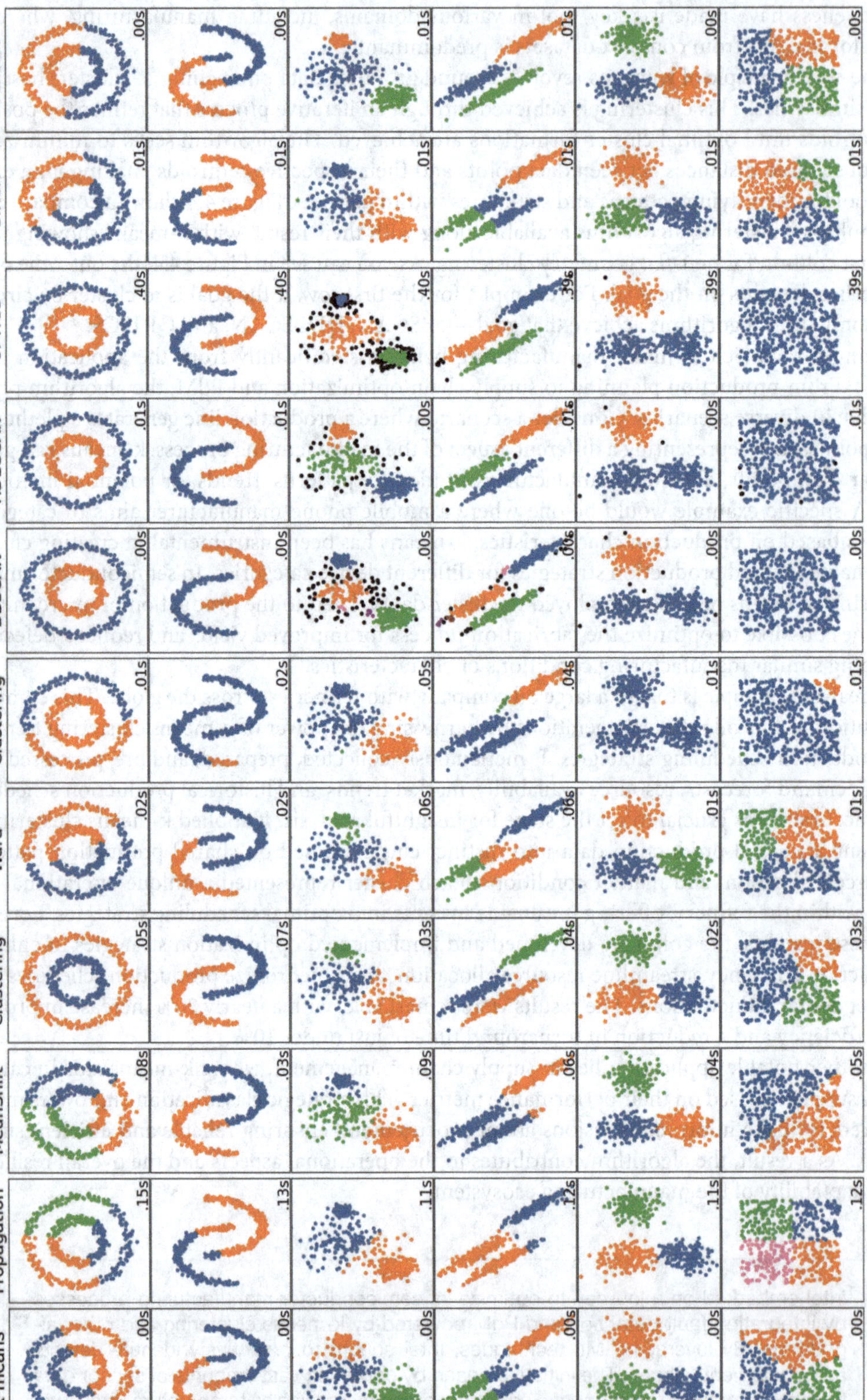

Figure 4.4 Comparison of multiple clustering algorithms.[4]

> and vibration levels. Through careful data preprocessing and feature engineering, Intel transformed raw sensor data into actionable insights through k-means clustering for eight clusters (i.e., k = 8). This played a pivotal role in the process of detecting failures, enabling Intel to categorize equipment behavior into distinct clusters representing normal and anomalous operational states. By identifying subtle deviations from normal behavior, the PdM system significantly reduced unplanned downtime and maintenance costs and increased equipment reliability.[5]

However, despite its robustness, k-means is not without challenges. Critical considerations include determining the optimal number of clusters (i.e., the k values), selecting appropriate initial centroid placements, and addressing sensitivity to outliers. Manufacturers must navigate these challenges thoughtfully to harness the full potential of the algorithm.

Hierarchical Clustering

Another unsupervised algorithm that finds many uses in manufacturing is hierarchical clustering, which has been widely used in manufacturing applications. One notable advantage is its ability to capture hierarchical relationships in the data, reflecting a nuanced understanding of the underlying structure and providing insights that go beyond the capabilities of the typical flat clustering algorithms like k-means. This algorithm is particularly advantageous when the underlying structure of the data is not well-defined and the number of clusters is not predetermined. This is crucial in scenarios where different levels of granularity are required for analysis, allowing manufacturers to explore relationships at various abstraction levels to understand intricate relationships between different facets of production. Another strength lies in its adaptability to different dissimilarity metrics and linkage methods. Manufacturers can tailor the algorithm to suit the specific characteristics of their data and the nature of the relationships they seek to uncover. This flexibility makes hierarchical clustering a versatile tool for diverse manufacturing datasets, from production planning to QC.

Hierarchical clustering also provides a visual representation of the data hierarchy through dendrograms. This visual aid helps in the interpretation of complex relationships, facilitating the decision-making process. Manufacturers can leverage a dendrogram to identify natural breaks or groupings within the data, guiding strategic choices in areas such as product categorization, supply chain optimization, and quality assurance. Figure 4.5 shows a sample dendrogram for different tools and components used in manufacturing. The earlier the join, the more similar the tools or components being compared, as depicted with the bolts and nails pair or the screwdriver and torx driver pair. However, drills are very dissimilar to all other tools or components in the data set!

Hierarchical clustering operates on the principle of recursively merging or dividing clusters until a dendrogram—a tree-like structure representing the data hierarchy—is formed. Two main approaches, *agglomerative* and *divisive*, define hierarchical clustering. The agglomerative method starts with individual data points as separate clusters and then merges them based on similarity (as in Figure 4.5), while the divisive method begins with a single cluster and progressively divides it into smaller, more distinct groups.

Imagine a manufacturing environment with multiple interconnected processes. Hierarchical clustering can be applied to group these processes based on similarities, creating a hierarchical tree that reflects the relationships to decide on, for instance, product mix. This hierarchical organization

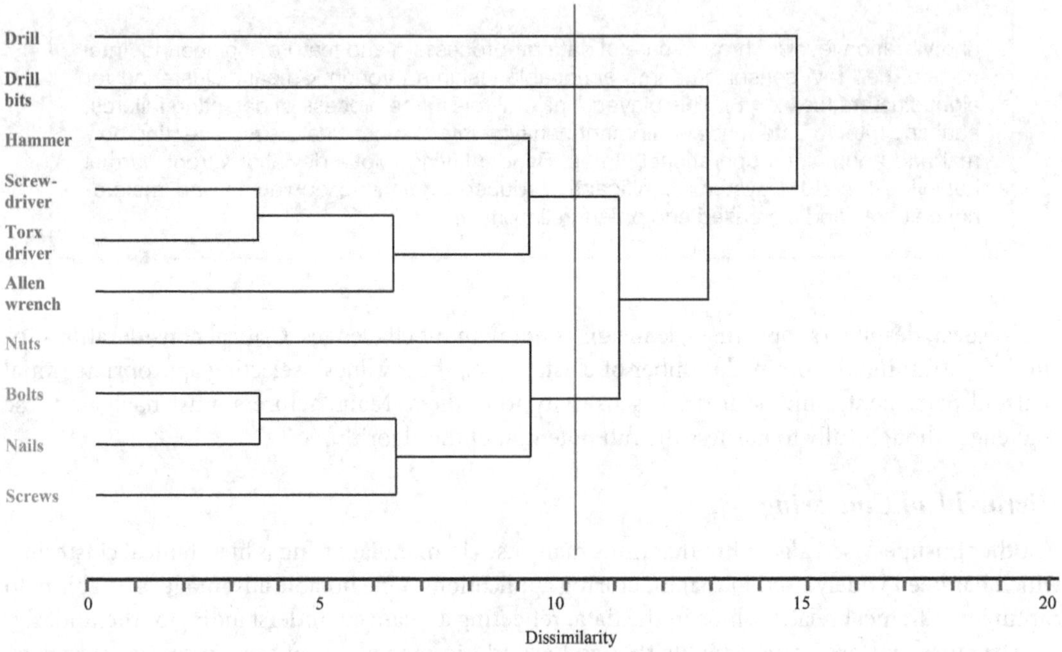

Figure 4.5 Sample dendrogram for hierarchical clustering.[6]

allows manufacturers to identify the individual steps in a process and the dependencies and influences between them.

> A global soft drink company employed hierarchical clustering for product mix decision making. By clustering its diverse range of beverage products based on factors such as flavor profiles, packaging sizes, and target demographics, it gained valuable insights into consumer preferences and market trends. This enabled the company to optimize its product portfolio, identify emerging market opportunities, and tailor its marketing strategies to specific consumer segments. The key performance indicators (KPIs) were very favorable—market shares saw more than a 10% increase, product sales saw nearly a 15% boost, and for overall profitability, the company saw a 10% improvement. This highlighted the success of the company's hierarchical clustering approach in product mix decision making.

In supply chain management, manufacturers can use hierarchical clustering to categorize suppliers based on multiple criteria such as reliability, cost-effectiveness, and geographical proximity. This assists in creating a tiered supplier structure, optimizing procurement strategies for each tier. In an automotive assembly line, hierarchical clustering can reveal the hierarchical structure of the production process, from individual component manufacturing to final assembly. Understanding these relationships enables targeted improvements, ensuring optimizations in one part of the process consider the broader impact on the entire production. In production planning, hierarchical clustering aids in classifying products based on features, allowing manufacturers to tailor production strategies for distinct product categories. An automotive manufacturer can use hierarchical clustering to group vehicles with similar specifications, streamlining production processes and adapting strategies to different product segments. QC is another critical area where hierarchical clustering proves valuable. By clustering defective products based on shared characteristics, manufacturers can identify patterns indicative of potential issues in production. This proactive approach enables timely interventions, reducing the likelihood of defects and enhancing overall product quality.

A large automobile manufacturer has strategically used hierarchical clustering to categorize its suppliers effectively. This approach enhances its supply chain management by facilitating optimized supplier relationships and ensuring consistent product quality across its vehicle lineup. This improved supplier segmentation accuracy by over 20%, reduced supply chain lead times by 18%, and increased overall supply chain efficiency by nearly 15%.[7,8]

While hierarchical clustering offers numerous advantages, it is not without challenges. One notable consideration is the sensitivity to outliers. The algorithm can be strongly influenced by extreme data points, potentially leading to skewed cluster formations. Pre-processing techniques or incorporating robust dissimilarity metrics can mitigate this challenge. Another consideration is the computational complexity, especially in large datasets. Hierarchical clustering can be computationally intensive, requiring substantial processing power and memory. Manufacturers must assess the scalability of the algorithm based on the size and complexity of their datasets.

Isolation Forests

Isolation forest, another unsupervised learning algorithm, is a novel and powerful anomaly detection algorithm that has gained prominence in manufacturing for identifying irregularities or anomalies within datasets. Anomalies, which represent deviations from the norm, can signify defects, faults, or unexpected events in the manufacturing process. Isolation forests offer an effective solution for detecting such anomalies by leveraging the principles of isolation and recursion. Figure 4.6 is an example of an isolation forest clearly showing the red point as anomalous or isolated, and the four green points as normal. The isolation forest algorithm stands out for its simplicity and efficiency in isolating anomalies. Unlike traditional methods that focus on building a profile of normal behavior, isolation forests take a different approach. They isolate anomalies by identifying features that make them stand out as *rare* or *isolated*. The fundamental idea is that anomalies are typically less frequent and can be isolated with fewer partitioning steps in a tree-like structure.

Consider a scenario where a manufacturing system produces a consistent output, and sudden anomalies, such as product defects or irregular machinery behavior, need to be swiftly detected. Isolation forests work by isolating anomalies since they require fewer steps to isolate them in the tree structure. This efficiency makes them particularly suitable for real-time anomaly detection in manufacturing environments. Isolation forests can be employed in PdM strategies where the algorithm identifies anomalies in machinery behavior, allowing for proactive intervention before significant failures occur. This minimizes unplanned downtime and improves OEE.

Isolation forests offer several advantages in the context of manufacturing anomaly detection. They excel at identifying anomalies across various dimensions, and their adaptability makes them suitable for applications ranging from QC to PdM. Isolation forests are also inherently scalable, making them well-suited for large and complex manufacturing datasets. The algorithm's reliance on random sampling and isolation trees maintains efficiency even as the dataset size increases. This scalability is particularly advantageous in manufacturing scenarios where extensive data collection is essential for robust decision making. Isolation forests require minimal input parameters, simplifying the implementation process. The algorithm's self-adjusting nature reduces the need

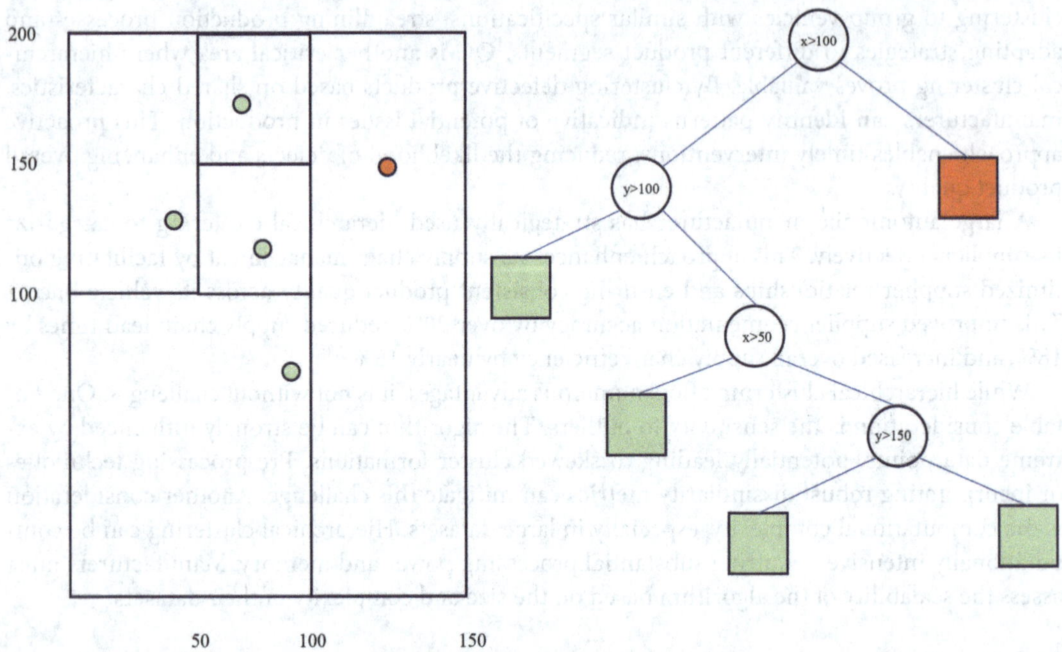

Figure 4.6 Sample anomaly detection using an isolation forest.[9]

for extensive hyperparameter tuning, making it accessible to practitioners with varying levels of expertise. This ease of use facilitates quick integration into manufacturing workflows, contributing to enhanced anomaly detection capabilities.

Isolation forests find diverse applications in manufacturing, contributing to improved QC and operational efficiency. In product quality assessment, manufacturers leverage the algorithm to identify anomalies in sensor data, indicating potential defects in real time. For example, anomalies in temperature or vibration patterns can signal equipment malfunctions or deviations in the production process. PdM is another critical area where isolation forests shine. By detecting anomalies in machine performance data, manufacturers can proactively address issues before they escalate, minimizing downtime and extending equipment lifespan. This proactive approach aligns with Industry 4.0 principles, emphasizing data-driven decision making for optimized operations. Supply chain management also benefits from the anomaly detection capabilities of isolation forests. Manufacturers can identify irregularities in inventory levels, shipment schedules, or supplier performance, enabling timely interventions and maintaining a resilient and efficient supply chain.

> A large chemical manufacturer implemented isolation forests in its chemical manufacturing plants to identify anomalies in process parameters and detect potential safety hazards. This initiative resulted in a more than 30% reduction in safety incidents and a nearly 25% improvement in process reliability. Another example involves a large pharmaceutical company that utilized isolation forests in their pharmaceutical manufacturing facilities to detect anomalies in drug production processes and ensure product quality. This implementation led to a significant decrease in batch rejection rates and a similar improvement in regulatory compliance.

While isolation forests offer compelling advantages, manufacturers must consider certain challenges in their implementation. One notable consideration is the potential sensitivity to the choice of the contamination parameter, the expected proportion of anomalies in the dataset. This parameter requires careful tuning to align with the specific characteristics of the manufacturing process. Another challenge is the reliance on random sampling, which might impact the algorithm's consistency across multiple runs. Manufacturers need to assess the trade-off between computational efficiency and result stability, particularly in scenarios where consistent anomaly detection is paramount. Table 4.2 shows a high-level comparison of each of the three algorithms in this section.

Table 4.2 A high-level comparison of unsupervised learning algorithms across different dimensions.

Feature	k-means	Hierarchical Clustering	Isolation Forest
Learning Curve	Easy to learn	Moderate learning curve	Easy to learn
Popularity	Widely used in machine learning	Moderate popularity	Moderate popularity
Flexibility	Limited for deep learning tasks	Moderate flexibility	Moderate flexibility
Community Support	Strong community support	Moderate community support	Moderate community support
Use Cases	Customer segmentation; image compression	Taxonomy creation; document clustering	Anomaly detection; fraud prevention
Pros	Simple and easy to implement; efficient for large datasets; works well with spherical clusters	No assumption about the shape of clusters; provides a dendrogram for visualization	Efficient for detecting anomalies; works well with high-dimensional data; robust to outliers
Cons	Sensitive to initial cluster centroids; may converge to local optima	Computationally intensive for large datasets; may not perform well with nonglobular clusters	Limited interpretability; may struggle with complex relationships

Reinforcement Learning

Reinforcement learning (RL) has emerged as a transformative approach in manufacturing, introducing dynamic decision-making capabilities and adaptive systems. Unlike supervised and unsupervised learning, RL focuses on an agent interacting with an environment and learning optimal strategies through trial and error. In manufacturing, where complex systems and dynamic environments prevail, RL offers unprecedented opportunities for optimizing processes, reducing costs, and enhancing overall efficiency.

Rooted in behavioral psychology and learning by consequence, RL mimics the human learning process. This learning paradigm, inspired by how humans learn from rewards and penalties, has evolved with computational breakthroughs. It is pivotal in reshaping how machines comprehend and respond to dynamic manufacturing environments. It enables machines to make decisions through continuous interaction with their surroundings. This is ideal in manufacturing, characterized by intricate processes, ever-changing variables, and the need for adaptive decision making.

Manufacturing, as an intricate web of processes, demands solutions that can navigate uncertainties, adapt to changing variables, and optimize decision making in real time. RL is perfect for this environment, offering a departure from conventional rule- or pattern-based systems and setting the stage for machines to dynamically adapt to the evolving demands of modern manufacturing. The journey of RL in manufacturing is not merely a technical progression; it signifies a paradigm shift, where machines learn, adapt, and optimize processes similar to their human counterparts. From autonomous robots orchestrating tasks on the shop floor to self-optimizing supply chain management systems, RL enables the learning of optimal strategies, enhances operations, and fine-tunes performance over time. The significance of RL extends beyond incremental improvements; it envisions a future where manufacturing systems autonomously respond to challenges, capitalize on opportunities, and continuously learn from their interactions with the environment.

RL brings several advantages to manufacturing, contributing to adaptive and intelligent systems. One key strength is its ability to handle sequential decision making, a common characteristic in manufacturing. RL shines in scenarios where actions at one stage influence outcomes in subsequent stages, enabling manufacturers to optimize complex workflows. Further, RL's ability to learn from changing environments makes it a valuable tool for real-time decision making in dynamic manufacturing settings. The algorithm's capacity to learn from experience allows it to respond to unseen events, adapt to fluctuations in demand, and optimize production schedules on the fly.

RL is based on the interaction between an agent and an environment. The agent takes actions within the environment, receiving feedback in the form of rewards or penalties based on the consequences of its decisions. Through continuous exploration and learning, the agent refines its decision-making policies to maximize cumulative rewards over time. A simple framework of RL is shown in Figure 4.7. It depicts the interactions between the agent and the environment and captures the actions, including the state and reward (or penalty). This learning paradigm aligns with manufacturing scenarios where adapting to uncertainties and optimizing resource allocation are critical for success.

RL finds diverse applications in manufacturing, demonstrating its versatility in optimizing various aspects of production. One use case is in building automated robots where the robots learn from the environment and the action they took with the rewards they received. Another use case is in PdM where RL algorithms analyze equipment sensor data to discover patterns indicative of potential failures. By preemptively scheduling maintenance based on learned insights,

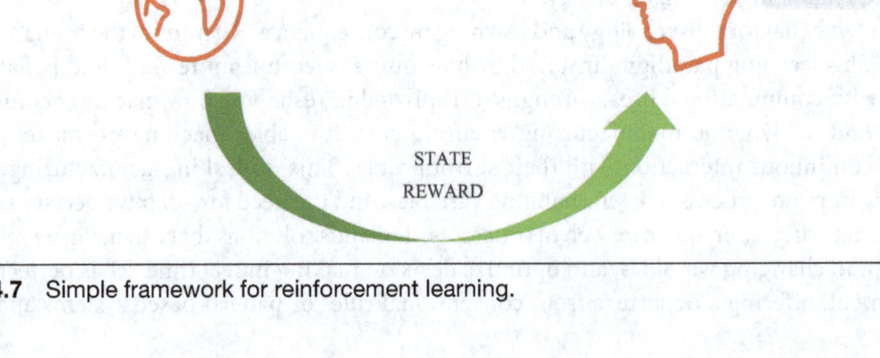

Figure 4.7 Simple framework for reinforcement learning.

manufacturers can reduce unplanned downtime and extend the lifespan of machinery. Supply chain optimization is another area where RL shines. From demand forecasting to inventory management and transportation route optimization, RL algorithms adapt to strategies based on changing market conditions, supplier performance, and unforeseen disruptions. This adaptability enhances the resilience of manufacturing supply chains, aligning with the industry's increasing focus on agility. RL can also support QC processes. By learning from historical data on product defects and variations, RL algorithms can optimize inspection procedures, ensuring higher product quality and reducing the likelihood of defects reaching the market. Finally, RL also supports the production process—especially those where robots are trained using RL.

> An automotive manufacturer deployed RL algorithms to optimize its robotic assembly line to improve efficiency, minimize downtime, and reduce errors in the assembly process. This effort improved production efficiency by almost 15%, reduced downtime by about 18%, and decreased the error rate by nearly 25%. Another case study is from a consumer products company. It implemented RL to optimize its inventory management system. The algorithms continuously learned from real-time data to adjust inventory levels, reorder points, and supply chain parameters dynamically. It resulted in a reduction in excess inventory of nearly 25%, increased on-time deliveries by nearly 15%, and improved inventory turnover ratio by almost 18%.

While RL presents promising opportunities, manufacturers must navigate certain challenges during its implementation. One key consideration is the significant computational resources required for training RL models, especially in complex manufacturing environments. This complexity and the need for extensive exploration of agents and environments may demand substantial computing power, impacting the feasibility of RL in resource-constrained settings. The interpretability of RL decisions also becomes a crucial consideration, particularly in applications where transparency and accountability are paramount. This is especially true when dealing with critical aspects such as safety protocols or compliance standards. Ensuring that RL algorithms provide interpretable outputs remains a priority for their widespread adoption in manufacturing. The intricate interplay between the promises and challenges of RL sets the stage for a comprehensive exploration of specific RL algorithms tailored to address the unique demands of manufacturing.

This section demonstrated the essence of ML, elucidating its fundamental principles and applications within the manufacturing value chain. From supervised learning (regression, SVM, tree-based algorithms), where models are trained on labeled data, to unsupervised learning (clustering—k-means, hierarchical, and isolation forests) uncovering patterns in unlabeled datasets, ML techniques empower manufacturers to extract valuable insights, optimize processes, and enhance decision making. Meanwhile, reinforcement learning introduces a paradigm where agents learn by interacting with an environment and receive feedback in the form of rewards or penalties. This iterative process equips the agent to optimize its actions, making reinforcement learning particularly adept in scenarios like game-playing, robotics, and autonomous systems.

AI ALGORITHMS IN MANUFACTURING

In the heart of modern manufacturing, a quiet revolution has been taking place—driven not by the noise of the machinery but by the subtle hum of AI algorithms within on-site data centers or those on the cloud. These algorithms, empowered by the prowess of highly sophisticated mathematics within lines of code, are rewriting the narrative of traditional manufacturing. They are turning raw information into actionable insights and transforming the shop floor into a perceptive space where machines learn, adapt, and optimize in real time.

> *"Some people call this artificial intelligence, but the reality is this technology will enhance us. So instead of artificial intelligence, I think we'll augment our intelligence."*
>
> —Ginni Rometty, Former CEO of IBM

The fusion of AI and manufacturing is not merely a technological evolution; it is a transformative trend reshaping the landscape of industry. Imagine a factory where every cog and gear, from product design to distribution, is managed by algorithms. In this age of intelligent manufacturing, AI algorithms are the digital artifacts that are the architects of efficiency and precision and are catalysts of innovation to reshape the future of production. In the past, factories were shrouded in uncertainty, and the fear of unexpected downtime made each worker a firefighter, working to extinguish the *downtime fire* by shaping the final product with hands and tools. Today, the landscape of manufacturing is no longer confined to the physical realm of gears and levers—it has evolved. Manufacturing now extends into the virtual domain, where algorithms, fueled by mathematical concepts and the principles of AI, navigate the complexities of production with unparalleled finesse.

The role of the workforce has shifted from manual labor to strategic oversight, where human expertise combines with the precision of AI to build factories of the future. It is evident that the true revolution lies not in the tools but in the intelligence that guides them. AI algorithms are the catalysts propelling this industrial renaissance, reshaping manufacturing processes, elevating precision, and heralding an era where efficiency is not just a metric but a dynamic learning entity. This section explores the various algorithms used within AI, the profound impact these algorithms have, and their role as transformative agents in modern manufacturing.

> *"While the popular view is that insights are the key benefit of artificial intelligence, in truth, AI creates value by improving the quality of decisions."*
>
> —Ajay Agrawal, Joshua Gans, and Avi Goldfarb

The AI Architectures of Intelligent Manufacturing

In the bustling realm of manufacturing, where all cogs and gears play a crucial role, neural networks (NNs) have emerged as the unsung heroes orchestrating a silent revolution. Imagine them as the master blueprints of a factory's brain, akin to a foreman with an encyclopedic knowledge of every process. These NNs, inspired by the intricacies of the human brain, bring a cognitive prowess to machines, enabling them to learn, adapt, and make decisions—transforming the manufacturing landscape into an intelligent and responsive ecosystem. NNs, at their core, mirror the neural architecture of the human brain, learning from experience to grasp the nuances of manufacturing by analyzing vast datasets. NNs, through layers of mathematical algorithms, have discerned patterns in data, enabling them to spot irregularities, defects, or opportunities for optimization with a precision that exceeds human capability. As the manufacturing landscape evolves, NNs and their various embodiments have guided the way toward intelligent manufacturing. They have become the accelerators, propelling manufacturing into the jurisdiction of adaptability and intelligence. The following sections describe some enhancements to NNs that are widely used within the industry.

Convolutional Neural Networks (CNNs)

Imagine a factory where machines have eyes—CNNs provide that vision. They discern patterns, inspect products with the precision of a hawk, and ensure that every creation meets the highest standards of quality. Like a seasoned quality inspector with an eagle eye for detail, CNNs have transformed the landscape by revolutionizing visual data analysis. CNNs have become the vigilant custodians of QC, ensuring that each product that rolls off the assembly line meets the highest standards with scrutiny unmatched by human eyes. CNNs excel in image recognition, making them indispensable in tasks such as defect detection, quality assurance, or process deviation. Their ability to dissect complex visual information enables them to identify product imperfections, ensuring that only perfect items proceed down the production line.

Imagine a CNC machine as a skilled artisan crafting components with the utmost precision. Much like this artisan, CNNs possess a unique spatial understanding that allows them to discern patterns and features in visual data. This spatial awareness is harnessed in manufacturing for additional tasks like product design and layout optimization. CNNs can analyze blueprints or product schematics, ensuring that every component is meticulously placed for optimal functionality and efficiency. In the manufacturing landscape, efficiency is paramount. CNNs excel in the art of feature extraction, a process akin to identifying and highlighting the critical components of a well-tuned machine. By extracting essential features from visual data, CNNs contribute to optimizing production processes. This ensures that manufacturing operations run seamlessly, identify bottlenecks, and streamline workflows to enhance overall efficiency. Two case studies, one from textiles and one from the aerospace industry, show how CNNs play an important role in their industry. Figure 4.8 shows a simplified representation of a CNN. It comprises two main elements—feature extraction and classification—with steps like convolution, pooling, and full connection before generating output. Many complexities and tuning parameters must be experimented with, evaluated, and selected before getting to the output. Figure 4.8 shows that CNNs are typically used in multiple-class detection (hence, three output values). Multiclass CNNs generate a probability for each class detected, with the highest probability being the most likely.

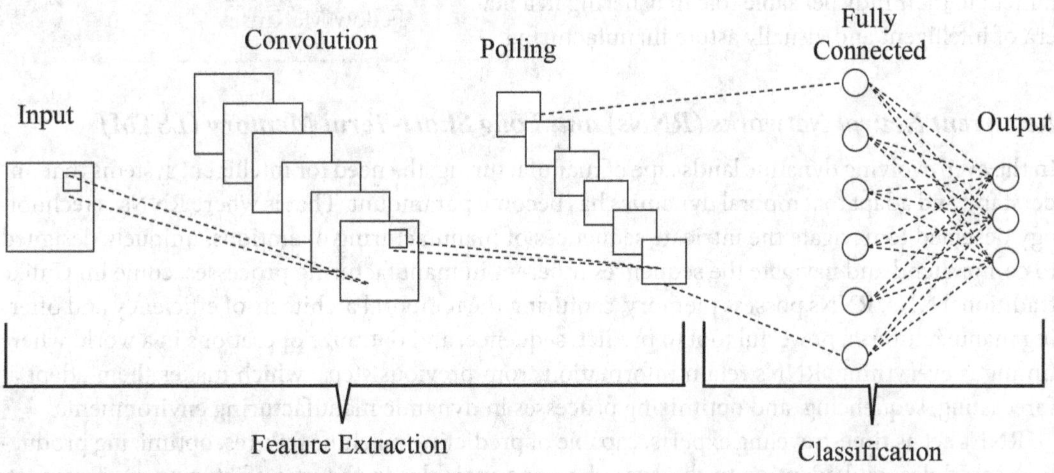

Figure 4.8 Simple representation of convolutional neural networks.[10]

A leading textile manufacturer implemented a CNN-based automated inspection system to detect defects in woven fabrics during production. The CNN model was trained on a dataset of thousands of high-resolution images of both defective and nondefective fabrics, allowing it to learn complex patterns and variations indicative of defects such as tears, stains, or weaving irregularities. Following the deployment of the fabric inspection system, this textile manufacturer improved defect detection accuracy to over 95%, much better than traditional manual inspection methods. It also reduced inspection time by over 50%. Finally, there was a reduction in defective products reaching customers and an almost 30% decrease in customer complaints. Overall, this technology in fabric inspection has enabled the textile manufacturer to enhance product quality, reduce manufacturing costs, and improve customer satisfaction.

In the aerospace industry, CNNs are employed in various applications, including the visual inspection of aircraft components, to ensure safety and compliance with industry standards. A leading aerospace manufacturer implemented a CNN-based inspection system to analyze images of aircraft components captured during the manufacturing process. Its CNN model was trained on a comprehensive dataset containing images of both pristine and defective components, enabling it to identify defects such as cracks, dents, or surface irregularities with high accuracy. This helped the aerospace manufacturer achieve significant improvements in KPIs related to quality assurance and manufacturing efficiency. Defect detection accuracy improved by over 96%, reducing the amount of manual structure inspection time by nearly 30% and allowing it to meet tight delivery deadlines. Finally, the system decreased rework and scrap by over 30%.[11]

In the fast-paced world of manufacturing, split-second decisions can make all the difference. CNNs are active decision makers; whether they identify a faulty component on the assembly line or assess the structural integrity of a product, CNNs make real-time decisions to prevent defects, minimize waste, and contribute to the overall reliability of the manufacturing process. As industries increasingly embrace the future of AI, the adoption of CNN algorithms stands as a testament to their indispensable role in ushering in a new era of intelligent and visually astute manufacturing.

". . . about 70% of companies might adopt some AI technologies by 2030, up from today's 33%, and about 35% of companies might have fully absorbed AI, compared with only 3% today."

—Jacques Bughin, Director McKinsey, Jeongmin Seong, Sr. Fellow McKinsey

Recurrent Neural Networks (RNNs) and Long Short-Term Memory (LSTM)

In the ever-evolving dynamic landscape of manufacturing, the need for intelligent systems that understand and adapt to temporal dynamics has become paramount. That is where RNNs, a technology designed to navigate the intricate sequences of manufacturing operations, uniquely designed to comprehend and navigate the sequences inherent in manufacturing processes, come in. Unlike traditional NNs, RNNs possess memory, capturing the temporal architects of efficiency and offering manufacturers a powerful tool to predict, sequence, and optimize operations in a world where timing is everything. RNNs retain information from previous steps, which makes them adept at forecasting, sequencing, and optimizing processes in dynamic manufacturing environments.

RNNs act as time-traveling experts, capable of predicting machine failures, optimizing production schedules, and adapting to the ever-changing variables in the manufacturing environment.

Their adaptive learning capability and capacity for continuous improvement position them as indispensable assets for manufacturers seeking not just to keep pace with industry rhythms but to lead the charge into a future where intelligent, anticipatory, and flexible systems define the new standard for manufacturing excellence. The narrative of RNNs in manufacturing is not just about algorithms; it is about orchestrating operational precision and future-ready adaptability that resonates throughout the production floor. Figure 4.9 shows a simplified architecture of an RNN, showing the different parameters used within RNN algorithms and capturing the temporal nature of the algorithm.

Manufacturing requires precise timing and sequencing in production. RNNs contribute a sense of rhythm by understanding and optimizing the sequence of production steps. This sequencing skill minimizes idle time, maximizes resource utilization, and ensures a production symphony that plays to the beat of efficiency. In the dynamic world of manufacturing, where change is constant, RNNs showcase their adaptive learning capabilities by dynamically responding to variations in demand and supply chain disruptions, offering a level of adaptability that aligns seamlessly with the unpredictable nature of manufacturing operations. RNNs position themselves well in allocating and optimizing resources. By analyzing past resource utilization patterns, they predict future demands, whether it is for the product or for the availability of resources on the production line. This helps to enhance operational efficiency, minimize waste, and ensure that every move in the manufacturing environment is strategic. Using a window into the historical performance of manufacturing processes, RNNs contribute to a culture of continuous improvement.

> A prominent textile manufacturer implemented an RNN-based demand forecasting system to optimize production planning and inventory management. The RNN model was trained on historical sales data, market trends, and external factors such as seasonal variations and economic indicators. By analyzing sequential data patterns, the RNN model learned to predict future demand for textile products with profound accuracy. This textile manufacturer observed significant improvements in KPIs: demand prediction accuracy increased to over 90%, inventory holding costs were reduced by over 21%, and on-time delivery rates increased by almost 15%.

An extension of an RNN is LSTM. Traditionally, RNNs have the problem of vanishing and exploding gradients, where they become extremely small or large. LSTMs effectively address this gradient challenge within RNNs. Picture the flow of information in an RNN as a stream, where

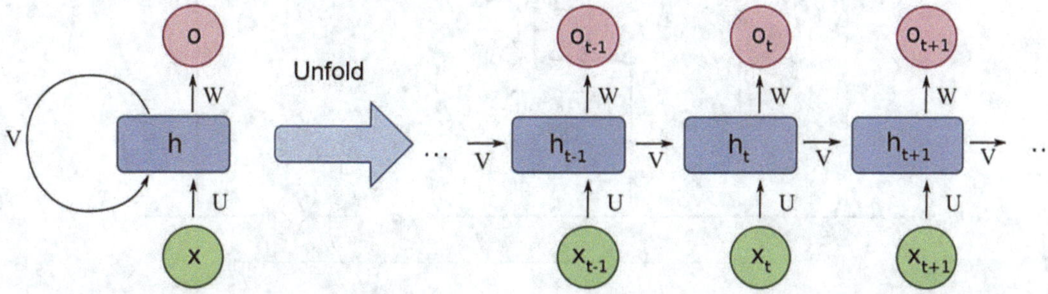

Figure 4.9 Architecture of recurrent neural network (x_t = input at time t; h_t = hidden state at time t; o_t = output at time t, U, V; W = weights).[12]

each item in the sequence affects the subsequent item. LSTMs, equipped with a more sophisticated architecture, introduce the concept of a *memory cell*, a crucial element that regulates the flow of information. This design allows LSTMs to capture and retain information over extended sequences, enabling them to discern long-term dependencies in manufacturing processes. In essence, LSTMs act as the custodians of historical context, ensuring that past information continues to influence future predictions, making them particularly powerful in scenarios where understanding temporal nuances is paramount. Figure 4.10 is a simplified architecture of a single LSTM cell, showing the different parameters used within LSTM algorithms. Think of each *h* in an RNN as a single cell and compare it to an LSTM cell, which is more intricate and complex than an RNN. It captures the temporal nature of the algorithm and the different operations performed within an LSTM cell.

Within the broader context of RNNs, LSTMs can be envisioned as specialized instruments that refine the overall architecture. While RNNs, in general, are adept at processing sequential information, LSTMs elevate this capability by addressing the challenges posed by long-term dependencies. Imagine the manufacturing floor as a complex composition of tasks, where some steps have enduring impacts on subsequent processes. LSTMs, by virtue of their architecture, excel at capturing these enduring patterns. They seamlessly integrate into the RNN framework, enriching the NN's ability to predict, sequence, and optimize manufacturing operations. In essence, LSTMs within RNNs bring depth and nuance to the solution, ensuring that manufacturing unfolds with precision and foresight.

> A leading pharmaceutical company implemented an LSTM-based drug discovery platform to accelerate the identification of potential drug candidates and optimize the drug development process. The LSTM model was trained on large datasets containing molecular structures, biological assays, and chemical properties, allowing it to learn complex relationships between molecular features and drug efficacy. This resulted in a high hit rate for identifying promising drug candidates, surpassing traditional screening methods and reducing the time and resources required for preclinical testing. Implementing LSTM reduced drug discovery time by nearly 28% and increased research productivity by almost 20%.

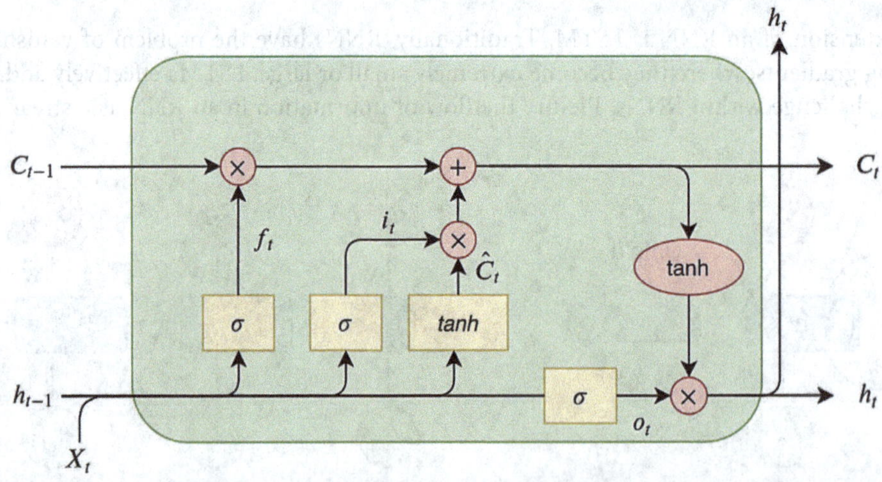

Figure 4.10 Architecture of a single LSTM cell (X_t = input time; h_t = output; C_t = cell state; f_t = forget gate; i_t = input gate; o_t = output gate; σ, *tanh*, x, + = operations to be performed within cell.)[13]

In summary, within the intricate complexity of manufacturing, RNNs and LSTMs emerge as indispensable partners—orchestrating efficiency, adaptability, and continuous improvement. Their unique temporal understanding makes them ideal for manufacturers aiming to keep pace with the rhythms of their industry and to lead intelligent and future-ready production.

Generative Adversarial Networks

In the realm of AI, generative adversarial networks (GANs) play an important role in manufacturing with their unique ability to create synthetic data. GANs are not merely algorithms but act as catalysts for an industrial renaissance. They operate on a simple yet profound principle: pitting two NNs—a generator and a discriminator—against each other. This results in the creation of artificial data that is indistinguishable from real data, unleashing a spectrum of possibilities for applications in manufacturing, especially in situations where large amounts of data are required but not available. In addition, GANs can generate new ideas, designs, and even products. Imagine a machine that not only follows instructions but also contributes innovative designs, sparking a creative renaissance on the factory floor. As we stand at the crossroads of machine intelligence and manufacturing, the significance of GANs becomes a sign guiding us toward uncharted territories of creativity and efficiency. This allows one to explore a transformative force and has the power to redefine product design, optimize supply chains, and usher in an era where the boundaries between the conceivable and the created dissolve. The GAN narrative in manufacturing transcends the technicalities; it is a story of how artfully generated data can become a catalyst for a new industrial revolution.

At the heart of GANs is a pair of NNs—the generator and the discriminator—where the generator composes synthetic data and the discriminator evaluates its authenticity. The generator refines its creations based on feedback from the discriminator, and this iterative process continues until the synthetic data becomes virtually indistinguishable from the real. In manufacturing, this process finds uses in tasks like image synthesis, where GANs can generate lifelike images of products that do not yet exist, aiding in the design and prototyping phases. The GAN-generated prototypes happen before a physical iteration is produced, accelerating the design cycle and enabling manufacturers to swiftly iterate and refine their products. GANs enable manufacturers to generate personalized and customized products by understanding consumer preferences.[14] From clothing designs to product configurations, GANs empower manufacturers to cater to individualized demands at scale. Figure 4.11 shows a block diagram for a GAN, showing how the generator and discriminator interact using the decision probability of how close the generated data is to the real data.

> A leading electronics manufacturer implemented a GAN-based virtual prototyping system to streamline product development and accelerate time-to-market for new devices. The GAN model was trained by using 3-D CAD models, material properties, and performance specifications of electronic components. The resulting prototypes enabled the exploration of design alternatives, the assessment of performance characteristics, and the identification of potential design flaws before physical prototyping. With this system, the electronics manufacturer reduced prototyping costs by almost 35% and shortened design iteration cycles by about 27%, allowing for faster and cheaper introduction of new products. Design accuracy increased by nearly 25%, and time-to-market for new electronic devices decreased by over 20%.
>
> *continued*

In the food and beverage industry, GANs are employed for various applications, including food image synthesis, recipe generation, and product design. A leading food and beverage company implemented a GAN-based product design platform to create photorealistic renderings of new food products and packaging designs. The GAN model was trained on a dataset containing images of food products, packaging materials, and branding elements, enabling it to generate realistic visualizations of product concepts and designs. The GAN solution reduced design iteration cycles by over 50% and shortened time-to-market for new products by nearly 32%, allowing the company to launch innovative products more quickly and efficiently. Consumer engagement increased by almost 25% as photorealistic product renderings captured the attention of consumers and enhanced brand perception. This increased the sales conversion rates by close to 22%.

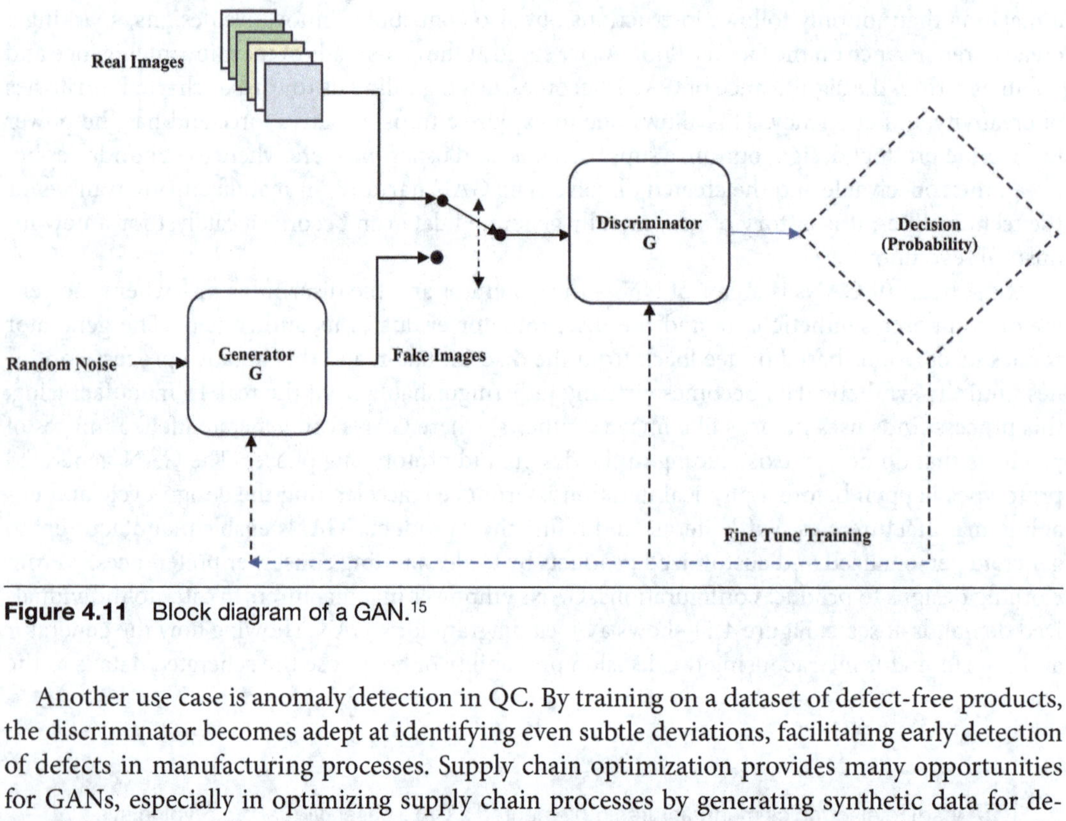

Figure 4.11 Block diagram of a GAN.[15]

Another use case is anomaly detection in QC. By training on a dataset of defect-free products, the discriminator becomes adept at identifying even subtle deviations, facilitating early detection of defects in manufacturing processes. Supply chain optimization provides many opportunities for GANs, especially in optimizing supply chain processes by generating synthetic data for demand forecasting. This enables manufacturers to simulate various scenarios and strategize for fluctuating market demands, reducing the risk of overstock or shortages. Since data always becomes a bottleneck, especially for testing, using GANs to create a virtual test environment by simulating real-world scenarios facilitates testing the robustness of manufactured products. These include anomaly data and supply chain execution data. GANs play a pivotal role by creating synthetic environments for testing, ensuring that products are resilient and adaptive under diverse conditions. Extending this to human-robot collaboration, GANs facilitate the training of robots for complex manufacturing tasks by generating diverse training datasets. This allows robots to learn in simulated environments, enhancing their adaptability and efficiency when deployed in actual manufacturing.

While the promise of GANs in manufacturing is immense, navigating the intricacies poses challenges. One significant hurdle is the potential bias in generated data. If the training data has inherent biases, GANs may replicate and even exacerbate them in the synthetic data they create. Additionally, ensuring the stability of GAN training remains a nuanced task because achieving equilibrium between the generator and discriminator can be delicate. The explainability of GAN-generated results is another challenge. Understanding how and why a GAN produces a specific output is often obscured, raising concerns in industries where interpretability and traceability are crucial. Security concerns also loom large, as GANs can be vulnerable to adversarial attacks where malicious actors manipulate the training data to compromise the model's performance. Moreover, implementing GANs in manufacturing demands a substantial computational infrastructure. Training for GANs can be resource-intensive, requiring powerful GPUs and significant time investments. This poses a barrier for smaller manufacturers with limited resources.

Despite these challenges, the potential benefits of GANs in manufacturing are compelling and warrant exploration and refinement. As researchers and practitioners delve deeper, addressing these challenges becomes integral to harnessing the full potential of GANs in reshaping the future of manufacturing.

Algorithms with Attention Mechanisms and Transformers

Within AI, *attention mechanisms* have recently garnered a lot of attention. Their inception marked a transformative period, introducing a profound shift in how machines perceive, understand, and respond to information. Often likened to a spotlight illuminating the most relevant facets of data, an attention mechanism is not just an algorithm. It is an architecture of refined intelligence, allowing machines to focus on crucial elements and ignore the extraneous. Conceived within the NNs landscape, an attention mechanism addresses a fundamental challenge—the need for machines to selectively attend to different parts of input sequences. This mechanism introduces an adaptive and learnable component to the NN architecture, dynamically adjusting its focus based on the context of the data at hand. Attention mechanisms bring us into uncharted territory where algorithms transcend rigid patterns and adapt with finesse, bringing forth a narrative of intelligent interaction and response.

An attention mechanism functions as a dynamic solution within NNs, enabling selective focus on different segments of input sequences. It operates through a two-step process:

- First, calculate attention weights for each element in the input
- Second, form a weighted sum based on these weights

The calculation of attention weights involves assessing the significance of each element in relation to the context of the task. Elements with higher weights are deemed more important and contribute more significantly to the overall computation. This adaptable mechanism infuses the NN with the ability to attend to specific elements during different processing phases. It introduces a form of memory, allowing the model to assign varying degrees of importance to different parts of the input sequence, enhancing its capacity to capture intricate patterns and dependencies. This dynamic and context-aware approach positions the attention mechanism as a pivotal tool in processing data and comprehending and responding to it with a depth akin to human cognition. The attention mechanism has played a pivotal role in revolutionizing natural language processing and machine translation, paving the way for the development of transformer architecture. Even with LSTM, traditional sequence-to-sequence models struggled with capturing long-range dependencies in sequential data, often leading to suboptimal performance in tasks like language translation.

The attention mechanism addressed this limitation by allowing the model to focus selectively on different parts of the input sequence when generating each element of the output sequence.

The *transformer architecture*, derived from the fusion of attention mechanisms and NNs, has emerged as a groundbreaking paradigm shift in sequence modeling. Unlike CNN or RNN, transformers avoid sequential processing in favor of parallelization. The attention mechanism enables the model to assign varying degrees of importance to different parts of the input sequence, allowing for more effective information capture across the entire context. This parallel processing, coupled with self-attention mechanisms, accelerates training. It also improves the model's ability to handle long-range dependencies, making transformers particularly potent in tasks requiring contextual understanding, such as language translation and sentiment analysis. The transformative impact of attention mechanisms on transformer architecture has extended beyond natural language processing. It is finding applications in diverse domains, including computer vision (both images and videos), speech processing, and, more recently, as the basis for generative AI. Figure 4.12 shows the transformer architecture that is based on an attention mechanism.

> *"Computers are able to see, hear, and learn. Welcome to the future."*
>
> —Dave Waters, Professor, Oxford University

The adoption of attention mechanisms and transformer architectures in manufacturing has ushered in a new era of efficiency and precision. In the realm of PdM, attention mechanisms play a pivotal role by allowing machines to selectively focus on critical components, predicting potential failures with unprecedented accuracy. This targeted approach minimizes downtime and enhances the overall reliability of manufacturing equipment. Transformer architectures are also revolutionizing QC processes, offering a robust solution for detecting defects in real time. By leveraging the inherent parallel processing capabilities of transformers, manufacturers can scrutinize intricate details across production lines, ensuring products meet stringent quality standards with unparalleled precision.

Beyond PdM and QC, the use cases of attention mechanisms and transformer architectures extend to optimizing supply chain operations. Attention mechanisms enhance demand forecasting by discerning patterns and anomalies in vast datasets, enabling manufacturers to align production schedules with market demands. Transformer architectures, with their ability to capture contextual dependencies, prove instrumental in route optimization and logistics planning, streamlining the movement of goods and reducing operational costs. As the manufacturing landscape continues to evolve, attention mechanisms and transformer architectures stand as indispensable tools, driving innovation and reshaping the industry's approach to efficiency and productivity.

In NLP (natural language processing), specifically, as it relates to language translation, the attention mechanism shines because it enables the model to selectively attend to different words in the input sequence, producing more contextually accurate translations. This simplifies the task of translating user and technical manuals in multiple languages, automatically localizing on-premise and cloud-based applications. Cross-geography web pages can be created in one region, automatically pushed out, and localized as region-specific products. As a part of computer vision, image captioning uses the power of attention mechanisms and transformer architecture and allows models to focus on specific regions of an image, aligning the generated captions more closely with the visually significant elements. This can help with focusing on areas where anomalies are detected. This can be extended to videos (e.g., training and events) with a voice-over on what the video is showing. Video-related diagnosis can highlight conditions before and after an event, whether it is production or safety related, show deviations from the normal production process, and, potentially, identify material shrinkages. For speech-to-text applications, attention mechanisms

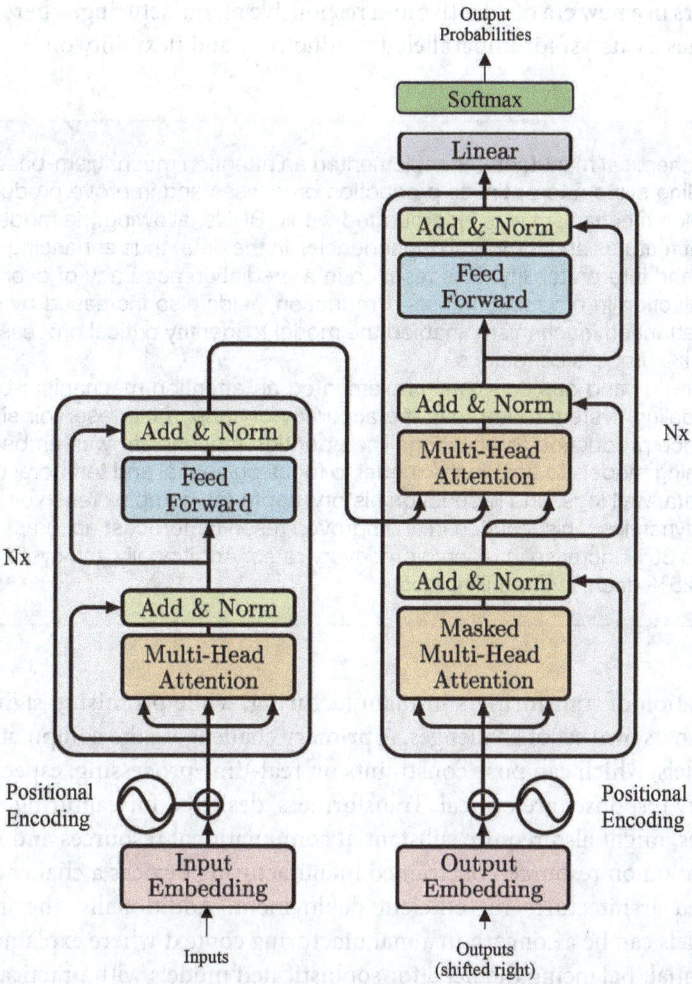

Figure 4.12 The transformer architecture.[16]

enhance accuracy by dynamically focusing on different segments of audio input, capturing the sentiment and nuances of the spoken language. Alternatively, using text-to-speech within applications allows for a multi-modal dissemination of information. Adding images and videos to this can extend the modality depending on the use case and user needs.

In the dynamic landscape of manufacturing, attention mechanisms and transformer architectures present a transformative synergy when integrated with intelligent autonomous systems. In the realm of self-driving cars, attention mechanisms are crucial. By leveraging attention mechanisms, *intelligent systems* can selectively focus on the critical aspects of the manufacturing process, adapting their focus to areas requiring immediate intervention or optimization. The transformer architecture empowers self-driving industrial transport systems and automated material handling vehicles to navigate and interact seamlessly within complex manufacturing environments. This integration opens novel avenues for efficiency, as these vehicles, equipped with attention mechanisms and transformers, can dynamically adjust their operations based on real-time production demands, collaborate with other machinery, and anticipate potential bottlenecks. This symbiotic relationship ushers in a new era of adaptive and responsive manufacturing, where attention-driven autonomy becomes a catalyst for unparalleled productivity and flexibility on the factory floor.

A leading chemical manufacturer implemented an attention mechanism-based predictive modeling system to optimize production processes and improve product quality. The attention mechanism was incorporated within RNNs, allowing the model to focus on relevant features and temporal dependencies in the data, thus enhancing predictive accuracy and interpretability. This resulted in a prediction accuracy of over 90% and a 17% reduction in production costs. Production yields also increased by over 22% since the attention mechanism enabled the model to identify critical process variables and optimize process conditions.

A leading oil and gas company implemented an attention mechanism-based reservoir modeling system to improve the accuracy of subsurface reservoir simulations and enhance production forecasting. The attention mechanism was embedded into deep learning models to enable the model to focus on spatial and temporal features in seismic data, well logs, and production history, capturing complex reservoir behaviors and fluid dynamics. This resulted in an improved reservoir forecast accuracy of almost 95% and a 30% increase in reservoir recovery rates. Additionally, the system contributed to a 25% decrease in drilling costs.

The incorporation of transformers in manufacturing, while promising significant advancements, comes with its own set of challenges. A primary challenge is the computational intensity of transformer models, which can pose constraints on real-time processing, especially in scenarios where low-latency responses are critical. Transformers, designed for capturing extensive contextual dependencies, might also require substantial computational resources and memory, making their implementation on resource-constrained manufacturing devices a challenge. These models require optimized architectures for efficient deployment. Additionally, the interpretability of transformer models can be a concern in a manufacturing context where explainability and transparency are essential. Balancing the need for sophisticated models with practical constraints and ensuring interpretable outcomes are key challenges manufacturers face when adopting transformer architectures. Further, adapting transformers to handle specific manufacturing nuances and ensuring seamless integration into existing workflows are ongoing challenges that necessitate careful consideration and customization. Like all other algorithms, whether in ML or AI, careful tuning and robust training data are mandatory to avoid biases and make the models more accountable, ethical, and responsible.

As we navigate the attention mechanism and transformer architecture in AI, the mechanism's role extends beyond mere algorithmic functionality. It becomes an indispensable component in the narrative of machines that not only process but also understand, learn, and adapt dynamically. The attention mechanism and transformers are an inspiration guiding us toward more human-like interactions between machines and data, a key mechanism shaping the trajectory of AI applications across industries. In the history of AI, attention

> *". . . an area of risk is making sure that your AI sensors or classifiers are in fact not biased and/or they're resilient. And one of the ways you make sure they're resilient and unbiased is you make sure that you have much more diversified data."*
>
> —Mark Maybury, CTO of Stanley Black & Decker

mechanisms and transformer architecture have emerged as a guiding force, redefining how machines engage with and respond to information generated within their environment. Their adaptive nature with data surpasses conventional algorithms, offering a nuanced approach to tasks ranging from language processing to image recognition. Challenges notwithstanding, these techniques draw us toward an era where the interactions between machines, information, and humans are marked by accuracy and a profound depth of understanding. As we explore the intricacies of attention in AI, we find ourselves not in the presence of mere algorithms but in the company of tools and techniques of intelligence, shaping the future of human-machine collaboration. Table 4.3 depicts a high-level comparison of these AI algorithms.

Table 4.3 A high-level comparison of AI algorithms across different dimensions.

Feature	CNN	RNN	LSTM	GAN	Attention Mechanisms & Transformers
Learning Curve	Moderate	Steep	Steep	Moderate	Moderate
Popularity	High	Moderate	High	Moderate	High
Flexibility	Low	Moderate	Moderate	High	High
Community Support	Strong	Moderate	Strong	Moderate	Strong
Use Cases	Image recognition; computer vision	Time-series prediction; NLP	Time-series prediction; NLP	Image generation; data synthesis	Machine translation; document summarization
Pros	Effective for spatial feature extraction	Sequential data processing; language modeling	Improved long-term dependency modeling	Generates realistic data; unsupervised learning	Captures long-range dependencies; self-attention mechanism
Cons	Limited contextual understanding	Vanishing/exploding gradient problem	Complex training; computational intensive	Mode collapse; training instability	Computational intensive; large memory requirements

AI TOOLS AND FRAMEWORKS IN MANUFACTURING

This exploration will dig into some intricacies of AI tools and frameworks in the manufacturing domain, unraveling their multifaceted applications and the pivotal role they play in revolutionizing the industry. The description unfolds beyond mere technological adoption, offering manufacturers a strategic lens through which they can harness the power of AI to elevate their operations, streamline processes, and achieve unparalleled competitiveness. It provides a narrative on the journey through the realm of AI tools and frameworks in manufacturing, illuminating the transformative potential that lies within their digital architecture.

Picture a manufacturing floor where cognitive computing tools seamlessly intertwine with data streams, deciphering intricate patterns and predicting maintenance needs before a single cog falters. Envision robotic process automation frameworks managing and streamlining repetitive tasks and liberating human talent for more intricate endeavors. Amidst the digital AI landscape, manufacturers have discovered not just a technological upgrade but a compelling storyline of adaptability, resilience, and foresight.

Scikit-Learn

Scikit-learn is a powerful and user-friendly open-source ML library in Python. It is a go-to tool for data analysis in manufacturing and has become indispensable for engineers, data scientists, and ML practitioners. The rich array of algorithms provides simple and efficient tools for data analysis and modeling. Its extensive set of algorithms simplifies model development and experimentation to unlock valuable insights from complex datasets. Offering a range of algorithms, scikit-learn is designed to be user-friendly while maintaining high-performance standards. Scikit-learn has become vital for manufacturers trying to optimize processes, reduce costs, and enhance overall efficiency.

Manufacturing enterprises, driven by Industry 4.0, are navigating a transformative journey toward smart and connected production systems. Scikit-learn provides a Pythonic gateway to this transformation, facilitating the deployment of ML models that can predict equipment failures, optimize supply chain logistics, and enhance QC processes. Its user-friendly interface and extensive documentation make it accessible to practitioners, fostering collaboration across interdisciplinary teams. As this section unfolds, we explore the technological intricacies of scikit-learn and some key use cases, exploring how its functionalities align with the unique challenges and opportunities presented by manufacturing.

Scikit-learn, as a comprehensive ML library, has found substantial application across various facets of manufacturing, ushering in a new era of data-driven decision making. One of its key technological strengths lies in the extensive set of algorithms in various Python libraries for numerical and scientific computing and data visualization. From traditional classifiers and regression models to more advanced techniques like clustering and dimensionality reduction, scikit-learn provides a versatile toolbox for addressing various manufacturing challenges. This diversity enables practitioners to choose algorithms that align with specific use cases, offering flexibility in solving problems ranging from PdM to QC. Moreover, scikit-learn excels in its support for feature engineering and preprocessing, which are crucial steps in preparing manufacturing data for analysis. Its rich suite of functionalities allows engineers to handle tasks such as data scaling, normalization, and handling missing values, ensuring that the input to ML models is optimized for accuracy. In manufacturing, where data can be heterogeneous and messy, these preprocessing capabilities become invaluable.

Scikit-learn is designed to seamlessly integrate into the Python ecosystem, making it well-suited for manufacturing environments that leverage Python as a primary programming language. This integration fosters collaboration between data scientists and engineers. Additionally, scikit-learn's compatibility with popular data science tools and frameworks further enhances its utility, providing manufacturing professionals with a familiar and efficient environment for model development and deployment. Further, the scikit-learn library's emphasis on model interpretability aligns with the transparency required in manufacturing processes. Understanding how an ML model arrives at a decision is crucial for gaining trust in its predictions, especially in applications where safety and compliance are paramount. Scikit-learn incorporates interpretability features that contribute to the explainability of models, ensuring that manufacturing stakeholders can comprehend and validate the insights generated.

In the era of Industry 4.0, manufacturing enterprises are increasingly investing in Internet of Things (IoT) infrastructure for real-time data collection. Scikit-learn's compatibility with IoT data pipelines positions it as a valuable tool for harnessing the sensor-generated data in manufacturing processes. Whether analyzing sensor readings for PdM or optimizing production based on real-time feedback, scikit-learn becomes an integral component in the journey toward intelligent and adaptive manufacturing systems. Scikit-learn finds applications in diverse fields, from data preprocessing and feature selection to classification, regression, clustering, and more. A common use case for scikit-learn includes PdM, where it plays a pivotal role in applications within manufacturing. By utilizing classification algorithms, such as random forests or SVMs, the library assists in forecasting equipment failures, enabling proactive maintenance interventions. This helps industries minimize downtime, optimize resource allocation, and extend the lifespan of critical machinery.

Other common use cases are QC and anomaly detection. In manufacturing, maintaining product quality is paramount. Scikit-learn's clustering algorithms, including k-means and DBSCAN, find applications in QC by identifying patterns in production data. Anomalies and defects can be detected efficiently, allowing for timely corrective measures and ensuring that only products meeting stringent quality standards reach the market. Using unsupervised learning algorithms, scikit-learn aids in fault detection within industrial processes.

> A leading automotive manufacturer implemented scikit-learn-based ML models for anomaly detection to mitigate potential issues that could impact product quality and operational efficiency. The manufacturer collected sensor data from production lines, vehicle components, and supply chain processes to train scikit-learn models, such as isolation forests and one-class SVMs. The outcome demonstrated a successful solution with a detection accuracy of over 95% and a reduction in defect rates of 23%. Additionally, the models contributed to a decrease in production downtime of over 15%.

Scikit-learn is also instrumental in supporting optimization in several areas. Scikit-learn's regression and optimization capabilities contribute to supply chain management in manufacturing. By analyzing historical data, scikit-learn assists in demand forecasting, helping manufacturers optimize inventory levels, streamline procurement processes, and enhance overall supply chain efficiency. This use case is crucial for meeting customer demand while minimizing excess inventory costs. Hand-in-hand with supply chain efficiency is process optimization and yield improvement. Scikit-learn's diverse algorithms support process optimization by analyzing complex relationships within manufacturing processes. This involves employing regression models to identify variables impacting yield and efficiency. Manufacturers can then fine-tune operational parameters to maximize output and minimize resource waste. Diving deeper into process optimization, you will find energy consumption optimization. The manufacturing sector often consumes a substantial amount of energy. Regression models within scikit-learn enable the development of algorithms to predict and optimize this consumption. By analyzing factors such as production schedules, weather conditions, and equipment efficiency, manufacturers can implement energy-efficient practices that contribute to the manufacturer's environmental, social, and governance (ESG) initiatives.

While scikit-learn is powerful, implementing ML models using the scikit-learn library in a manufacturing environment includes several challenges and considerations that organizations must carefully address. One of the primary challenges lies in the quality and readiness of manufacturing data. Inconsistent or incomplete datasets can adversely impact the performance of scikit-learn models. Manufacturers need to invest significant effort in thorough data preprocessing, addressing issues like missing values and outliers, to ensure the reliability of the models. Interpreting ML models is another significant consideration. While scikit-learn provides a plethora of powerful algorithms, some of them—particularly the complex ones—generate less interpretable models. This lack of interpretability can pose challenges in industries where explaining model decisions is crucial for regulatory compliance and transparency. Scalability is also a concern when dealing with large manufacturing datasets. The ability of scikit-learn models to efficiently handle extensive data volumes warrants careful evaluation. Manufacturers must assess the suitability of scikit-learn for specific use cases where scalability is paramount.

Integration with existing manufacturing systems is a crucial aspect of scikit-learn implementation. Compatibility with data storage solutions, real-time data streaming, and seamless communication with other software and hardware components must be ensured to avoid disruptions in ongoing operations. Continuous model maintenance is vital in dynamic manufacturing environments. Processes evolve over time, and models need regular updates to remain accurate and relevant. Establishing robust processes to update models based on model drift, new data, changing business and market needs, and changing operational dynamics is essential for sustained success.

Security and privacy concerns are heightened in manufacturing, where sensitive information related to proprietary processes and intellectual property is at stake. Implementing robust security measures to protect data from unauthorized access and potential cyber threats is imperative for maintaining trust and compliance with regulations. To ensure this, having the necessary skill set and training is paramount and must be addressed from the start, posing a significant challenge. Successful implementation of scikit-learn models requires a skilled workforce proficient in AI, ML, and data science. Manufacturers face challenges when acquiring or upskilling personnel to effectively leverage scikit-learn's capabilities. Ongoing training and fostering a culture of data-driven decision making are critical to overcoming this challenge. Its simplicity and extensive documentation make it an excellent choice for quick prototyping and exploration, especially in scenarios where deep learning complexity is unnecessary. Users should consider the nature of their data and the specific requirements of their tasks when considering scikit-learn for their projects.

TensorFlow

In the intelligent manufacturing landscape, TensorFlow commands the spotlight as a player of progress. Born from a collaboration within Google Brain, TensorFlow is as an open-source platform that organizes the complex interplay of AI, ML, and manufacturing. This section explains the intricate layers of TensorFlow's influence on the manufacturing stage, from its foundational technology and versatile applications to the challenges encountered in its implementation.

TensorFlow is a versatile open-source AI and ML framework that empowers manufacturers to create and deploy AI and ML models. Its scalability and adaptability make it a cornerstone in various industrial applications. TensorFlow resonates across the manufacturing landscape, promising to transform the traditional patterns of production and acting as a catalyst, steering manufacturing enterprises into a new era where predictive intelligence, efficiency, and innovation converge seamlessly. TensorFlow's genesis lies in the pursuit of democratizing AI and ML, empowering both novices and experts to compose the solutions needed for intelligent manufacturing. Its open-source nature invites collaboration, echoing the spirit of cooperation for advancements in various areas of manufacturing.

This section illuminates the technique of TensorFlow's architecture by delving into the technological nuances. Its modular and scalable design lays the foundation for a powerful AI/ML symphony that is capable of orchestrating tasks ranging from PdM to QC. TensorFlow also supports both training and deployment, emphasizing its role as the architect unifying diverse ML tasks.

TensorFlow boasts a design that helps build intelligence seamlessly into AI/ML models. Its adaptability and power in handling complex AI/ML tasks make it an indispensable tool for modern manufacturers. At its core lies a flexible and modular architecture, embodying the essence of adaptability for diverse manufacturing needs. This foundational modular and scalable design makes TensorFlow powerful and enables manufacturers to craft tailored solutions that resonate with the unique nuances of their production processes. The ability to seamlessly integrate with existing manufacturing systems positions TensorFlow as a catalyst for efficiency, driving a wave of innovation in the industry.

TensorFlow's versatile architecture, rich array of tools, and extensive library of prebuilt modules and application programming interfaces (APIs) facilitates the implementation of complex algorithms, reducing the learning curve. TensorFlow's design philosophy revolves around accessibility, empowering users to leverage its capabilities without compromising on sophistication. In manufacturing, scalability and flexibility are nonnegotiable, as are the needs to handle vast datasets and intricate models. TensorFlow's architecture, built on a dynamic computational graph of NN structures, facilitates the efficient development and deployment of AI/ML models. This accessibility and flexibility become a compelling reason for manufacturers to embrace TensorFlow, providing them with a technological compass to navigate the intricate landscape of intelligent manufacturing and future-proofing their operations.

Manufacturers have found that TensorFlow's design acts as a catalyst for innovation, enabling seamless integration into existing workflows and offering transformative solutions to age-old challenges. Its compatibility with various hardware configurations ensures manufacturers can harness its power without overhauling their infrastructure.

A leading manufacturer specializing in automotive components implemented TensorFlow-based ML models to optimize its robotic assembly lines. The company collected sensor data from robotic arms, conveyor belts, and other equipment to train TensorFlow models for PdM and real-time performance monitoring. The models increased the detection of equipment failures with a precision of over 90%, led to a reduction in maintenance costs of 40%, and increased equipment uptime by 30%, ensuring the smooth operation of assembly lines and maximizing production output. With optimized robotic operations, OEE improved by almost 20%, and the defect rate decreased by nearly 15%.

For PdM, TensorFlow's prowess becomes evident by predicting equipment failures, optimizing maintenance schedules, and minimizing downtime. QC ascends to new heights as TensorFlow empowers automated visual inspections, ensuring products meet stringent standards with unprecedented precision. Process optimization takes center stage as TensorFlow refines manufacturing processes through predictive analytics, enhancing efficiency and resource utilization. Supply chain management experiences a transformative wave with TensorFlow through optimizing logistics and demand forecasting and bolstering decision-making processes. TensorFlow facilitates seamless interactions between workers and AI-powered systems, fostering efficiency and safety on the shop floor. The integration of TensorFlow in manufacturing lays the foundation for intelligent production lines, orchestrating synchronized operations and adaptive manufacturing strategies.

While TensorFlow is a technological marvel, its integration into manufacturing unearths challenges and vital considerations. Tailoring models for manufacturing processes demands a nuanced approach, striking a balance between model complexity and computational efficiency. Interoperability hurdles arise as manufacturers seek to seamlessly integrate TensorFlow with diverse and esoteric legacy systems, demanding meticulous planning for harmonious coexistence. Privacy and security concerns and demand for skilled expertise loom large, emphasizing the need for skilled people not just to build TensorFlow models but also to incorporate robust cybersecurity measures and privacy-preserving techniques. Adapting to TensorFlow's dynamic landscape and ensuring scalability aligns with real-time processing requirements add layers to the manufacturing narrative. A strategic, informed approach becomes imperative for a successful TensorFlow manufacturing journey.

Keras

Keras is an open-source deep learning library built with high-level NN APIs written in Python. It has gained prominence in manufacturing applications due to its simplicity and versatility. It also seamlessly integrates popular and powerful NN frameworks like TensorFlow, Theano, and Microsoft Cognitive Toolkit, making complex NN architectures more accessible to a broader audience. Keras has democratized deep learning, empowering both seasoned ML experts and novices to effortlessly craft intricate NNs. Keras has also facilitated rapid prototyping and experimentation. Whether embarking on image recognition endeavors, natural language processing challenges, or predictive analytics, Keras offers an intuitive interface that accelerates the model development process. It bridges complexity and simplicity in deep learning, connecting the complexity of cutting-edge NN architectures with the simplicity sought by practitioners. Envisioned as an open-source tool to facilitate fast experimentation with NNs, Keras transcends traditional barriers, fostering an environment where creativity and problem-solving converge seamlessly. This section explores the multifaceted landscape of Keras, unraveling its technological underpinnings, applications across manufacturing, challenges encountered in real-world implementations, and the promise it holds for the future of AI in manufacturing.

Keras provides manufacturing industries with a simplified interface for building, training, and deploying deep learning models. It provides a high-level abstraction that shields users from the complexities of low-level operations, enabling developers to focus on building, training, and deploying NNs effortlessly. Its compatibility with frameworks like TensorFlow allows manufacturers to leverage advanced features for developing complex NN architectures. Keras facilitates the creation of CNNs, RNNs, and other architectures crucial for tasks like image recognition, PdM, and QC. Keras supports both CPU and GPU acceleration, offering manufacturers flexibility in choosing hardware configurations based on their computational requirements. This adaptability is crucial for addressing the diverse computing infrastructures present in manufacturing environments. The library's modular design enables seamless experimentation with different NN configurations, fostering innovation and adaptation to specific manufacturing challenges. Manufacturers also benefit from Keras's extensive documentation and large community of developers, ensuring ongoing support and updates.

Keras was developed with a user-centric philosophy and provides an abstraction layer atop popular deep learning and powerful frameworks. Its design focuses on enabling rapid experimentation and iteration, fostering a conducive environment for innovation and breakthroughs in

AI. One of the standout features defining Keras is that the framework is designed with a modular building-block approach, allowing users to assemble NN layers like Lego blocks. This modularity enhances flexibility and facilitates the creation of complex architectures with minimal code. Keras enables users to construct NNs through its expressive API, removing the intricate details of gradient calculations and optimization while still offering room for customization.

In recent years, advancements in the deep learning landscape have led to updates and enhancements in the Keras library. For instance, the transition to TensorFlow 2.x as the default backend has improved performance, added eager execution capabilities, and tightened integration with TensorFlow's ecosystem. The continuous evolution of Keras, supported by active community contributions, reflects its commitment to staying at the forefront of technological innovations in the field of deep learning.

In QC, manufacturers use Keras-powered models for image classification to identify defects in products during production. PdM benefits from Keras's ability to create effective models for analyzing sensor data, predicting equipment failures, and optimizing maintenance schedules. Keras is also employed in demand forecasting, helping manufacturers predict market trends, optimize inventory levels, and enhance supply chain management. Its versatility in handling various types of data makes it a valuable tool for addressing multiple use cases across the manufacturing value chain.

Keras is also instrumental in enhancing industrial automation processes, particularly in robotics and PdM (see the case study in the TensorFlow section about the leading manufacturer specializing in automotive components). That company integrated its TensorFlow with Keras-based AI models to optimize its robotic assembly lines. The company gathered sensor data from robotic arms, conveyor belts, and other equipment to train Keras models for PdM and real-time performance monitoring.

Keras, known for its user-friendly and adaptable deep learning framework, has made significant strides in transforming manufacturing processes. In PdM, Keras models are employed to analyze historical equipment data. Utilizing RNNs or LSTM networks, Keras enables accurate predictions of potential equipment failures. This proactive approach helps manufacturers minimize unplanned downtime, enhance operational efficiency, and optimize maintenance costs.

Keras is instrumental in supply chain and process optimization within manufacturing while offering a robust solution for fine-tuning parameters in complex processes. As a part of demand forecasting, Keras models leverage historical data to predict future demand patterns with greater accuracy. As with earlier benefits of demand forecasting, this enables manufacturers to optimize inventory management, preventing excess stock or stockouts and ensuring a streamlined supply chain. NN models created with Keras excel at learning intricate patterns and relationships within the data, leading to improved process efficiency and resource utilization. In the realm of sustainability, Keras finds application in energy consumption optimization within manufacturing facilities. Deep learning models created with Keras analyze energy usage patterns, identify inefficiencies, and support the implementation of strategies to reduce energy consumption. This contributes to environmental sustainability goals and cost-saving initiatives.

Implementing Keras in manufacturing environments comes with its own set of challenges and considerations. Manufacturers must carefully address the interpretability of complex deep learning models created with Keras. Deep NNs, often employed in Keras, are inherently complex, making it challenging to interpret their decisions. In critical manufacturing processes, where transparency and explainability are essential, addressing the black-box nature of deep learning models becomes crucial. Balancing the accuracy of predictions with the interpretability of models is an ongoing consideration. Manufacturers need to ensure that decisions made by these models align with operational requirements and regulatory standards.

Another significant challenge is the need for substantial computing resources, especially when dealing with large datasets and complex NN architectures. Training deep learning models in Keras demands significant computational power, and manufacturers may need to invest in high-performance hardware or cloud-based solutions to overcome this obstacle. Scaling up a Keras model, especially when dealing with large datasets common in manufacturing, adds its own challenge. Manufacturers must evaluate the efficiency of Keras models in handling extensive data volumes and ensure that the chosen architecture aligns with scalability requirements.

Another key consideration is the need for substantial labeled data for training effective Keras models. Obtaining and annotating large datasets can be expensive, resource-intensive, and time-consuming. Manufacturers need to strategize data acquisition and labeling processes to ensure the robustness of their models. Additionally, maintaining data quality and relevance over time poses an ongoing challenge, requiring continuous efforts in data management. In scenarios where more data simply cannot be obtained, manufacturers might opt for generating synthetic data. Continuous monitoring, updating, and retraining of models are critical to ensuring their relevance and accuracy in dynamic manufacturing settings. Manufacturing processes evolve, and the models developed with Keras need to adapt to these changes effectively. Regular monitoring and updates, if necessary, may be required to keep Keras models accurate and aligned with evolving operational dynamics. As models are built and updated, security concerns surrounding the deployment of AI models, including those built with Keras, merit careful consideration. Manufacturers must implement robust cybersecurity measures to safeguard sensitive manufacturing data. This includes protecting models against adversarial attacks that could manipulate input data to deceive the AI system, potentially leading to quality issues or process disruptions.

While Keras offers immense potential for innovation in manufacturing, addressing these challenges and considerations is essential for successful and responsible integration. Manufacturers need to align their AI strategies with organizational goals, invest in suitable infrastructure, and establish robust data governance practices to harness the full capabilities of Keras in enhancing manufacturing operations.

PyTorch

PyTorch is a dynamic and versatile deep learning framework that has become an indispensable asset in manufacturing, transforming the landscape of AI and ML applications. Renowned for its flexibility and ease of use, PyTorch empowers manufacturers to harness the potential of NNs, facilitating groundbreaking advancements in various facets of production. At its core, PyTorch leverages tensors, the fundamental building blocks of NNs. Tensors are multidimensional arrays that efficiently represent data in PyTorch, providing the mathematical foundation for operations and transformations within NN architectures. The seamless integration of tensor computations into the framework's design ensures that PyTorch can efficiently handle complex mathematical operations, making it well-suited for the computational demands of deep learning in manufacturing. PyTorch provides a combination of dynamic computational graphs with the ability to make on-the-fly adjustments and tensor-based operations. These form the technological backbone that empowers PyTorch to deliver a versatile, expressive, and powerful deep learning environment for a myriad of applications, including those within the manufacturing domain.

PyTorch's technological underpinnings align with the diverse needs of manufacturing. Its robust support for NN architectures, including CNNs, RNNs, and LSTMs, propels applications such as PdM, QC, and process optimization. In PdM, PyTorch demonstrates its prowess by leveraging RNNs to analyze time-series data from machinery. The technology identifies patterns indicative of potential failures, enabling preemptive maintenance and reducing unplanned downtime. QC processes benefit from PyTorch's CNNs, ensuring precise defect detection in manufacturing lines. Here PyTorch's image classification capabilities empower manufacturers to identify and rectify defects with unprecedented accuracy. PyTorch's versatility unfolds across several other use cases, each contributing to operational excellence and strategic advancement. The dynamic nature of PyTorch lends itself to process optimization, enabling adaptive manufacturing through reinforcement learning models. PyTorch's reinforcement learning capabilities enable systems to learn and adjust strategies autonomously, optimizing efficiency and resource utilization. Supply chain management benefits from PyTorch's predictive analytics by optimizing logistics and demand forecasting. The collaboration between human workers and PyTorch-driven AI systems enhances operational efficiency and safety in manufacturing environments.

> A large chemical manufacturer embraced PyTorch-based AI models to optimize its production processes. The company aggregated sensor data from various stages of the production process, including reactors, distillation columns, and storage tanks, to train PyTorch models for PdM and real-time performance monitoring. This reduced equipment failure rates, lowered maintenance costs by 28%, and boosted equipment uptime by 17%. It also led to a substantial OEE improvement along with a decrease in defects.

While PyTorch offers a wealth of possibilities, deploying it in manufacturing presents challenges. Integration with existing systems, data privacy concerns, and the need for specialized expertise in NN architectures are among the key considerations manufacturers must address. Additionally, ensuring the interpretability and explainability of PyTorch models becomes crucial for sectors with stringent regulatory requirements. Successfully navigating these challenges requires a holistic approach involving collaboration between data scientists, engineers, stakeholders, and domain experts. However, the benefits of PyTorch in manufacturing far outweigh the challenges, offering an avenue for innovation and efficiency.

AutoML Tools

In recent years, the surge in data availability and the increasing complexity of ML models have underscored the need for streamlined and automated processes. Automated ML (AutoML)—including AI—represents a groundbreaking evolution in the realm of AI/ML, offering a revolutionary approach to democratizing and expediting the application of AI/ML in diverse industries. AutoML accomplishes this by encapsulating the end-to-end ML pipeline into an automated framework. AutoML tools streamline the model development process for the diverse needs of the manufacturing sector and make AI/ML accessible to non-experts. AutoML tools, such as Google AutoML and H2O.ai, accelerate the adoption of AI in manufacturing by simplifying complex tasks. AutoML is a transformative approach that allows the automation of the end-to-end process of applying AI/ML to real-world problems. AutoML platforms encapsulate complex algorithms, hyperparameter tuning, and model selection for the wide variety of issues encountered in manufacturing.

AutoML leverages diverse technologies to automate and optimize the ML pipeline. The common technologies building the foundations for AutoML include hyperparameter optimization in which algorithms and techniques automatically tune model parameters for optimal performance. Hyperparameters significantly impact the performance of ML models. AutoML employs advanced optimization algorithms, such as Bayesian optimization, grid search, and random search, to automatically tune hyperparameters, ensuring models achieve optimal accuracy and efficiency. Feature engineering is a critical aspect of model development. AutoML tools utilize algorithms to automatically identify, create, select, and generate relevant features from raw datasets. This reduces manual efforts, enhances model interpretability, and improves model generalization and accuracy.

AutoML combines predictions from multiple models to improve overall performance and robustness. Techniques such as *bagging and boosting* help aggregate diverse models, mitigating overfitting and improving overall predictive performance. Some flavors of AutoML allow the user to experiment with various strategies, starting with examining all models and selecting only those that seem to work best. AutoML frameworks and platforms such as Google AutoML, H2O.ai, and the Tree-based Pipeline Optimization Tool provide automated model selection, training, and deployment. The end-to-end ML offered by these frameworks also facilitates hyperparameter tuning, making the entire process more accessible to users with varying expertise.

With the increasing importance of model interpretability within manufacturing, AutoML technologies integrate explainable AI methods. Techniques such as local interpretable model-agnostic explanations and Shapley additive explanations help users understand model predictions, promoting trust and transparency. In manufacturing, where precision and reliability are paramount, understanding the rationale behind model predictions becomes crucial. This transparency fosters trust in automated systems and empowers manufacturing professionals to validate and fine-tune models based on comprehensible insights.

AutoML can be applied across various previously discussed manufacturing scenarios, including PdM, QC, demand forecasting, and anomaly detection. Its automated nature accelerates model development, making it particularly advantageous for organizations aiming to deploy ML solutions rapidly. AutoML plays a pivotal role in PdM by analyzing historical data. AutoML algorithms identify patterns indicative of potential breakdowns, allowing manufacturers to schedule maintenance proactively and minimize unplanned downtime. Along the same line of thinking, AutoML models can also be trained to identify defects or deviations (i.e., anomalies) in real time during the production process. Integrating computer vision and ML, these models ensure the early detection of issues, minimizing the production of faulty units and reducing waste. In addition, AutoML aids in automating QC processes, ensuring that only products meeting quality standards are shipped to customers, thereby improving customer satisfaction. Furthermore, AutoML supports the optimization of complex manufacturing processes. By analyzing sensor data, production metrics, and other relevant parameters, AutoML models can identify opportunities for process improvement. These processes can be fine-tuned to enhance efficiency and resource utilization.

AutoML is also instrumental in optimizing supply chain operations. AutoML contributes to accurate demand forecasting using historical sales data, economic indicators, and external factors, thus helping manufacturers anticipate market trends and plan production accordingly. AutoML models also provide valuable insights for inventory management and production planning. Manufacturers can make data-driven decisions to minimize lead times, reduce costs, and streamline logistics. Further, AutoML models assist in forecasting energy consumption within manufacturing facilities. By analyzing historical energy usage data and incorporating external factors, these models predict future energy demands. This enables manufacturers to implement energy-efficient practices and reduce operational costs. Human resources are key in manufacturing, and AutoML aids in human resource planning by predicting workforce requirements based on production schedules, demand forecasts, and historical patterns. This assists manufacturers in optimizing staffing levels, ensuring the right workforce is available when needed.

> An example of the successful use of AutoML, not just for business benefits but also for sustainability, is that of Epiroc. Epiroc, a large mining, infrastructure, and natural resources company, used AutoML to predict steel quality. Epiroc wanted to reduce CO_2 emissions during heat treatment in its processes. They used Azure Machine Learning, specifically its open-source (MIT license) solution accelerator called ESML (Enterprise Scale ML), to get the AI factory on Azure. The pilot program reduced 3.35 tons of CO_2 from propane annually, reduced electricity usage by 90 MWh, and reduced quality rejections by 30%. The quality rejection saved resources like coal and gas needed to produce the steel.
>
> Another example is from LG, which has implemented AutoML Vision Edge in its manufacturing intelligence solutions. These detect in-line anomalies within the manufacturing of LCD screens, optical films, and automotive fabrics. Defect detection accuracy was improved by 6% and the time to build the models (design and train) was reduced from one week to a few hours.[17]

While AutoML brings about remarkable advancements in democratizing ML, it is not without its challenges. Striking a balance between automation and domain expertise remains a crucial balancing act to consider, ensuring that automated processes align with the intricate needs of manufacturing. The challenge lies in the complexity of choosing an appropriate algorithm and feature set, as the computerized nature of AutoML might not always capture the intricacies of specific manufacturing contexts. Additionally, challenges related to interpretability, model explainability, and potential biases may exist in automated decision making. This poses a challenge in environments where model explanations and understanding the decision-making process of AI models are essential to ensure alignment with industry standards and regulatory requirements. Further, organizations must carefully navigate these issues to ensure the ethical and responsible use of AutoML in manufacturing settings.

In addition, selecting appropriate algorithms and handling large-scale data efficiently are persistent challenges in AutoML adoption. Data quality and preprocessing represent critical considerations in AutoML for manufacturing. Inconsistent or incomplete datasets may lead to inaccurate model outputs, emphasizing the need for thorough data cleaning and preprocessing strategies. Addressing these challenges and carefully considering the specific needs of manufacturing processes is pivotal in maximizing the benefits of AutoML adoption. Despite these challenges, the promise of democratizing AI/ML and empowering manufacturing processes positions AutoML as a valuable asset for the industry's future, offering a gateway to enhanced decision making, productivity, and innovation.

Table 4.4 shows a high-level comparison of the five frameworks.

Table 4.4 A high-level comparison of the five frameworks across different dimensions.

Feature	Scikit-learn	TensorFlow	Keras	PyTorch	AutoML
Learning Curve	Easy to learn; user-friendly	Steeper learning curve; more options	User-friendly; high-level API	Moderate learning curve; flexible	Automated; reduces need for in-depth ML expertise
Popularity	Widely used in ML	Dominant in deep learning and neural nets	Popular, especially for beginners	Gaining popularity; widely adopted	Automation is increasing in popularity
Flexibility	Limited for deep learning tasks	Highly flexible for deep learning	Built on TensorFlow; offers flexibility	Dynamic computational graph; flexible	Automated model selection; hyperparameter tuning
Community Support	Strong community support	Extensive community	Good community support	Growing community; active development	Increasing community engagement
Use Cases	General-purpose ML tasks	Deep learning; neural networks; NLP	Neural networks; quick prototyping	Deep learning; research-oriented	Automation of ML model development across industries
Pros	Easy to use for ML beginners	Versatile and scalable	Simple API; seamless with TensorFlow	Dynamic computational graph	Reduces manual effort; suitable for non-experts
Cons	Limited support for deep learning	Steeper learning curve	May lack some advanced features	Less production-ready than TensorFlow	Limited customization; might not suit complex tasks

SUMMARY AND KEY INSIGHTS

This chapter on common AI/ML algorithms and tools for use in manufacturing delved into the ML components of supervised, unsupervised, and reinforcement learning. The intent was not to cover all the algorithms, just the key ones, such as CNNs, RNNs (including LSTM), GANs, transformers, and attention mechanisms. This was followed by a discussion of five robust platforms that have shaped the industry's technological landscape: scikit-learn, TensorFlow, Keras, PyTorch, and AutoML. Despite their strengths, each component of an ML and AI footprint presents its unique challenges and considerations. From technology nuances to manufacturing-specific use cases, we navigated through the complexities of integrating these tools into industrial workflows.

The types of classes of algorithms used across the manufacturing value chain are displayed in Figure 4.13. Note that the graphic does not delve into detailed algorithms like CNN, LSTM, etc., but instead presents the high-level algorithms that form the foundation of these advanced analytics, AI, and ML approaches.

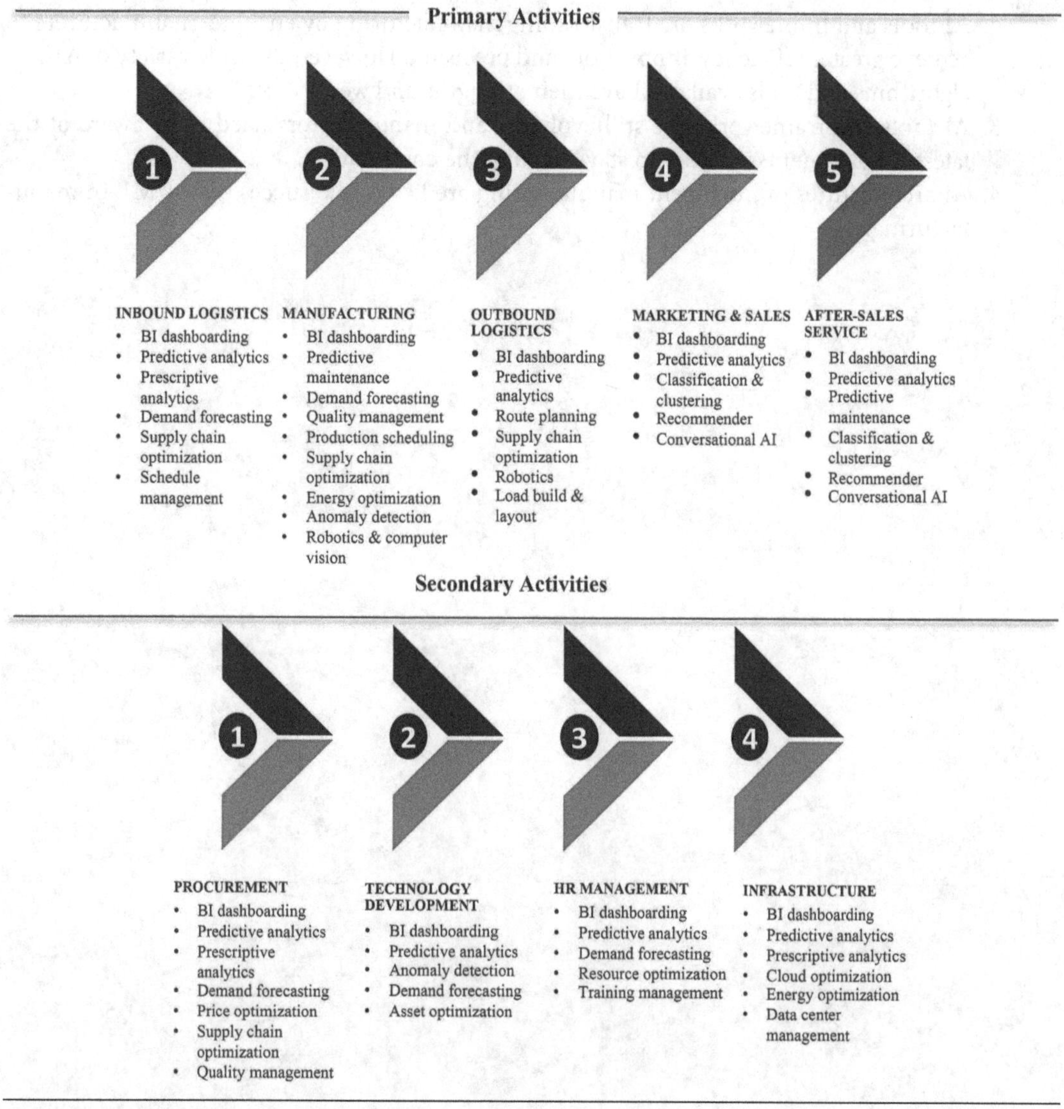

Figure 4.13 Algorithm classes across the manufacturing value chain.

This comprehensive chapter served as a valuable guide for industry professionals, researchers, and enthusiasts aiming to harness the power of AI/ML in manufacturing. It sets the stage for the rest of the book, in which we detail different phases of the manufacturing value chain. The future of manufacturing is bright, with promising opportunities to transform how manufacturers redefine their processes.

Some of the key takeaways from this chapter include:

1. AI/ML is rapidly transforming the manufacturing industry and is expected to play a pivotal role in defining which companies will come out ahead. AI tools and frameworks are being used in a variety of manufacturing applications, including PdM, QC, and process optimization.

2. AI tools and frameworks are transforming manufacturing by enabling manufacturers to achieve greater efficiency, innovation, and precision. However, the wide variety of AI/ML algorithms and tools available have their strengths and weaknesses.
3. AI tools and frameworks are still evolving, and manufacturers need to be aware of the latest developments in order to stay ahead of the competition.
4. AI architectures of intelligent manufacturing are key to the success of AI/ML in manufacturing.

5

PRODUCT DESIGN AND DEVELOPMENT IN MANUFACTURING

AI-Enhanced Product Design and Development at Stellar Electronics: Revolutionizing Manufacturing

Stellar Electronics is a leading player in consumer electronics. Their product design and development had long relied on traditional methods that were time-consuming and often prone to inefficiencies. Recognizing the opportunity for improvement, Rachel Miller, the head of product development, sought to incorporate AI into the design process to streamline workflows and enhance innovation. Partnering with Alex Thompson, Stellar's chief data officer, the team began by integrating generative artificial intelligence (AI) and machine learning (ML) models to expedite product design and prototyping.

Using AI-powered design tools, Rachel's team was able to generate multiple design iterations based on customer preferences, market trends, and historical product performance. This approach allowed them to cut the design cycle time by 30% while improving product quality and alignment with market demands. The AI system automatically generated design variations, simulated performance, and predicted customer response, enabling the team to focus on refining and selecting the most promising options.

The AI tools also revolutionized the prototyping phase. Using ML algorithms, John Lee, Stellar's senior manufacturing engineer, simulated various manufacturing processes and material choices, optimizing for cost and efficiency. This reduced the need for physical prototypes by 25%, saving both time and resources. Stellar's AI system accurately predicted production challenges, which allowed the team to address potential bottlenecks before entering full-scale manufacturing.

Additionally, Samantha Hayes, the director of innovation, leveraged advanced analytics to gather real-time feedback from pilot tests and customer insights. By incorporating this feedback into the AI models, the team refined product features and improved customer satisfaction rates by 20% in recent launches. The AI-driven design process enabled rapid iteration and data-backed decision making, ultimately helping Stellar bring innovative products to market faster and with greater precision.

Thanks to these AI-driven advancements, Stellar Electronics transformed its product development workflow, making it more agile, efficient, and customer-focused. The results were faster time to market and a notable increase in customer satisfaction and product innovation across its manufacturing processes.

"Innovation is the specific instrument of entrepreneurship. The act that endows resources with a new capacity to create wealth."

—Peter Drucker, management consultant, educator, and author

In the dynamic landscape of manufacturing, the process of product design and development is the foundation of innovation and competitiveness. It plays a pivotal role in determining the success or failure of a company's endeavors. Basically, product design and development involves the conceptualization, creation, and refinement of tangible goods and processes, ranging from consumer electronics to industrial machinery and supply chain processes. This multifaceted process involves numerous stakeholders, including designers, engineers, marketers, and supply chain professionals, collaborating to transform ideas into market-ready products. With the advent of advanced analytics, business intelligence (BI), and AI/ML technologies, product design and development has undergone a profound transformation, ushering in a new era of possibilities and challenges.

As manufacturing organizations strive to remain agile and responsive to evolving market demands, the role of data-driven decision making in product design and development has become increasingly prominent. Advanced analytics and BI empower manufacturers to leverage data insights from various sources, like customer feedback, market trends, and competitor analysis, to guide strategic product decisions. By harnessing the power of data visualization, predictive and prescriptive modeling, and trend analysis, manufacturers can gain valuable insights into consumer preferences, market dynamics, and emerging trends.

> *"We used AI to develop a system that advises our chemists on the next set of recipes to try as they iterate toward a final solution. We've found dramatic reductions, in many cases, with that approach; dramatic means reducing those R&D loops [by] 70% or 80%."*
> —Chris Couch, SVP, CTO at Cooper Standard

The integration of AI/ML technologies into the product design and development process holds additional immense potential for driving innovation and efficiency. AI algorithms, such as generative design and predictive modeling, enable manufacturers to explore a vast array of design possibilities, optimize product performance, and enhance manufacturing processes. Companies like General Electric leverage AI-powered tools to accelerate product prototyping and optimize product designs for performance, cost, and sustainability.[1] Rolls-Royce Civil Aerospace is using additive manufacturing to make intricate aircraft engines, thereby accelerating production.[2] Chemical and life science companies are using sophisticated AI algorithms to quickly discover new chemicals and pharmaceutical products. By automating repetitive tasks and uncovering hidden insights within data, AI/ML technologies enable manufacturers to streamline the product development life cycle, reduce time to market, and foster innovation.

Despite the transformative potential of advanced analytics, BI, and AI/ML in product design and development, manufacturers still face challenges in navigating the complexities of data management, interoperability, and organizational alignment. Data silos, legacy systems, and disparate workflows hinder the seamless integration of data-driven processes into the product development life cycle, limiting the effectiveness of analytics-driven decision making. Moreover, concerns surrounding data privacy, security, and regulatory compliance add another layer of complexity, requiring manufacturers to implement robust data governance frameworks and security protocols to safeguard sensitive information and mitigate risks. The convergence of advanced analytics, BI, and AI/ML technologies presents both opportunities and challenges for manufacturers in product design and development. By embracing a data-driven approach, harnessing the power of AI/ML, and addressing challenges effectively, manufacturers unlock new avenues for innovation, enhance operational efficiency, and gain a competitive advantage in today's rapidly evolving manufacturing landscape.

This chapter serves as a pivotal guide for manufacturers, offering insights into the intricate processes of transforming conceptual designs into tangible products efficiently and effectively. In modern manufacturing, production design and development integrates cutting-edge technologies such as AI, ML, advanced analytics, and BI into the fabric of production processes. This chapter explains how AI and ML algorithms can optimize design iterations, enhance predictive maintenance (PdM) strategies, and improve quality control (QC) measures, fostering innovation and efficiency. Advanced analytics techniques are employed to glean actionable insights from vast datasets, facilitating informed decision making at every stage of product development. BI tools provide real-time visibility into key performance indicators (KPIs), enabling proactive adjustments to production plans and resource allocation. By embracing these technologies, manufacturing practitioners can unlock new avenues for creativity, agility, and competitiveness in the ever-evolving landscape of production design and development.

FUNDAMENTALS OF PRODUCT DESIGN AND DEVELOPMENT

We now get into a comprehensive exploration of the fundamentals of product design and development, probing into the intricacies of a product life cycle. Product design and development encapsulates a multifaceted journey spanning from ideation to commercialization. This life cycle typically comprises several interconnected stages, each crucial for ensuring the success of a new product. The process typically begins with ideation, where concepts are generated based on market needs, consumer preferences, and technological feasibility. Following ideation, the design phase involves translating concepts into tangible product designs, incorporating aesthetics, functionality, and manufacturability. Prototyping then allows for the creation of physical or digital models to validate design concepts and gather feedback from stakeholders. Figure 5.1 shows a sample of the steps involved in new product design and introduction.

Key principles and methodologies serve as guiding frameworks in product design and development, offering structured approaches to problem solving and innovation. Design thinking, for example, emphasizes understanding your customers, creativity, and iterative prototyping to address user needs effectively. Agile development promotes flexibility and collaboration, enabling cross-functional teams to adapt quickly to changing requirements and market dynamics. Lean manufacturing focuses on eliminating waste and optimizing processes to streamline production and enhance efficiency. By integrating such methodologies, organizations can foster innovation, reduce time to market, and deliver products that resonate with customers.

> *"The design process—particularly for the prototyping phase—required several months in the past. And for this, artificial intelligence is able to support R&D because you can now virtualize your product development with AI."*
>
> —Daniele Petecchi, Head of Data Management and AI at Pirelli

In modern product design and development, organizations face a myriad of challenges and opportunities. Rapid advancements in technology, evolving consumer

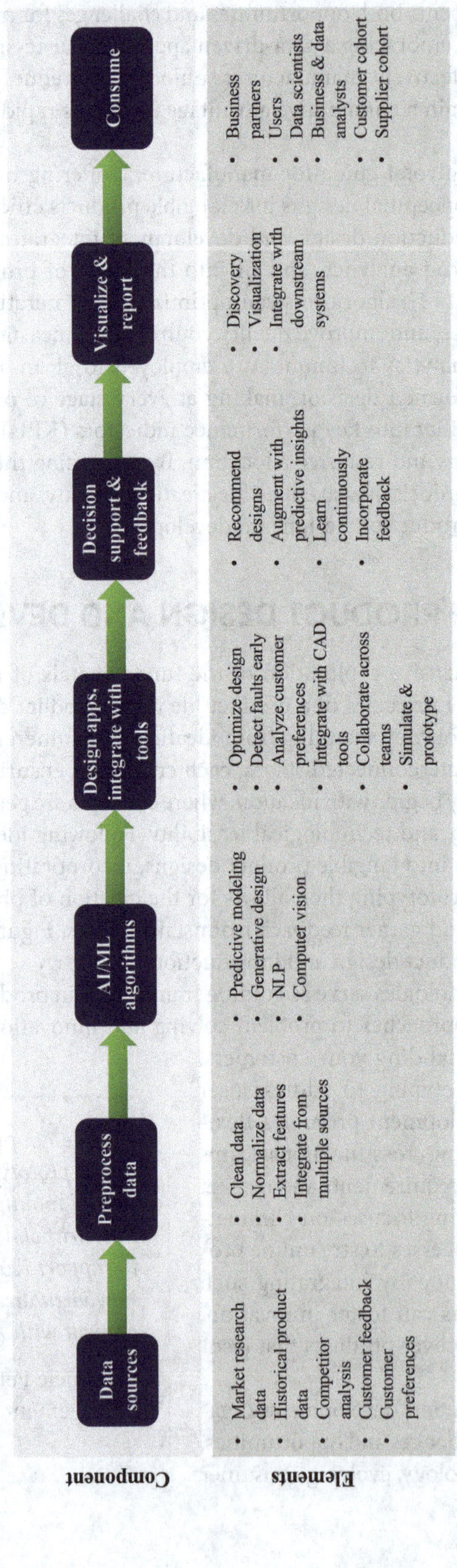

Figure 5.1 Sample process for new product design and introduction.

preferences, and increased competition necessitate agility and innovation. Successful examples include Apple's iterative approach to product development, evident in the continuous refinement of its iPhone lineup.[3] Conversely, failures such as the Samsung Galaxy Note 7 battery explosion highlight the importance of rigorous design review and testing along with quality assurance in preventing catastrophic product defects.[4] Further, globalization and supply chain disruptions underscore the need for resilient and adaptable manufacturing processes to mitigate risks and capitalize on emerging market trends.

Figure 5.2 depicts processes within traditional product design and development. Visually, the size of the icons for computer versus human resources represents the manual versus automated processes, with most process steps either fully or partially manual. The conceptualization phase is primarily manual, with technical tool support generating product ideas through brainstorming. Similarly, conducting market research and feasibility studies is primarily manual. The items in concept development, tooling, pilot and full-scale production, and distribution are similar, with a high level of human engagement. The other processes—detailed design, planning, and post-launch evaluation—have a higher level of computer involvement and support manual efforts to achieve their goals.

The adoption of emerging technologies such as AI, ML, and advanced analytics presents new opportunities for enhancing product and process design and development. Think of these three technologies as the legs of a stool that work together to keep the stool stable. AI-driven design

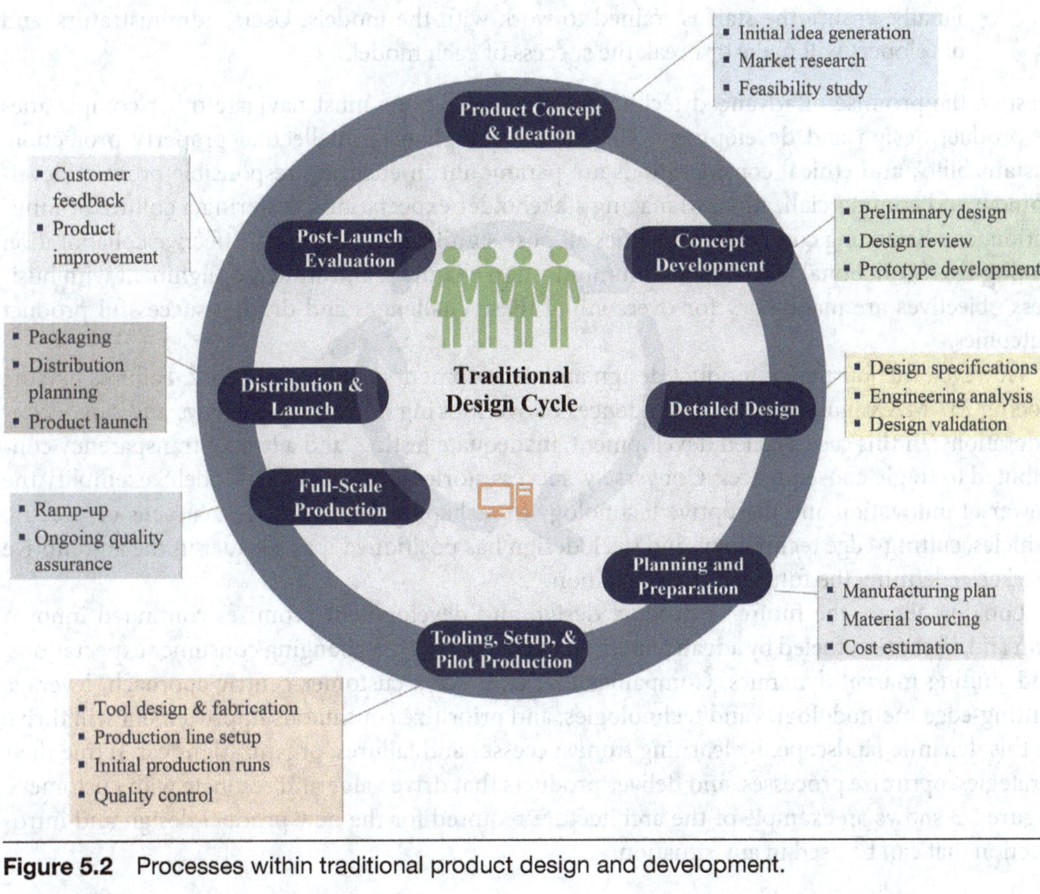

Figure 5.2 Processes within traditional product design and development.

tools can automate repetitive tasks, optimize designs based on performance data, and generate innovative solutions that align with user preferences. ML algorithms can analyze vast datasets to identify patterns, predict market trends, and optimize product features for maximum impact. Advanced analytics provides valuable insights into customer behavior, market dynamics, and product performance, enabling data-driven decision making throughout the product life cycle.

However, when companies embark on an AI/ML or advanced analytics journey and build models to meet specific needs, there are certain things they have to keep in mind:

- First, a governance framework must be established to ensure these models adhere to well-defined governance practices, including ensuring they work in conjunction to support the company's mission, not just a specific use case.
- Second, checks and balances need to be put in place to ensure each model is frequently monitored to ensure there is no model drift. There must be processes to:
  - Identify drift
  - Determine when and how to verify that the drift is unacceptable
  - Establish what to do when a drift is observed—refine the model, update it, or stop using it—all critical decisions
- Third, the company has to implement all operational processes (also called Ops—e.g., AIOps, MLOps) to ensure smooth operations of the models in the company's standard operating procedures (SOPs). This needs to happen before each model is put into production.
- Finally, ensure the staff is trained to work with the models. Users, administrators, and developers will make or break the success of each model.

Despite the promise of advanced technologies, organizations must navigate other complexities in product design and development. Regulatory compliance, intellectual property protection, sustainability, and ethical considerations are paramount in ensuring responsible product development and commercialization. Managing stakeholder expectations, fostering a culture of innovation, and balancing competing priorities all pose significant challenges. Effective collaboration among cross-functional teams, clear communication channels, and strategic alignment with business objectives are mandatory for overcoming these challenges and driving successful product outcomes.

However, the journey of product design and development is not without risk. Failures like the Boeing 737 MAX illustrate the consequences of overlooking critical design flaws and safety considerations. In this case, rushed development, inadequate testing, and a lack of transparency contributed to tragic consequences. Conversely, success stories such as Tesla's Model S exemplify the power of innovation and disruptive technology in reshaping industries. Tesla's focus on electric vehicles, cutting-edge technology, and sleek design has positioned it as a leader in the automotive market, redefining the future of transportation.

Looking ahead, the future of product design and development promises continued innovation and disruption fueled by advancements in technology, ever-changing consumer expectations, and shifting market dynamics. Companies that embrace a customer-centric approach, leverage cutting-edge methodologies and technologies, and prioritize continuous improvement will thrive in this dynamic landscape. By learning from successes and failures, organizations can refine their strategies, optimize processes, and deliver products that drive value and resonate with customers. Figure 5.3 shows an example of the architecture required for the new product design and introduction that can be used in any situation.

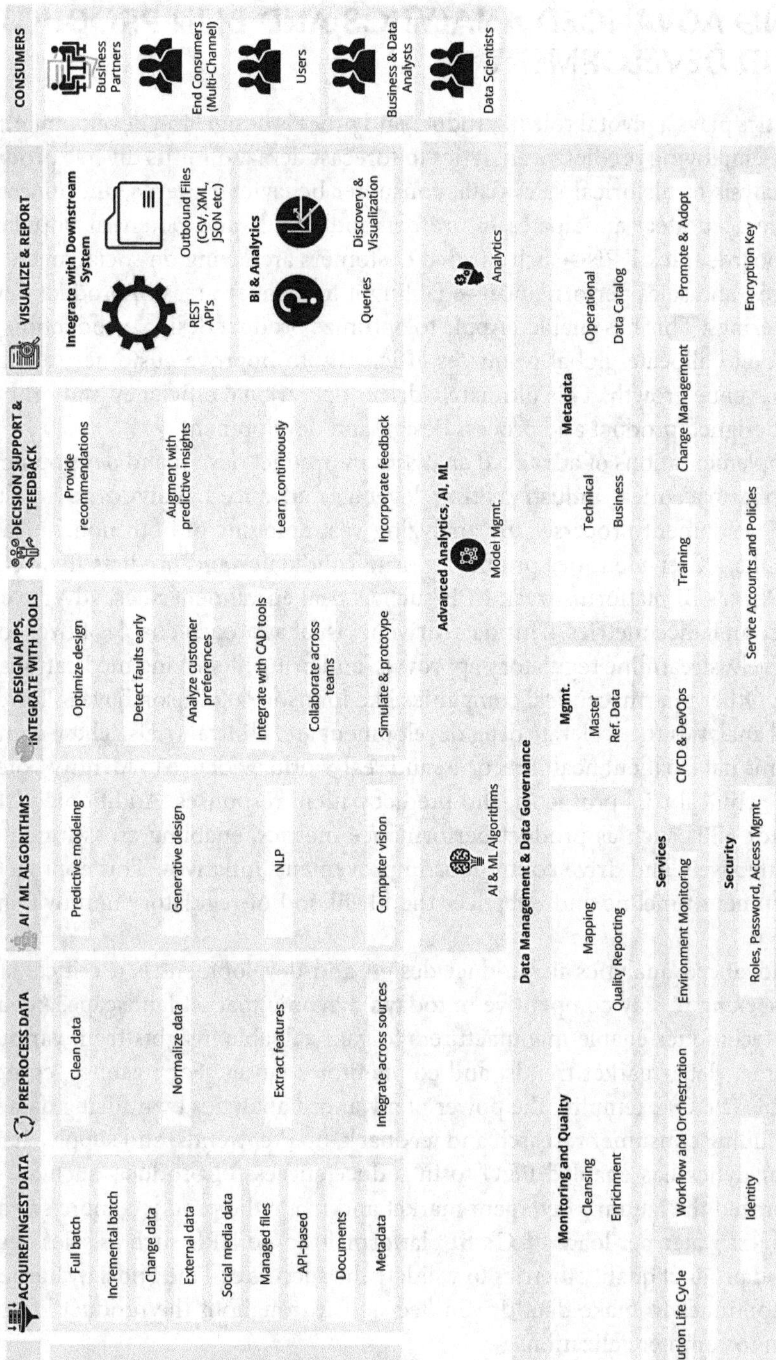

Figure 5.3 Example architecture for new product design and introduction.

LEVERAGING ADVANCED ANALYTICS AND BI IN PRODUCT DESIGN AND DEVELOPMENT

Predictive analytics plays a pivotal role in product and process design and development. Apple exemplifies this by employing predictive analytics to forecast demand for its diverse product lineup. Through the analysis of historical sales data, consumer behavior patterns, and macroeconomic indicators, Apple accurately anticipates future demand trends and potential popular features. Apple's BI dashboards track KPIs—such as what customers are saying on social media, customer satisfaction scores, and sales performance—enabling it to identify areas in product development for customer offerings. This has enabled Apple to optimize product design, production schedules, and inventories, and allocate global resources efficiently to improve customer satisfaction and drive sales and revenue growth. This ultimately drives operational efficiency and profitability, all through a well-designed product and process design and development.

Successful implementations of advanced analytics in product design and development are also evident in the pharmaceutical industry. Pfizer leverages advanced analytics to accelerate drug discovery and development processes. By analyzing vast amounts of both nonclinical data and clinical trial results, Pfizer identifies promising drug candidates and predicts their efficacy and safety profiles. Pfizer's BI platforms track KPIs such as trial enrollment rates, adverse event rates, and regulatory compliance metrics. This data-driven, visual approach enables Pfizer to prioritize resources effectively, streamline regulatory approvals, and bring life-saving medications to market more efficiently. Other pharmaceutical companies like Johnson & Johnson have followed suit and utilize advanced analytics to accelerate drug development and clinical trials. These companies are analyzing genomic data, patient health records, and real-world evidence to identify potential drug targets, optimize clinical trial protocols, and predict patient responses. Additionally, they use BI platforms to track KPIs such as product performance metrics, enabling cross-functional teams to collaborate effectively and drive continuous improvement initiatives. This approach shortens the drug development timeline and enhances the likelihood of regulatory approval and market success.

Leveraging advanced analytics in product design and development is a critical strategy for manufacturers seeking to stay competitive in today's dynamic market landscape. Advanced analytics tools and techniques enable manufacturers to gain valuable insights from various sources, including customer data, market trends, and competitor analysis. For instance, companies like Procter & Gamble (P&G) exemplify the power of advanced analytics by utilizing data from various sources, including consumer research and feedback, market trends, and supply chain dynamics. Predictive analytics has enabled P&G to introduce successful products such as Tide Pods, which revolutionized the laundry detergent market and made the products more sustainable because they used less water per load. P&G's BI platforms tracked KPIs such as sales performance, market share, and product quality metrics to validate the success of Tide Pods. By harnessing such insights, P&G continues to make data-driven decisions throughout the product life cycle, from concept ideation to commercialization.

Companies in the automotive sector also exemplify the effective use of advanced analytics in product design. They leverage advanced analytics to analyze vast amounts of data collected from vehicles, like sensor data, telemetry data, and customer feedback. By analyzing this data, they gain insights into vehicle performance, user behavior, and preferences, which contribute to their product design decisions. They use predictive analytics to forecast demand for specific vehicle configurations, allowing the company to optimize production schedules and inventory management. Additionally, they utilize advanced analytics to identify potential design flaws or quality issues early in the development process, thus improving overall product quality and lowering the risk of costly recalls and warranty expenses. By tracking KPIs such as customer satisfaction scores and vehicle performance indicators, they can identify areas for improvement and prioritize feature enhancements to meet customer expectations. Their use of BI tools enables them to monitor real-time data on vehicle performance, battery life (for EV or hybrid cars), and customer complaints. This allows them to quickly address issues and improve overall quality through a well-designed product and related processes.

> *". . . essentially, what comes in is a wish for a car, and what comes out is drawings and code. In that area, there are, of course, a lot of possibilities for data analytics and AI, both during the development process and also as part of the actual functions in the car—so intelligent functions and personalized functions and so on."*
>
> —Anders Sjögren, Senior Technical Leader at Volvo Cars

However, the journey toward leveraging advanced analytics in product design and development is not without its challenges. The failure of Google Glass serves as a cautionary tale, highlighting the importance of understanding market needs and user preferences. Despite significant investment in technology and design, Google Glass failed to gain widespread consumer adoption due to concerns of privacy, usability, social acceptance, and product features (e.g., short battery life). This underscores the importance of conducting thorough market research, user testing, and iterative prototyping to validate product concepts and address potential barriers to adoption.

The strategic adoption of advanced analytics in product design and development offers manufacturers unprecedented opportunities for innovation and growth. By leveraging predictive analytics, companies can gain valuable insights into market trends, optimize product features, and accelerate time to market, thereby driving competitive advantage and customer loyalty. However, success requires a holistic approach that integrates data-driven decision making with market insights, user feedback, and iterative design processes to deliver products that meet the needs and expectations of customers. This has to be followed by validation processes and a commitment to quality and safety.

UTILIZING BI IN PRODUCT DESIGN AND DEVELOPMENT

In contemporary manufacturing, BI is the foundation for driving innovation and maintaining competitiveness in product design and development. By harnessing advanced analytics tools and techniques with support from BI, manufacturers gain profound insights from an array of data sources, facilitating well-informed decision making across the product life cycle. This section explores the transformative role of BI in reshaping product design and development within the manufacturing sector.

Manufacturing firms employ BI methodologies to dissect data from diverse origins, spanning customer feedback, market trends, and operational metrics, among others. Such analyses furnish invaluable inputs for shaping product design strategies, enabling companies to align their offerings with evolving consumer preferences and market dynamics. BI empowers manufacturers to discern patterns, anticipate consumer behaviors, and refine product features. Further, BI cultivates a culture of innovation within manufacturing ecosystems. By providing insights into emerging technologies, market opportunities, and competitive landscapes, BI empowers companies to embark on the development of novel products and solutions that resonate with evolving consumer demands. BI-driven innovation empowers manufacturers to outpace competitors and uphold market leadership in a swiftly evolving business milieu.

Exemplary applications of BI in manufacturing can be seen in large industrial manufacturers who harness BI tools to analyze vast amounts of operational data collected from their equipment and machinery deployed worldwide. By leveraging BI dashboards and analytics platforms, they gain real-time visibility into equipment performance, maintenance needs, and usage patterns. They also integrate BI into the engineering processes to identify inefficiencies, eliminate bottlenecks, and accelerate time to market for new products. This granular understanding of equipment and machine health enables these companies to optimize product design and development processes, enhance product reliability, and deliver superior value to customers.

To further illustrate the impact of BI on product design and development within the manufacturing sector, let's probe into the case of a large aircraft manufacturer. It employs BI tools to analyze data generated from its supply chain, aircraft assembly lines, and maintenance operations to monitor production, identify optimization opportunities, and provide input into new product and process design. By leveraging BI dashboards and predictive analytics, the company has gained insights into component forecasts, inventory levels, and meeting customer delivery timelines. This data-driven approach has enabled it to design its products efficiently, optimize inventory, and reduce costs.

However, the journey toward leveraging BI in product design and development has challenges and pitfalls. One notable example of BI implementation gone awry is the case of a global electronics HVAC company that manufactures building automation systems. It embarked on a BI initiative to optimize its product portfolio and pricing strategies. However, due to inadequate data governance practices and siloed data sources, the BI project failed to deliver actionable insights, leading to wasted resources and missed opportunities. This cautionary tale underscores the importance of establishing robust data governance frameworks and ensuring data quality and integrity in BI initiatives.

The strategic deployment of BI in product design and development heralds promising prospects for manufacturing entities aiming to foster innovation and sharpen their competitive edge. Leveraging insights derived from diverse data sources, manufacturers can make informed decisions, optimize operational workflows, and craft products that resonate with consumer needs. Realizing the full potential of BI requires a holistic approach encompassing technological investment, organizational alignment, and a culture of data-driven decision making.

Figure 5.4 depicts the processes within modern product design and development. As you can see, compared to Figure 5.2, all the processes are at least semi-automated (as indicated by the icon size for computer and human resources), with a significantly higher contribution from computer resources compared to manual resources. Although semi-automated, most of the *hard* work is done by the models! These models provide guidance to human resources on actions to incorporate in operations. The manual intervention is to ensure checks and balances are in place through incorporating *human-in-the-loop*. This allows humans to quickly react and manually override poor outcomes from the modern tools getting pushed downstream.

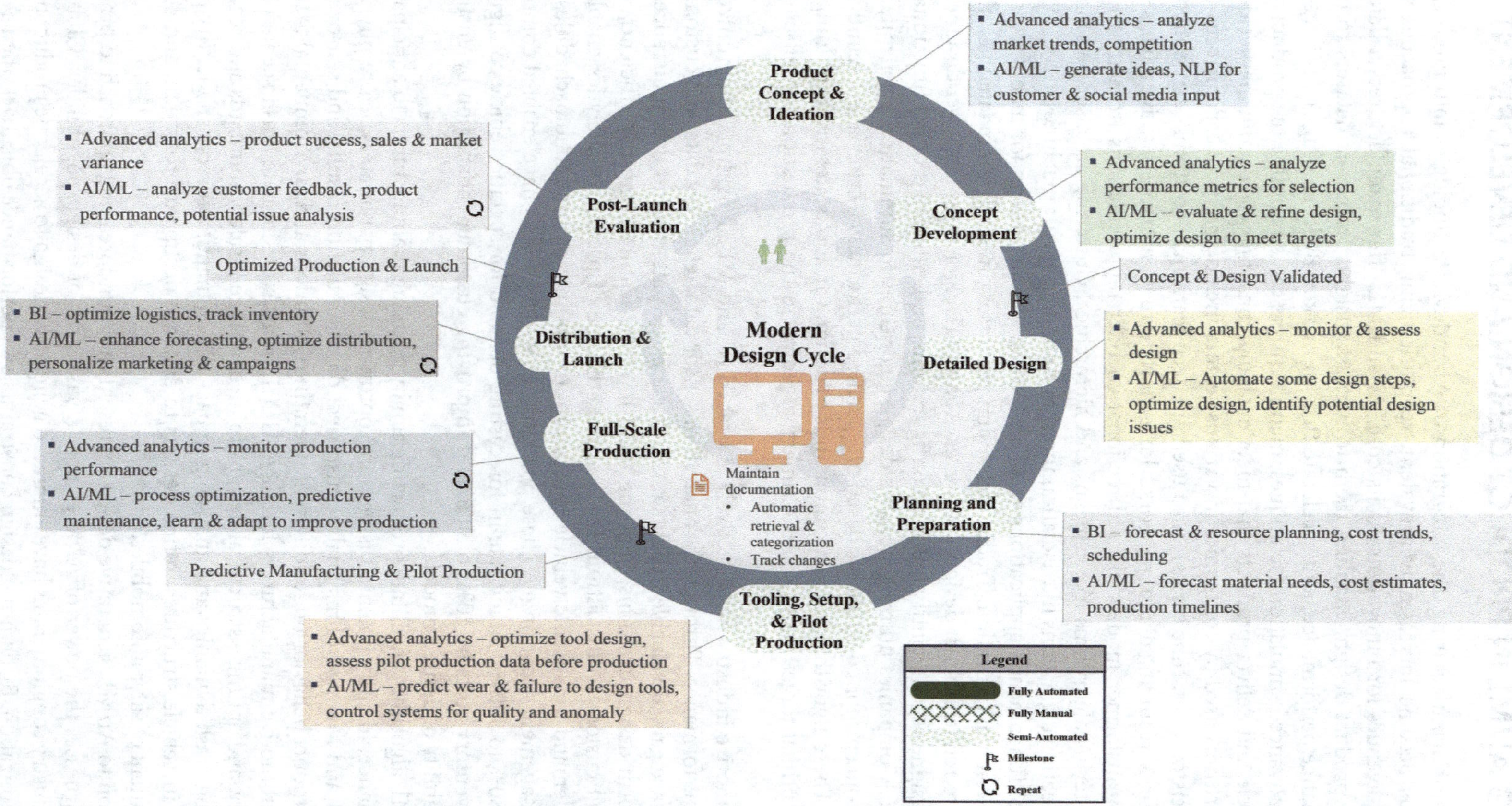

Figure 5.4 Processes within modern product design and development.

UTILIZING AI/ML IN PRODUCT DESIGN AND DEVELOPMENT

In the realm of manufacturing, the integration of AI and ML technologies into product design processes has become increasingly prevalent, revolutionizing traditional approaches and unlocking new avenues for innovation. At its core, AI and ML offer manufacturers a data-driven approach to product design, enabling them to harness insights from vast datasets to optimize designs, enhance performance, and accelerate time to market. By leveraging AI, ML, generative design, and advanced simulation techniques, manufacturers can streamline the product development life cycle and deliver products that meet evolving consumer needs and market demands. Typical algorithms used by companies in various sectors include predictive algorithms (e.g., regression, decision trees, time series). Also used are deep learning algorithms (e.g., convolutional neural networks, recurring neural networks, long short-term memory, generative adversarial networks, and transformer models like bidirectional encoder representations from transformers). Further, clustering algorithms like k-means and hierarchical clustering are also used along with anomaly detection algorithms like isolation forest and support vector machines.

One of the fundamental applications of AI and ML in product design for manufacturing is in the field of design optimization. Traditional design processes often rely on manual iterations and simulations to refine product designs. However, AI and ML algorithms empower designers to explore vast design spaces, identify optimal configurations, and enhance product functionality and performance. Companies utilize generative design algorithms to automatically generate innovative design concepts for components based on specified performance criteria and manufacturing constraints. These algorithms simulate thousands of design variations, iteratively refining designs to achieve optimal target metrics and ultimately leading to the production of well-engineered products.

The automotive industry has effectively leveraged AI and ML algorithms to optimize vehicle design and performance. Through the analysis of sensor data, telemetry data, and customer feedback, they have gained valuable insights into vehicle usage patterns and user preferences. These insights inform design decisions, allowing them to refine features, improve efficiency, and enhance overall customer satisfaction. For instance, the use of predictive analytics enables them to forecast demand for specific vehicle configurations, optimizing production schedules and inventory management accordingly. These predictive analytics models have enabled companies to achieve an 80% to 90% accuracy rate in forecasting demand for specific vehicle configurations, reducing inventory costs by 20% to 30%, and improving customer satisfaction by ensuring the timely availability of desired models.

AI and ML algorithms facilitate product design by enabling manufacturers to anticipate product performance, identify potential issues, and optimize design parameters before physical prototyping. AI and ML algorithms analyze historical data, simulation results, and sensor readings to predict product behavior under different operating conditions and environmental factors. For instance, automotive manufacturers leverage AI/ML models to simulate crash scenarios, assess vehicle safety, and optimize structural designs to meet stringent safety standards and regulations. By simulating crash dynamics and analyzing impact forces, manufacturers can identify potential weaknesses in vehicle structures, refine designs to enhance occupant protection, and accelerate the development of safer automobiles.

In addition to AI/ML, generative design represents a groundbreaking approach to product design that harnesses the power of AI to explore a vast array of design possibilities. Global heavy machine industry conglomerates have embraced generative design to accelerate product prototyping and optimization. By defining design constraints and performance objectives, their AI-driven design tools generate innovative solutions that optimize product performance, cost, and sustainability. For example, leveraging generative design to redesign aircraft components followed by additive manufacturing has resulted in significant weight reductions, fuel savings, and supply chain performance without compromising structural integrity and manufacturing speed. The adoption of generative design tools has led to substantial cost savings and environmental benefits.

The consumer electronics industry has mastered the art of data-driven innovation by using AI and ML. Through the analysis of consumer behavior patterns and market trends, these companies can anticipate future demand and identify potential product features that resonate with consumers. Using BI dashboards and enabling data-driven decision making at every stage of product development, they track KPIs—such as customer satisfaction scores, social media sentiment, and sales performance. This data-driven approach has enabled them to optimize product designs, production schedules, and global resources.

Large global pharmaceutical companies leverage AI, ML, and advanced analytics to accelerate drug discovery and development processes. As mentioned earlier, Pfizer uses vast amounts of data obtained with AI and ML (and advanced analytics) to identify promising drug candidates. Pfizer and other pharmaceutical companies use natural language processing tools in their pharmacovigilance initiatives. This, along with BI, helps them track adverse event rates and regulatory compliance metrics, enabling the company to prioritize resources effectively and streamline regulatory approvals. This digital approach has enabled Pfizer and others to bring life-saving medications to market faster and more efficiently, demonstrating the profound impact of AI and ML on product design and development. Moreover, their advanced analytics platform has significantly accelerated the drug discovery process, reducing time to market for new medications and improving patient outcomes. This has translated into revenue growth and improved regulatory approval rates for new drugs.

Further, the integration of AI and ML technologies into product design processes presents manufacturers with unique opportunities to optimize designs for sustainability and environmental impact. Consumer goods manufacturers have exemplified this by leveraging AI, ML, and advanced analytics to introduce eco-friendly product innovations.

> P&G's use of predictive analytics enabled the successful introduction of Tide Pods, which revolutionized the laundry detergent market. By harnessing AI and ML technologies, P&G continues to make data-driven decisions that drive sustainability and innovation in product design and development. Additionally, P&G's use of these tools has led to a significant reduction in warm or hot water usage per load with Tide Pods, resulting in significant environmental benefits and cost savings. P&G's goal is to help customers attain a 75% increase in cold-water washing by 2030, which will avoid having almost four million tons of CO_2 released into the atmosphere. In addition, P&G is working toward moving all Tide packaging to 100% recyclable materials by 2030.[5]

SUMMARY AND KEY INSIGHTS

The adoption of AI and ML in product design processes necessitates a holistic approach to data governance, model monitoring, and operational management. Companies must establish robust governance frameworks to ensure that AI and ML models align with business objectives and regulatory requirements. Regular monitoring of AI models is essential to detect and mitigate model drift, ensuring that models remain accurate and reliable over time. Further, companies must implement intelligent operational processes, such as AIOps and MLOps, to ensure smooth integration and operation of AI models within SOPs. By prioritizing staff training and collaboration among cross-functional teams, companies can maximize the value of AI and ML in product design and development while mitigating potential risks and challenges.

The strategic adoption of AI and ML technologies in product design within manufacturing offers unprecedented opportunities for innovation, efficiency, and sustainability. Through the analysis of vast datasets and the application of AI, ML, predictive analytics, generative design, and advanced simulation techniques, manufacturers can optimize designs, enhance performance, and accelerate time to market. Case studies from industries such as heavy manufacturing, automotive, consumer electronics, pharmaceuticals, and consumer goods demonstrate the transformative impact of AI and ML on product design and development. By embracing a digital approach and prioritizing collaboration, companies can unlock new avenues for growth and competitiveness.

This chapter provided a glimpse into the future landscape of product design in the manufacturing sector, where AI and ML promise to catalyze transformative change. From AI-based predictive analytics and automation to enhanced supply chain visibility and optimization, manufacturers are poised to unlock new frontiers of efficiency, sustainability, and innovation. As we embark on this journey, harnessing the power of data-driven insights and emerging technologies will be imperative for manufacturers to thrive in an increasingly dynamic and competitive marketplace.

Some of the key takeaways from this chapter include:

1. Product design and development is the foundation of innovation and competitiveness in manufacturing, with key principles and methodologies serving as frameworks to guide product design and development.
2. Rapid advancements in technology, evolving consumer preferences, and increased competition necessitate agility and flexibility in product design and development. This mandates the involvement of numerous stakeholders, including designers, engineers, marketers, and supply chain professionals.
3. AI and ML can optimize design iterations, enhance PdM strategies, and improve QC measures. Similarly, advanced analytics techniques are used to glean actionable insights from vast datasets. BI tools provide real-time visibility into KPIs, enabling proactive adjustments to production plans and resource allocation.
4. By embracing cutting-edge technologies such as AI, ML, advanced analytics, and BI, organizations can foster innovation, reduce time to market, and deliver products that resonate with customers.

6

LEVERAGING ADVANCED ANALYTICS, BI, ML, AND AI IN PROCUREMENT

AI-Driven Procurement Transformation at Stellar Electronics: Enhancing Manufacturing Efficiency

At Stellar Electronics, the procurement process for manufacturing had long been plagued by unpredictable supplier performance, fluctuating material costs, and inventory inefficiencies. Recognizing the need for innovation, Laura Gomez, the chief procurement officer, spearheaded a project to integrate artificial intelligence (AI) and machine learning (ML) into the company's procurement strategy. Her goal was to create a more data-driven and agile supply chain capable of responding to real-time market dynamics.

Working closely with James Yu, the senior data scientist at Stellar, Laura's team implemented an ML platform that leveraged predictive analytics to optimize supplier selection and manage risks. By analyzing vast datasets—including historical supplier performance, pricing trends, geopolitical risks, and quality control (QC) records—the AI model identified the most reliable and cost-effective suppliers for each component. This improved supplier performance by 15% and helped reduce procurement costs by 10% within the first six months.

Furthermore, the AI system was integrated with Stellar's material requirement planning (MRP) software. This allowed the company to better forecast material needs and dynamically adjust procurement schedules. With real-time insights, Stellar could avoid costly last-minute purchases and reduce the risk of overstocking raw materials. The predictive analytics model forecasted demand fluctuations, improving material availability by 20% while cutting inventory holding costs by 18%.

To address price volatility, Sarah Malik, the director of operations, worked with the AI team to implement automated negotiation algorithms. These algorithms analyzed historical pricing patterns and external market factors to recommend optimal purchasing windows. This allowed Stellar to secure raw materials at favorable prices, reducing overall material costs by 12% annually.

Thanks to the synergy between procurement and AI-driven insights, Stellar Electronics has drastically improved its procurement efficiency, ensuring that manufacturing operations run smoothly with minimal disruptions and optimized costs.

"With AI already being deployed in consumer applications and increasingly applied in multiple business capabilities, the number of leaders utilizing AI to transform procurement and sourcing is anticipated to dramatically increase in the coming months and years."

—Matthias Gutzmann, Founder of Digital Procurement World

Raw material procurement is the fundamental process of acquiring essential inputs for manufacturing operations. It involves sourcing, purchasing, and managing various materials necessary for production, ranging from metals, electronics, plastics, and chemicals to natural resources. It also includes sourcing shared services such as consulting, payroll, human resources, and janitorial. Raw materials are the foundational elements upon which manufacturing industries build their products, making procurement a critical aspect of the supply chain. Services support the manufacturing processes and make sure that they work smoothly.

Efficient procurement processes are vital for the seamless functioning of manufacturing operations. Streamlined procurement ensures the timely availability of high-quality materials in the correct format, at the right place, and at the right time to minimize production delays and disruptions. Efficient procurement contributes to the company's financials, like cost reduction, by optimizing purchasing practices, negotiating favorable terms with suppliers, and effectively managing inventory levels. By prioritizing efficiency in procurement, manufacturers can enhance operational performance, maintain product quality, and gain a competitive edge in the market. Each procurement dollar saved, whether through inventory management or payment terms, directly impacts the balance sheet and the profit and loss statement of the company.

The emergence of advanced analytics, business intelligence (BI), ML, and AI has revolutionized raw material procurement. These technologies offer manufacturers powerful tools to optimize procurement processes, improve decision making, and drive operational excellence. Advanced analytics and BI provide actionable insights by analyzing procurement data, supplier performance metrics, and market trends. By leveraging these insights, manufacturers can identify cost-saving opportunities, streamline supplier relationships, and enhance supply chain visibility. ML and AI algorithms further enhance procurement efficiency through activities such as automating routine tasks, predicting demand patterns, and optimizing supplier selection. Predictive analytics algorithms enable manufacturers to forecast raw material requirements accurately, anticipate market changes, and optimize inventory management strategies. AI-driven supplier management systems facilitate supplier selection, risk assessment, deviation from contracts, and contract negotiation, enabling manufacturers to make informed decisions and mitigate supply chain risks effectively.

In the following sections, we explore the realm of raw material procurement, exploring strategies and best practices to leverage advanced analytics, BI, ML, and AI technologies to optimize procurement processes. Through real-world experience, case studies, and practical insights, we demonstrate how manufacturers can utilize these technologies to enhance procurement efficiency, mitigate risks, and drive sustainable value creation in the manufacturing supply chain. Further, we uncover the transformative potential of data-driven procurement in the modern manufacturing landscape.

UNDERSTANDING DATA SOURCES AND INTEGRATION

Raw material procurement in the manufacturing sector relies heavily on accessing and integrating data from various sources to make informed decisions. These data sources encompass a wide range of information crucial to procurement professionals to ensure the timely acquisition of high-quality raw materials at optimal prices. Among these sources are supplier databases, market data, and internal enterprise resource planning (ERP) and manufacturing execution system (MES) software, each playing a vital role in the procurement process.

Supplier databases serve as repositories of information about potential suppliers, their capabilities, performance history, and pricing structures. By analyzing data from these databases, manufacturers can identify reliable suppliers, assess their track records, and negotiate favorable terms. Market data, on the other hand, provides insights into prevailing market trends, commodity prices, supply-demand dynamics, labor costs and availability, geopolitical factors affecting raw material availability, and weather data that can cause friction in the flow of raw material to the manufacturer. Accessing this data allows procurement teams to anticipate price fluctuations, mitigate supply chain risks, optimize sourcing strategies accordingly, and refine planning parameters to address these challenges. Internal ERP and MES systems house critical operational data, including inventory levels, production schedules, and financial records. Integrating data from ERP systems with other sources enables manufacturers to align procurement activities with production needs, minimize inventory costs, and streamline order fulfillment processes. However, despite the wealth of information available from these diverse sources, integrating and harmonizing disparate datasets pose significant challenges. Figure 6.1 shows a sample of the steps involved in raw material procurement.

> *"The key here is that supplier data is available where the sourcing decisions are made. This means that supplier risk and performance data has to be made available in the e-sourcing tool of the buyer so that when the supplier proposals come in, this data is taken into account to achieve a more optimized and data-based sourcing decision."*
>
> —Nico Bac, former Senior Director of Procurement at Procter & Gamble, Founder of Digital Procurement Now

One major challenge is the heterogeneous nature of data formats, structures, and standards across different sources. Supplier databases, market reports, and ERP systems often employ proprietary formats and terminology, making it difficult to establish a common data framework for integration. Data inconsistency, duplication, and incompleteness further complicate the harmonization process, leading to inaccurate insights and decision making. Further, not all data is captured at the same level of temporal granularity, so marrying such data with others becomes challenging. To address these challenges, manufacturers must implement robust strategies for data collection, cleansing, and integration to ensure data quality and consistency across the procurement process. First, establishing data governance frameworks is essential to defining standardized data models, formats, and validation rules for all data sources. This ensures data integrity and facilitates seamless integration across systems.

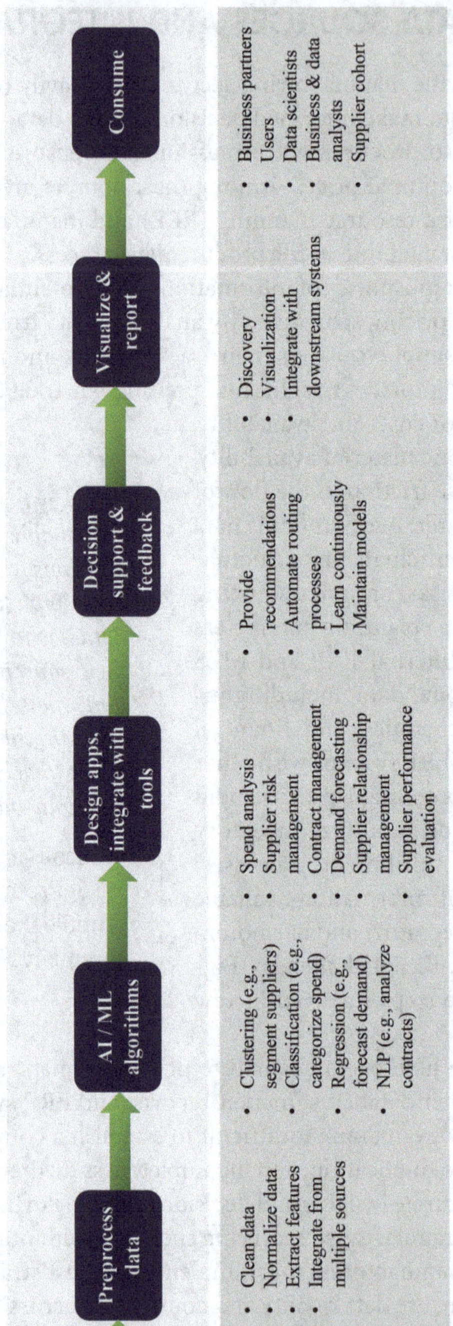

Figure 6.1 Sample of steps involved in raw material procurement.

Second, investing in data cleansing and enrichment technologies helps identify and rectify inconsistencies, errors, and redundancies in the dataset. Automated data cleansing tools can normalize data formats, correct inaccuracies, and enrich datasets with missing information. Adopting robust data integration platforms and middleware solutions enables manufacturers to establish seamless connectivity between disparate systems and applications. These platforms facilitate data transformation, mapping, and synchronization, ensuring smooth data flow and real-time visibility into procurement activities. However, technology by itself will not solve these problems; there must be supporting processes as well. As new technology is introduced and processes are refined, attention to educating practitioners via robust change management is critical. Remember, introducing new and improved technology without addressing the underlying processes will result in the same undesirable results manifesting themselves more quickly.

Third, leveraging advanced analytics and ML algorithms can be used to enhance data processing capabilities. Data quality is key to generating high-quality predictive insights and actionable intelligence in procurement decision making. By analyzing data such as historical procurement data, market trends, and supplier performance metrics, manufacturers can forecast demand, optimize inventory levels, and mitigate supply chain risks proactively. Fostering collaboration and data sharing among internal stakeholders and external partners promotes transparency and agility in the procurement process. By establishing data-sharing agreements, collaborative platforms, and common data quality rules, partners can exchange relevant and quality information with suppliers, customers, and industry partners, enabling better-informed decisions and fostering innovation.

Finally, ongoing monitoring, validation, and refinement of data quality and integration processes are critical to maintaining data consistency and relevance over time. Implementing performance metrics, data quality checks, and continuous improvement initiatives, along with a robust data governance framework, ensures that the procurement data ecosystem remains robust, reliable, and responsive to evolving business needs. Figure 6.2 depicts the processes within traditional procurement. As you can see, most process steps are either fully or partially manual. This is depicted by the size of the icons for computing and human resources. Remember, these are representative steps and sub-steps.

So, understanding data sources, processes, and integration is paramount for optimizing raw material procurement in the manufacturing sector. By harnessing data from multiple sources, like supplier databases, market reports, internal ERP systems, etc., manufacturers can make data-driven decisions to enhance procurement efficiency, mitigate risks, and drive competitive advantage. However, integrating and harmonizing diverse datasets requires careful planning, governance, investment in technology, and, of course, human resources. By implementing strategies for data collection, cleansing, and integration, manufacturers can ensure data quality and consistency across the procurement process. Figure 6.3 shows an example of an architecture for raw material procurement that you can use to develop your solution.

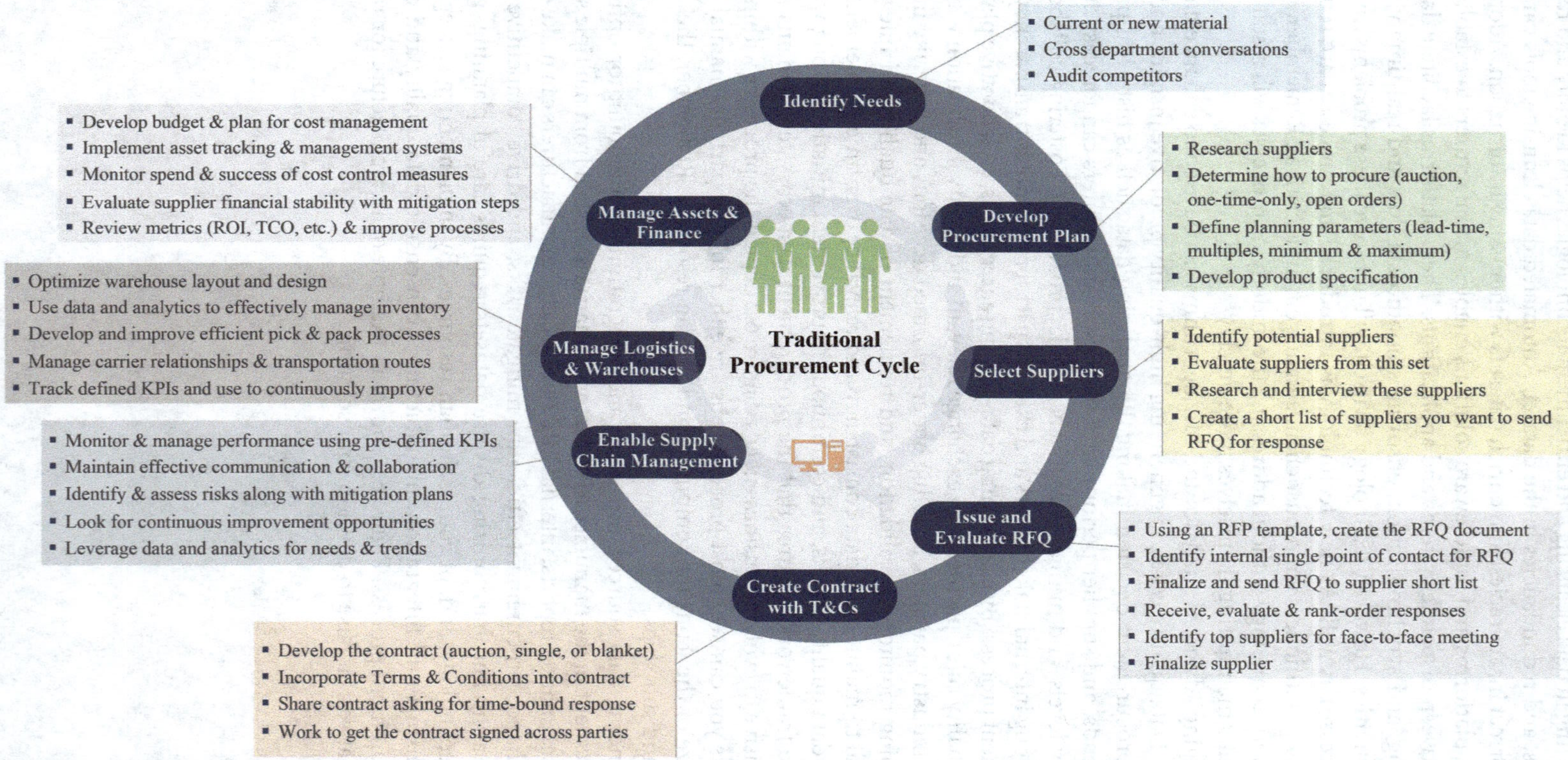

Figure 6.2 Processes within traditional procurement.

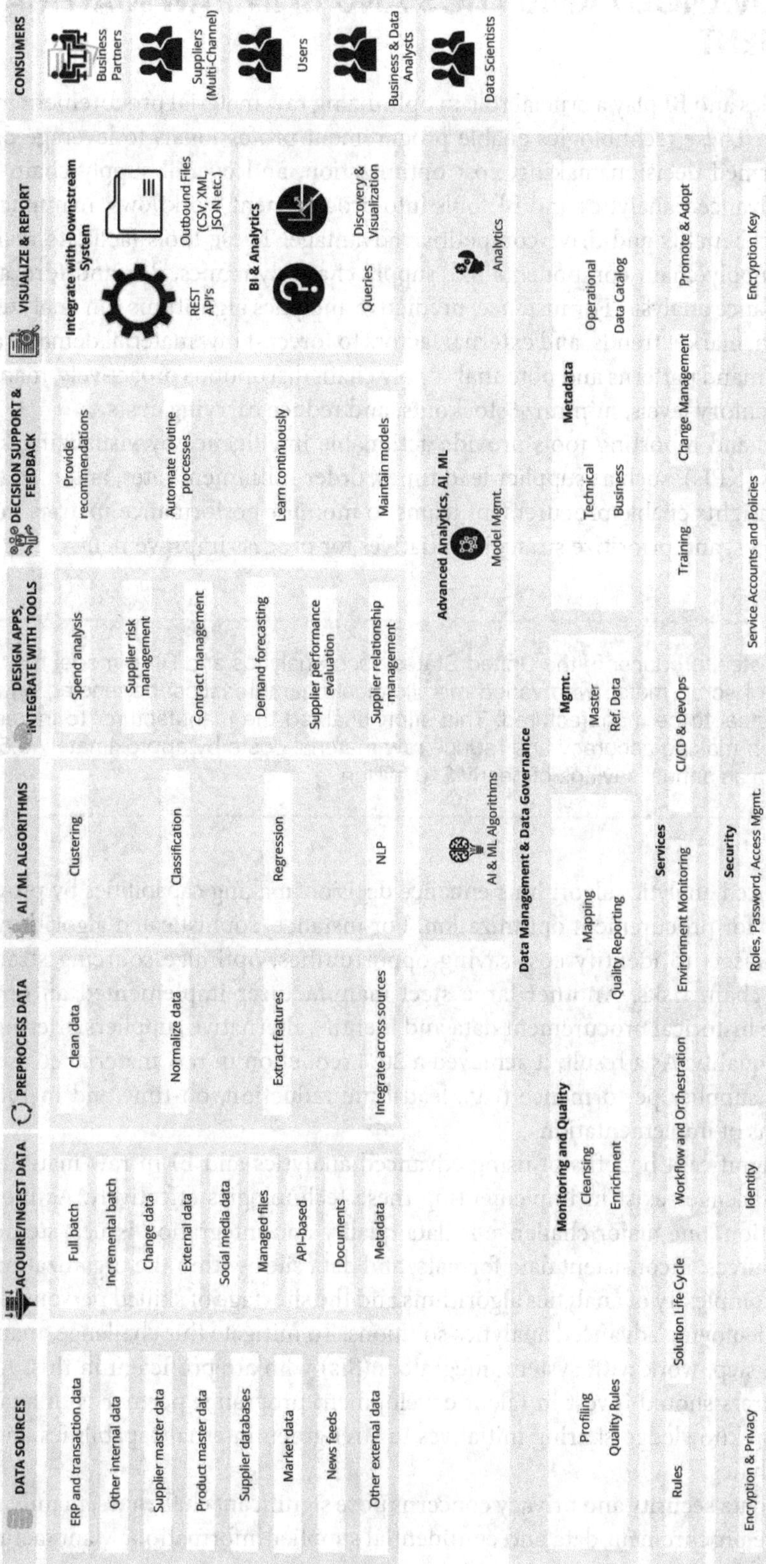

Figure 6.3 Example architecture for raw material procurement.

ROLE OF ADVANCED ANALYTICS AND BI IN RAW MATERIAL PROCUREMENT

Advanced analytics and BI play a crucial role in optimizing raw material procurement in the manufacturing sector. These technologies enable procurement professionals to leverage data-driven insights for informed decision making, cost optimization, and overall supply chain efficiency. By integrating advanced analytics and BI tools into procurement workflows, manufacturers can unlock numerous benefits and drive competitive advantage. These tools facilitate real-time visibility into key supply chain components like supply chain dynamics, demand forecasting, and supplier performance analysis. For instance, predictive analytics algorithms can analyze historical procurement data, market trends, and external factors to forecast raw material demand accurately. By identifying demand patterns and potential supply chain disruptions proactively, manufacturers can optimize inventory levels, minimize stockouts, and reduce carrying costs.

BI dashboards and reporting tools provide actionable intelligence by visualizing key performance indicators (KPIs), such as supplier lead times, order fulfillment rates, and material quality metrics. These insights enable procurement teams to monitor performance metrics in real time, identify bottlenecks, and prioritize strategic initiatives for process improvement.

> A leading steel producer in the United States used analytics and BI to model the use and cost of scrap metal (with varied metallurgical characteristics) depending on the grade of steel to be manufactured. This effort enabled the manufacturer to increase its decision-making accuracy and reduce raw material costs by approximately 0.5%, resulting in an annual savings of nearly $10 million.[1]

Further, advanced analytics algorithms enhance decision-making capabilities by providing prescriptive insights for procurement optimization. For instance, sophisticated algorithms can analyze complex datasets to identify cost-saving opportunities, optimize sourcing strategies, and mitigate supply chain risks. Another large steel manufacturer implemented an optimization model to analyze historical procurement data and identify alternative suppliers offering competitive pricing and quality. As a result, it achieved a 20% reduction in raw material costs and a 30% improvement in supplier performance (e.g., lead-time reduction, on-time and in-full delivery) within six months of implementation.

Despite the significant benefits of using advanced analytics and BI in raw material procurement, several challenges exist in implementing these technologies effectively. As mentioned in the previous section, one major challenge is data quality and integration issues, stemming from disparate data sources, inconsistent data formats, and data silos within the organization. Another challenge is the complexity of analytics algorithms and the shortage of skilled personnel capable of developing and deploying advanced analytics solutions. To mitigate this challenge, manufacturers should, as a first step, work with system integrators (SIs) who are proficient in this. At the same time, manufacturers should invest in talent development programs, partner with analytics vendors, and develop knowledge-sharing initiatives to strengthen internal capabilities and expertise in advanced analytics.

Additionally, data security and privacy concerns pose significant challenges, mainly when dealing with sensitive procurement data and confidential supplier information. Manufacturers, along with their SIs, must implement robust cybersecurity measures, access controls, and encryption protocols to safeguard data integrity and protect against cyber threats and data breaches. These should be embedded within standard operating procedures (SOPs) to minimize the risk of exposing sensitive data. Resistance to change and organizational inertia are real problems and can hinder the adoption and utilization of advanced analytics and BI tools among procurement professionals. To address this challenge, manufacturers should invest in strong change management initiatives, stakeholder engagement, and training programs to foster a data-driven culture and promote user adoption of analytics solutions.

AI AND ML APPLICATIONS IN RAW MATERIAL PROCUREMENT

AI and ML are revolutionizing raw material procurement in manufacturing by enabling predictive analytics, optimization, and automation of sourcing processes. These technologies enable manufacturers to analyze vast volumes of data, identify patterns, and make data-driven decisions to optimize raw material sourcing strategies and enhance supply chain efficiency. Several AI and ML algorithms are instrumental in raw material procurement, including:

- *Demand forecasting*: AI and ML algorithms analyze historical sales data, market trends, and external factors to predict future demand for raw materials accurately. Improved demand forecasting can optimize inventory levels, reduce stockouts, and minimize excess inventory holding costs. For example, a large global chip manufacturer utilizes AI-driven demand forecasting to enhance its procurement processes. By leveraging advanced AI algorithms, it accurately predicts the demand for various components and raw materials required for its semiconductor manufacturing. This enabled it to optimize inventory costs by 20% (freeing up working capital), reduce lead times by 25%, and improve supplier performance to ensure the timely availability of essential materials by 25%.
- *Supplier risk assessment*: AI-driven risk assessment models analyze supplier performance data, market intelligence, and geopolitical factors to evaluate supplier reliability and mitigate supply chain risks. These models identify high-risk suppliers and enable proactive risk management strategies to ensure business continuity. For instance, a global manufacturing conglomerate utilizes AI-driven demand forecasting to enhance its procurement processes. By leveraging advanced AI algorithms, it has improved its risk detection accuracy by 25%, reduced supplier-related disruptions by 30%, and increased procurement cost savings by 30%.
- *Price optimization*: ML algorithms analyze historical pricing data, market dynamics, and demand-supply patterns to optimize raw material prices and negotiate favorable contracts with suppliers. By leveraging price optimization models, manufacturers can achieve cost savings and improve profit margins. A global consumer products company has effectively leveraged advanced AI algorithms for price optimization in procurement. It analyzes historical pricing data, market trends, supplier performance, and external factors, such as economic indicators and raw material prices. This AI-driven approach enables it to make data-informed decisions on procurement strategies, negotiate more favorable prices with suppliers, and accurately forecast future price trends. This has led to an annual procurement cost reduction of 12%. It has also reduced the supplier price variance by 15% while improving supplier performance by 10% through better quality and reliable delivery.
- *Supply chain optimization*: AI-powered supply chain optimization models optimize sourcing decisions, transportation routes, and inventory management strategies to minimize

lead times, reduce transportation costs, and enhance overall supply chain efficiency. By dynamically adjusting procurement and logistics operations based on real-time data, manufacturers can respond rapidly to market changes and customer demands. The previously mentioned consumer products company also leverages AI to predict demand, manage inventory, optimize supplier selection, and streamline procurement processes. It reduced inventory holding costs by 15% and reduced supplier lead times by 25%, leading to a 25% decrease in stockouts by identifying and mitigating issues in the supply chain.
- *Quality control*: ML algorithms analyze sensor data, production records, and quality inspection results to detect defects, identify root causes, and improve product quality. By implementing AI-driven QC systems, manufacturers can reduce rework, scrap rates, and warranty claims, enhancing customer satisfaction and brand reputation. One example of a manufacturing company that has been implementing AI for QC in procurement is Flex Ltd. By analyzing data from suppliers, production lines, and historical quality issues, Flex can predict potential defects and proactively address them before they impact production. Flex has reduced its test time by about 30%, improved its product yield by 97%, and achieved an efficiency improvement of over 30%. The return on initial investment was achieved in under one month, with a return on investment (ROI) of 388%.[2]

> *"Interest in AI for procurement has increased dramatically, and we expect to see a rise in use. In our survey, 58% of procurement leaders said they already are implementing or plan to implement AI in the next 12 months."*
>
> —Kaitlynn Sommers, Senior Director Analyst at Gartner Supply Chain Practice

Despite the transformative potential of AI and ML in raw material procurement, manufacturers face several challenges. First, as referenced earlier, data silos created by fragmented data sources and legacy systems hinder data integration and AI/ML initiatives. Manufacturers must implement data integration strategies to consolidate and harmonize disparate data sources, enabling seamless data flow and analysis. Data silos also impact data quality. Inaccurate, incomplete, or outdated data can undermine the effectiveness of AI and ML algorithms. Manufacturers must prioritize data quality management practices, including data cleansing, validation, and governance, to ensure the reliability and relevance of procurement data. Biases inherent in AI and ML algorithms can lead to unfair or suboptimal procurement decisions, perpetuating inequities or overlooking valuable opportunities. Manufacturers must assess and mitigate algorithmic biases through transparent algorithm design, diverse training data, and ongoing monitoring and validation. Resistance to change and lack of organizational readiness can impede the adoption of AI and ML technologies in procurement processes. Manufacturers must invest in change management initiatives, training programs, and stakeholder engagement to foster a culture of innovation and collaboration. Finally, regulatory requirements and data privacy concerns pose significant challenges to the adoption of AI and ML, particularly in highly regulated industries. Manufacturers must ensure compliance with data protection regulations, such as the General Data Protection Regulation (GDPR) or the California Consumer Protection Act (CCPA), and implement robust security measures to safeguard sensitive procurement data.

To mitigate these challenges, manufacturers must implement several strategies. They have to establish robust data governance frameworks to define data standards, policies, and procedures for data management, quality assurance, and security. Manufacturers can go a long way in mitigating these challenges by fostering collaboration with technology vendors, industry partners, and academic institutions to access expertise, resources, and best practices in the implementation of AI and ML. They must embrace a culture of continuous improvement and innovation by encouraging experimentation, promoting learning, and adapting to evolving technologies and market dynamics. All these are moot until manufacturers invest in workforce development initiatives to build AI and ML capabilities within the organization, including training programs, knowledge sharing, and talent acquisition. Ultimately, manufacturers must incorporate ethical considerations into AI and ML initiatives, ensuring fairness, transparency, and accountability in procurement decision-making processes. By adopting these strategies, manufacturers can harness the power of AI and ML to optimize raw material procurement processes, drive operational efficiency, and gain a competitive edge in the marketplace.

ADVANCED ANALYTICS, BI, ML, AND AI IN COST OPTIMIZATION

Advanced analytics, BI, AI, and ML play crucial roles in optimizing raw material procurement costs across diverse manufacturing sectors. These technologies enable manufacturers to analyze internal and external data to identify cost-saving opportunities. This helps them make informed decisions to streamline procurement processes. In raw material procurement cost optimization, various advanced analytics techniques and algorithms are employed, including predictive modeling, optimization algorithms, and anomaly detection.

ML-based predictive modeling is a key component of advanced analytics in raw material procurement cost optimization. By analyzing large amounts of historical data and forecasts, manufacturers have predicted future raw material prices and availability. This enables them to optimize sourcing strategies, negotiate favorable contracts with suppliers, and mitigate procurement risks. Techniques such as time series analysis, regression analysis, long short-term memory, and forecasting models are used to develop predictive models for raw material prices and demand.

Optimization algorithms are utilized to optimize sourcing decisions and minimize procurement costs. These algorithms optimize supplier selection, inventory management, transportation logistics, and production planning to achieve cost savings while meeting production targets. Techniques such as linear programming, integer programming, and constraint optimization algorithms are applied to solve complex optimization problems in raw material procurement. Anomaly detection techniques are employed to identify irregularities or deviations in procurement data that may indicate inefficiencies or fraudulent activities. By detecting anomalies in supplier invoices, purchase orders, and inventory levels, manufacturers can identify cost-saving opportunities and improve procurement processes. Techniques such as statistical analysis, ML algorithms, natural language processing (NLP), and pattern recognition are used to detect anomalies and outliers in procurement data.

In the automotive industry, predictive modeling and optimization algorithms are used to enhance sourcing decisions for critical components, reduce procurement costs, and improve supply chain efficiency. In the consumer goods sector, anomaly detection techniques are employed to identify discrepancies in supplier invoices and eliminate overpayments, leading to cost savings and process efficiencies. Further, BI tools and dashboards provide manufacturers with real-time visibility into procurement KPIs, performance metrics, and cost trends. BI platforms aggregate and visualize procurement data from multiple sources, enabling stakeholders to monitor key metrics, identify areas for improvement, and make data-driven decisions. Dashboards display metrics such as supplier performance, procurement costs, inventory levels, and lead times, allowing manufacturers to track performance and optimize procurement processes accordingly.

Despite the benefits, implementing advanced analytics, BI, AI, and ML in raw material procurement cost optimization comes with several challenges. These challenges include data integration, data quality, limited model trustworthiness, explainability of model results, change management, and regulatory compliance. Manufacturers must address these challenges by investing in data governance, data cleansing, algorithm transparency, organizational training, and compliance measures to ensure the success of their procurement optimization initiatives.

> A large global manufacturer of industrial, safety, and consumer products utilizes advanced analytics and AI to optimize raw material procurement costs. By analyzing historical data and market trends, it achieved a 10% reduction in raw material costs through better negotiations with suppliers and demand forecasting accuracy.

Other examples of large companies using AI in their procurement process include a global technology, engineering, and software company that leverages BI tools and ML algorithms for cost optimization in procurement. Through data-driven insights and predictive analytics, the company has realized a 12% reduction in procurement costs and a 20% improvement in inventory turnover. Similarly, a U.S.-based technology and manufacturing company utilizes advanced analytics and AI in its procurement processes to optimize sourcing decisions and mitigate supply chain risks. By analyzing supplier performance data and market intelligence, it has achieved a 15% reduction in raw material costs and a 10% increase in profit margins. Another manufacturer of electronics and HVAC equipment uses BI tools and ML algorithms for cost optimization in raw material procurement. Through real-time data analysis and predictive modeling, the company has reduced procurement cycle times by 20% and achieved a 15% improvement in supplier performance metrics. Finally, an industrial automation and digital transformation technologies company leverages advanced analytics and AI to optimize inventory management and forecast demand for raw materials. By implementing predictive analytics models, it has achieved a 10% reduction in inventory holding costs and a 25% improvement in on-time delivery rates for critical components.

Advanced analytics, BI, AI, and ML are instrumental in optimizing raw material procurement costs across diverse manufacturing sectors. By leveraging predictive modeling, optimization algorithms, anomaly detection, and BI tools, manufacturers can achieve cost savings, improve supply chain efficiency, and gain a competitive advantage in the marketplace. However, addressing challenges such as data integration, quality, bias, and compliance is essential to maximizing the effectiveness of these technologies in procurement cost optimization.

Figure 6.4 illustrates the processes with modern procurement. As you can see, compared to Figure 6.2, most of the processes are automated—again, depicted by the size of the icons for computing and human resources. As the figure shows (along with the legend), 50% of the process steps are fully automated with the aid of advanced analytics, BI, AI, and ML. Also, for the semi-automated steps (4, 6, and 8), the text in bold black font indicates the manual steps within, while the others (in green font) are automated. Only step 3 is fully manual, making the entire procurement process much simpler than the traditional cycle.

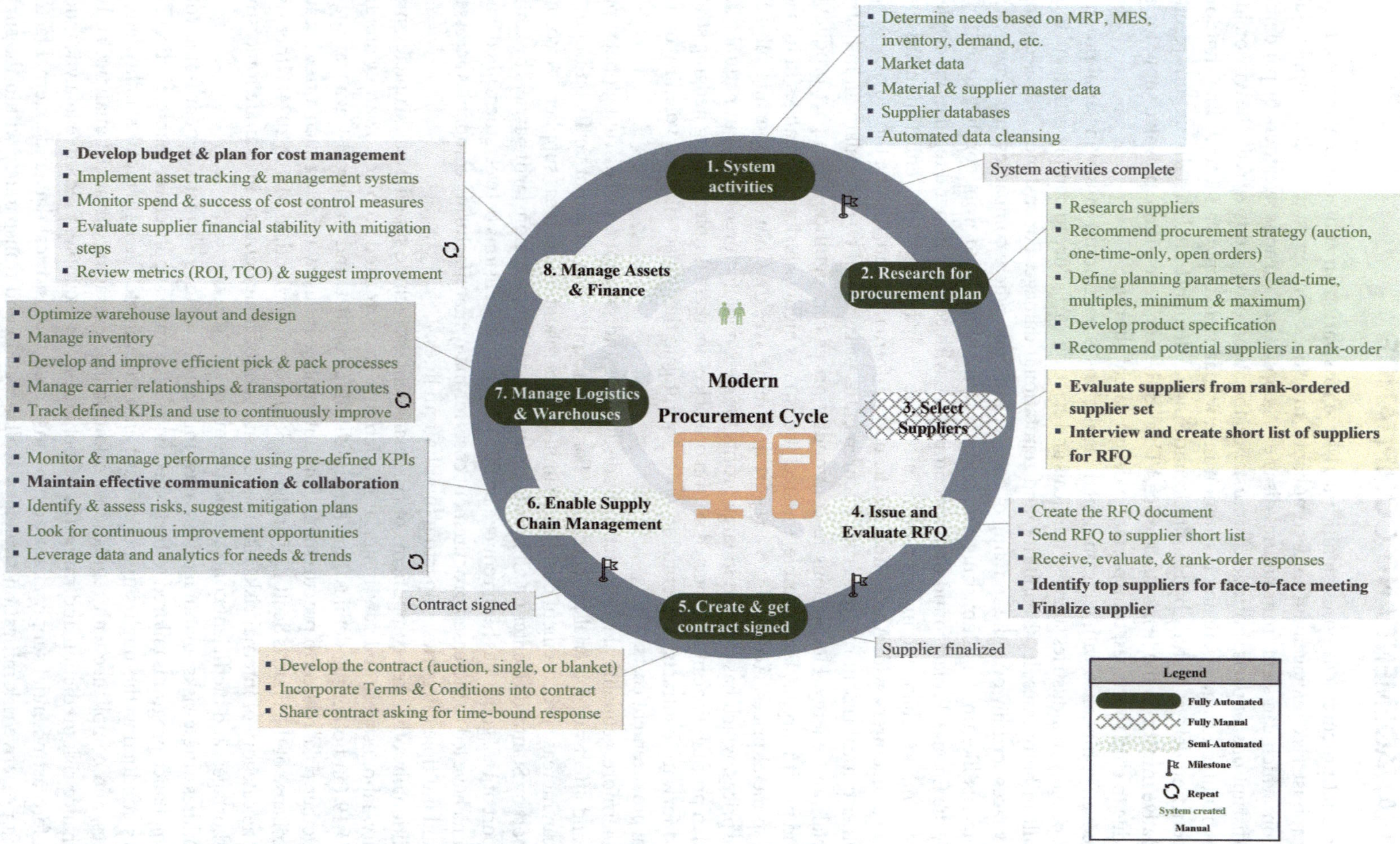

Figure 6.4 Processes within modern procurement.

RISK MANAGEMENT AND COMPLIANCE

Raw material procurement poses significant risks across various manufacturing sectors, necessitating robust risk management and compliance measures. One primary risk is supply-chain disruption, which can stem from natural disasters, geopolitical tensions, weather- or pandemic-caused disruptions, or parts shortages. For instance, the automotive industry relies heavily on just-in-time manufacturing, where any disruption in raw material supply can lead to line shutdown and production delays. A study by Hendricks and Singhal demonstrated that disruptions resulted in a negative impact on the firm's stock price.[3] This has become even more prevalent now, for example, due to chip shortages, transportation delays, talent shortage, etc.

Moreover, compliance with environmental regulations is paramount in raw material procurement, especially in industries such as chemicals and electronics. Failure to comply with regulations can result in hefty fines, legal liabilities, and reputational damage. For instance, the electronics industry faces scrutiny over the sourcing of conflict minerals, where companies must ensure that raw materials like tin, tantalum, tungsten, and gold are not sourced from conflict zones.[4,5,6] A study by the Enough Project found that noncompliance with conflict mineral regulations can lead to a loss of market share and investor confidence. The study highlights the efforts of many electronics manufacturers, with Apple showing the best performance and Toshiba the worst.[7]

Another critical risk in raw material procurement is price volatility, particularly in industries like metals and energy. Fluctuations in commodity prices can significantly impact production costs and profit margins. For example, the steel industry is highly sensitive to changes in iron ore and scrap metal prices. According to a report in Mining Weekly, China is supporting sub-$100 iron ore prices, which has resulted in large profits for steel manufacturers.[8] It is but natural that a reversal in profitability can be expected as iron ore prices increase. Implementing hedging strategies and forward contracts is essential in order to mitigate the impact of price volatility on raw material procurement costs.

Furthermore, geopolitical risks, such as trade disputes and sanctions, can disrupt global supply chains and affect raw material procurement. For instance, the imposition of tariffs on imported goods can increase procurement costs for manufacturers reliant on overseas suppliers. According to a report by Stanford's Center on China's Economy, since the use of non-tariff barriers by China, the import of U.S. goods has been reduced by 50%.[9] Such trade tensions between the United States and China have led to increased procurement costs for many manufacturing companies across the globe and multiple categories, impacting their bottom lines.

Additionally, other key components of risk management include supplier bankruptcies, price volatility, labor issues, and raw material availability. Risks related to black swan events like COVID-19 (and other pandemics), the Russia-Ukraine conflict, and the Israel-Gaza war spawn several other risks related to procurement. The automobile industry is facing such a risk because of the collapse of the Francis Scott Key bridge in Baltimore, Maryland, in March 2024. Other risks fall in the category of internal risks like inadequate vendor management, inaccurate forecasting, talent shortages, and inefficient contract management.

To address these risks, manufacturers must implement robust risk management frameworks and compliance programs tailored to their specific industry and supply chain characteristics. This includes conducting thorough supplier assessments to evaluate their financial stability, ethical practices, and compliance with regulations. Additionally, leveraging advanced analytics and AI can enhance risk prediction and mitigation capabilities. For example, predictive analytics models can forecast potential supply chain disruptions based on historical data and external factors, enabling proactive risk mitigation strategies. Collaboration and transparency within the supply chain are crucial for effective risk management. Establishing strong partnerships with suppliers and fostering open communication channels can help proactively identify and address potential risks. Investing in technologies that enhance supply chain visibility, such as blockchain and Internet of Things (IoT) sensors, can improve traceability and transparency, reducing the risk of noncompliance and unethical sourcing practices.

DATA GOVERNANCE AND SECURITY

Just like risk management and compliance, data governance and security are of paramount importance when implementing advanced analytics, BI, AI, and ML in raw material procurement across diverse manufacturing sectors. With the increasing reliance on data-driven decision making, ensuring the integrity, confidentiality, and availability of data is essential to mitigate risks and safeguard sensitive information. According to a recent survey, in 2023, the global average cost of a data breach was $4.45 million across all industries.[10] Similarly, according to another report by IBM, data breaches in manufacturing cost an average of nearly $5 million per incident, highlighting the importance of robust data governance and security measures.[11]

One key aspect of data governance is establishing clear policies and procedures for data collection, storage, and usage. This includes defining data ownership and stewardship, as well as assigning roles and responsibilities within the organization to ensure accountability and compliance with regulatory requirements such as GDPR and CCPA. Implementing data quality controls and standards is essential to maintaining the accuracy and reliability of data used in advanced analytics, AI, and ML algorithms. Ensuring data security is critical to prevent unauthorized access, tampering, or theft of sensitive information. Implementing encryption, access controls, and multifactor authentication mechanisms are effective ways to protect data that is at rest or in transit. Additionally, conducting regular security audits and vulnerability assessments helps to identify and mitigate potential security risks in AI and ML systems used for raw material procurement. Establishing tools and processes ensuring protection against malware and ransomware is critical—not just in manufacturing but in all industries. In addition, having strong protection procedures prevents the company from being held hostage by ransomware attacks, such as the one that shut down the Colonial Pipeline for five days in 2021.

Manufacturers must also consider the security implications of third-party data sources and AI/ML vendors. Evaluating the security posture and compliance certifications of vendors and conducting due diligence on data-sharing agreements are essential steps in mitigating risks associated with third-party data usage. Implementing data anonymization and de-identification techniques can help protect sensitive information when sharing data with external partners or suppliers. The emergence of edge computing and IoT devices in manufacturing introduces new challenges for data governance and security. Edge devices rapidly generate vast amounts of data at the source, requiring robust data governance frameworks to ensure data integrity and compliance with privacy regulations. Implementing secure communication protocols and endpoint security measures is crucial to protect edge devices from cyber threats and unauthorized access.

Transparency and accountability are essential to building trust and confidence among stakeholders. Ensuring trustworthy and ethical data, along with the use of explainable AI, is essential for gaining buy-in from stakeholders. Providing explanations for algorithmic decisions—especially those of AI algorithms, which often appear as black boxes—and monitoring algorithm performance for bias and fairness can help mitigate ethical concerns and ensure responsible use of AI and ML.

Terex, a global manufacturer of lifting and material processing products, has enumerated its information and cybersecurity policy in its 2023 annual report. They have adopted the National Institute of Standards and Technology cybersecurity framework, which is supported by the best practices in the industry. This enables them to embed security into all their procurement activities, including those that utilize AI. This has resulted in Terex being honored as one of America's Most Responsible Companies by Newsweek. It has also resulted in receiving a perfect score in the Human Rights Campaign Equality Review Index.[12,13] Another case study is that of Applied Materials, a global leader in materials engineering solutions. One of Applied Materials' suppliers had a cybersecurity incident that impacted Applied Materials to the tune of $250 million in quarterly revenue. This put more pressure on them to incorporate cybersecurity not just in their procurement processes but also in their key suppliers.[14]

These case studies showcase how companies prioritize data governance and security along with environmental, social, and governance (ESG) principles in their AI- and ML-driven processes. Unfortunately, though, most initiatives start after an incident has occurred; few are proactive.

MANAGING PROCUREMENT PROCESSES

Manufacturers employ various dimensions to effectively manage procurement processes and ensure the seamless acquisition of raw materials. Major key parameters fall into the following categories: cost, risk, supplier, inventory, and procurement processes. By effectively managing these parameters, manufacturers can optimize procurement processes, minimize risks, and enhance the overall efficiency and sustainability of their supply chains. Some details on each of the key parameters, with each having its own set of measurable key performance indicators (KPIs), are:

1. **One of the primary metrics is *cost*:** This involves evaluating the total cost of ownership—not just focusing on the unit cost but on the costs associated with procuring raw materials, including purchase price, shipping costs, taxes, duties, and inventory holding costs. Manufacturers aim to minimize costs while maintaining product quality and supplier reliability. The goal of this is to measure the amount of money saved through activities such as actual procurement, negotiated discounts, reduced purchase prices, and optimized sourcing strategies. Advanced analytics and BI provide actionable insights by analyzing procurement data, supplier performance metrics, and market trends. ML and AI algorithms enable manufacturers to evaluate supplier performance based on criteria such as on-time delivery, quality of goods or services, and responsiveness. By leveraging these insights, manufacturers can monitor supplier performance, identify top-performing suppliers, and foster long-term strategic partnerships. These insights also help identify cost-saving opportunities and negotiate favorable terms with suppliers to achieve significant cost reductions.
2. **Cost is followed by *risk* to the supply chain:** Components of risk management include identifying and mitigating potential risks associated with procurement activities, such as supply chain disruptions, geopolitical instability, supplier bankruptcies, and price volatility. Risks related to black swan events like COVID-19 (and other pandemics), the Russia-Ukraine conflict, the Israel-Gaza war, and the collapse of the Francis Scott Key bridge in Baltimore spawn several other risks related to procurement. Automobile manufacturers,

among others, faced the risks associated with chip shortages. Other risks fall into the category of internal risks, for example, inadequate vendor management, inaccurate forecasting, talent shortages, and inefficient contract management. Manufacturers must have a strong risk management and evaluation process in place that is periodically executed to identify gaps in the process. Use risk assessment parameters to proactively manage risks and ensure business continuity.

3. **Cost and risk are key drivers for *supplier selection*:** Manufacturers assess suppliers based on criteria such as product quality, reliability, financial stability, geographic proximity, and compliance with regulatory requirements. Evaluating suppliers helps ensure that they align with the organization's quality standards and business objectives. Within the United States, one dimension in supplier selection is supplier diversity. This tracks the percentage of procurement spend allocated to diverse suppliers, including minority-owned, women-owned, veteran-owned, and small businesses. Promoting supplier diversity fosters inclusion and supports economic development within diverse communities. Supplier performance plays a major role in this. It evaluates the performance of suppliers based on criteria such as on-time delivery, quality of goods or services, and responsiveness to inquiries or issues. Monitoring supplier performance ensures that vendors meet contractual obligations and maintain high standards of service.
4. **Based on the outcome of the supplier selection, building strong *supplier relationship management* (SRM) is key to procurement:** SRM assesses the strength and effectiveness of relationships with key suppliers based on factors such as communication, collaboration, and strategic alignment. It involves building and nurturing relationships with key suppliers to foster collaboration, drive innovation, and mitigate supply chain risks. Parameters such as communication, performance feedback, and contract negotiations contribute to effective SRM.
5. **It is critical to measure and track *contract compliance*:** Primarily, this involves measuring the degree to which procurement activities adhere to contract terms and conditions, including pricing agreements, delivery schedules, demand variance ranges, and quality specifications. Ensuring contract compliance minimizes risks associated with noncompliance and contractual disputes.
6. **Advanced analytics tools help track *procurement cycle time*:** These tools will measure the average time taken to complete procurement processes, from requisition to purchase order issuance. By identifying bottlenecks and streamlining workflows, manufacturers can reduce cycle times, enhance efficiency, and meet production timelines more effectively. Lead time refers to the time taken for suppliers to deliver raw materials after an order is placed. Manufacturers monitor lead times and delivery performance to ensure timely replenishment of inventory and to minimize production disruptions. Analyzing this data quickly highlights issues in lead and delivery time across suppliers. It also highlights trends toward stockouts or excess inventory.
7. **Track sustainability, including *environmental and social impact* metrics related to procurement activities:** These metrics have a significant impact on manufacturers. They must consider ethical and sustainability parameters when selecting suppliers and procuring raw materials. This includes metrics such as carbon emissions, waste reduction, and ethical sourcing practices, which supports sustainable supply chain practices and garners goodwill from customers and the community. Other factors, such as ethical sourcing practices, labor conditions, and compliance with social responsibility standards, influence procurement decisions and supplier relationships. About a decade ago (in the

2010s), this was referred to as corporate social responsibility. In recent years, this has been known as ESG.

8. **Calculate the *ROI* generated from procurement activities:** This calculation should include cost savings, process efficiencies, and value-added contributions to the organization. Procurement ROI measures the financial benefits derived from procurement efforts and demonstrates the department's contribution to organizational goals.
9. ***Inventory management, QC, and assurance* are the final items in managing the procurement process:** By analyzing inventory turnover ratios, organizations can assess the efficiency of their inventory management practices. A high inventory turnover ratio indicates efficient utilization of raw materials and optimal inventory levels, minimizing holding costs and reducing the risk of obsolete inventory. Manufacturers use parameters such as reorder points, safety stock levels, and economic order quantities to balance inventory levels and demand variability. Manufacturers establish QC measures to ensure that procured raw materials meet specified quality standards and regulatory requirements. Parameters such as product specifications, inspection criteria, and quality assurance processes help maintain consistent product quality and reliability.

By tracking metrics in these dimensions, organizations can assess the performance of their procurement functions, identify areas for improvement, and drive strategic decision making to optimize procurement processes and achieve business objectives. BI, along with the company's data, is instrumental in supporting the management of procurement processes.

FUTURE TRENDS AND OPPORTUNITIES

As the manufacturing landscape evolves, the integration of AI and ML technologies in raw material procurement presents promising opportunities and future trends across various sectors. One significant trend is the adoption of AI/ML-based predictive analytics to forecast demand and optimize inventory levels. By analyzing historical data and market trends, manufacturers can anticipate raw material requirements more accurately, reducing stockouts and excess inventory. A research report by a large systems integrator indicated that companies using predictive analytics in procurement achieved a 20% reduction in inventory holding costs and a 15% improvement in on-time delivery rates.

Furthermore, AI and ML algorithms offer opportunities for automating routine procurement tasks, such as supplier selection, negotiation, and contract management. By leveraging NLP and ML, manufacturers can streamline procurement processes, improve efficiency, and reduce manual errors. For example, in 2021, according to Gartner, utilizing AI/ML was expected to reduce the manual processing and review of contracts by 50%.[15] AI-based procurement has resulted in up to 40% faster transactions at a 30% lower cost.[16]

Another emerging trend is the use of AI-driven predictive maintenance (PdM) in raw material procurement for equipment and machinery maintenance. By analyzing sensor data and equipment performance metrics in real time, manufacturers can detect potential failures or maintenance needs before they occur, minimizing downtime and optimizing production schedules. For instance, a report by McKinsey indicated that PdM can reduce maintenance costs by 10–40%, reduce downtime by 50%, and increase equipment lifetime by 20–40%.[17] A sound PdM system highlights the real need for maintenance components and improves procurement.

AI and ML technologies enable manufacturers to enhance supply chain visibility and traceability, addressing growing consumer demand for transparency and sustainability. By leveraging blockchain technology and IoT sensors, companies can track raw material origins, verify supplier certifications, and ensure compliance with ethical and environmental standards. For example, Nestlé partnered with IBM Food Trust to implement blockchain-based traceability solutions to its Zoégas coffee brand, thereby enabling consumers to track the journey from the farm to their coffee mug and improving trust in the brand.[18]

AI-powered risk management solutions offer opportunities to mitigate supply chain disruptions and identify potential risks proactively. By analyzing external factors such as geopolitical events, natural disasters, and supplier financial health, manufacturers can anticipate and mitigate risks before they impact production. For instance, AutoVentive and Ford Motor Company implemented an AI and big data analytics inventory monitoring system to minimize risk, enabling real-time monitoring of supplier performance. As a result, Ford reduced its supply chain disruptions significantly and improved overall supply chain resilience.[19]

The emergence of AI-driven market intelligence solutions provides manufacturers with valuable insights into parameters such as raw material pricing and availability trends, competitor strategies, and market dynamics. By utilizing NLP to analyze unstructured data from various sources—including news articles, social media, upcoming regulatory changes, and industry reports—companies can make informed decisions and gain a competitive edge in raw material procurement. For example, a global chemical company implemented an AI-driven market intelligence platform that improved pricing accuracy by 20% and reduced procurement costs by 15%.

AI-powered optimization algorithms offer opportunities to optimize procurement decisions and maximize cost savings. By considering multiple factors, such as those mentioned in the previous section, manufacturers can identify the most cost-effective sourcing strategies and negotiate favorable contracts with suppliers. For example, a company that provides agricultural solutions and industrial products implemented an AI-driven optimization solution for its raw material procurement processes that reduced procurement costs by 10% and improved supplier performance metrics by 15%.

Overall, the future of AI and ML in raw material procurement holds immense potential for driving efficiency, agility, and sustainability across different manufacturing sectors. By embracing these technologies and leveraging real-time data analytics, manufacturers can gain competitive advantages, mitigate risks, and unlock new opportunities for growth and innovation in the dynamic global marketplace.

SUMMARY AND KEY INSIGHTS

This chapter has meticulously navigated the complex realm of adopting advanced analytics, BI, ML, and AI within raw material procurement in manufacturing. We examined key domains ranging from data sourcing and integration to advanced analytics, risk management, and future trends. Beginning with a deep dive into understanding data sources, we explored how manufacturers amalgamate information from diverse channels, such as supplier databases, market insights, and internal ERP systems, to inform their procurement strategies.

Predictive analytics, automation, and demand forecasting have become indispensable tools in trimming procurement costs, enhancing efficiency, and bolstering competitive advantage. Moreover, this chapter scrutinized the critical imperative of risk management, spotlighting the relentless efforts of manufacturers to navigate supply chain disruptions, compliance hurdles, and price fluctuations. Drawing insights from case studies, we discerned how proactive measures such as PdM, supplier diversification, and real-time risk monitoring fortify resilience and safeguard against potential vulnerabilities.

Some of the key takeaways from this chapter include the following:

1. Data-driven procurement helps procurement managers achieve efficiency, agility, and sustainability in manufacturing.
2. Advanced analytics, BI, ML, and AI play major roles in improving procurement processes, mitigating risks, and driving sustainable value creation.
3. Fostering collaboration and data sharing among internal stakeholders and external partners can promote transparency and visibility in procurement.
4. Manufacturers are increasingly adopting AI- and ML-based predictive analytics to forecast demand and optimize their inventory levels.

7

INCORPORATING ADVANCED ANALYTICS, BI, ML, AND AI IN MANUFACTURING AND PRODUCTION

AI-Powered Manufacturing and Production at Stellar Electronics: Redefining Efficiency

Stellar Electronics' manufacturing processes have historically relied on manual planning and reactive maintenance, leading to inefficiencies, downtime, and quality control (QC) challenges. Recognizing the potential for improvement, Tom Willis, the VP of manufacturing, partnered with Olivia Chen, Stellar's director of data analytics, to integrate artificial intelligence (AI) and machine learning (ML) into their production lines, revolutionizing how the company operates.

The first major initiative involved implementing AI-driven predictive maintenance (PdM) systems across the production floor. By installing Internet of Things (IoT) sensors on critical machinery, the AI models analyzed real-time data to detect patterns and predict equipment failures before they occurred. This proactive approach resulted in a 40% reduction in unplanned downtime and extended the lifespan of key machinery by 25%, significantly lowering maintenance costs. Carlos Rivera, the plant manager, remarked that the shift from reactive to PdM had been a game-changer, allowing the plant to run more efficiently with fewer disruptions.

In addition, AI-powered QC systems were introduced to enhance production accuracy. Olivia Chen's team deployed ML algorithms capable of inspecting products in real time, identifying defects with over 95% accuracy. This resulted in a 20% reduction in defective units and lowered the rate of product returns, which contributed to improved customer satisfaction and cost savings in rework and warranty claims. By automating this critical aspect of the production line, Stellar was able to scale production without sacrificing quality.

The company also utilized AI for production scheduling and optimization. By analyzing variables such as workforce availability, machine capacity, and material supply, Stellar's AI-driven system optimized production schedules to minimize idle time and maximize throughput. This resulted in a 15% increase in production efficiency and a significant reduction in lead times. Jessica Moore, the operations manager, highlighted how the AI system helped streamline workflows and ensure that production aligned with demand fluctuations more effectively than ever before.

By embracing AI and ML in their manufacturing and production operations, Stellar Electronics has not only improved efficiency but also reduced costs, minimized downtime, and enhanced product quality. These technological innovations have positioned the company as a leader in advanced manufacturing, with a flexible and future-ready production system.

"Human forecasters often struggle with predicting rapid changes in demand, especially when faced with complex time-series signals."

—Devadrita Nair, Postdoctoral Associate at MIT and Maria Jesus Saenz, Director of the MIT Supply Chain Transformation Lab

In the modern-day landscape of manufacturing, the integration of advanced analytics, business intelligence (BI), ML, and AI has emerged as a foundation for driving operational excellence and sustaining competitive advantage. This chapter probes into the pivotal role of these cutting-edge technologies in revolutionizing manufacturing and production processes, ushering in an era of unparalleled efficiency, quality control, and productivity gains.

In today's fiercely competitive, rapidly changing market environment, manufacturers are under constant pressure to optimize production operations, among other things, while maintaining stringent quality standards and meeting ever-evolving customer demands. The significance of data-driven insights cannot be overstated in this endeavor. These insights provide manufacturers with actionable intelligence to make informed decisions, identify bottlenecks, and streamline processes for maximum efficiency. Efficient production operations lie at the heart of achieving and sustaining competitive advantage in the manufacturing sector. By leveraging advanced analytics, BI, ML, and AI, manufacturers can unlock new levels of efficiency, agility, and responsiveness in their production processes. From PdM and demand forecasting to real-time process monitoring and adaptive control systems, these technologies empower manufacturers to optimize resource allocation, minimize downtime, optimize inventory across multiple levels, and enhance overall operational performance.

The advent of Industry 4.0 has ushered in a paradigm shift in manufacturing, where interconnected systems and smart technologies enable seamless integration and automation across the production value chain. Advanced analytics, BI, ML, and AI are the foundation of this digital transformation, providing manufacturers with the tools and capabilities to realize the full potential of data-driven insights for optimizing manufacturing and production. By harnessing the power of these technologies, manufacturers can unlock new frontiers of efficiency, quality, and innovation to strengthen their competitive position in the global marketplace. PdM, QC, supply chain optimization, and adaptive manufacturing provide limitless possibilities for manufacturers.

In this chapter, you will embark on a journey to explore the transformative impact of advanced analytics, BI, ML, and AI on manufacturing and production processes. By probing into real-world experience and industry insights, you will gain a comprehensive understanding of how these technologies are reshaping the manufacturing landscape, driving benefits such as improved operational efficiency, QC, and overall productivity gains. Furthermore, by unraveling the significance of data-driven insights and efficient production operations in maintaining competitive advantage, this chapter aims to equip readers with actionable strategies and best practices for leveraging advanced analytics, BI, ML, and AI in optimizing manufacturing and production processes. As you navigate through the chapter, you will discover the transformative potential of these technologies and unlock new avenues for innovation, growth, and success in the dynamic world of manufacturing.

UNDERSTANDING MANUFACTURING AND PRODUCTION PROCESSES

As mentioned in Chapter 1, manufacturing and production processes form the backbone of industrial operations across various sectors, encompassing a myriad of activities aimed at transforming raw materials into finished goods. They also play a large role in the economic activities of various countries. This chapter explores the intricate workings of manufacturing processes, focusing on key components such as planning, scheduling, and execution. We examine the challenges and complexities inherent in manufacturing operations, drawing insights from real-world examples and industry metrics to reveal the multifaceted nature of modern production environments.

Manufacturing operations involve a series of interconnected processes aimed at efficiently converting raw materials into finished products. Central to these operations are planning, scheduling, and execution, which collectively dictate the flow of activities and material within a production facility. Planning involves activities like defining production objectives, allocating resources, and determining the optimal sequence of tasks to meet demand forecasts and production targets. Scheduling includes assigning specific tasks to resources, machines, and personnel within predefined timeframes to ensure smooth workflow and minimize bottlenecks. Execution contains the actual manufacturing processes, including machining, assembly, and QC, where raw materials are transformed into final products through a series of operations (see Figure 7.1).

> *"... [We] rolled out an AI application to do our supply and demand planning. Before, it was spreadsheets and those kinds of things. The moment we started using the AI application, we saw ... 30% increases in our forecasting accuracy."*
>
> —Sowmya Gottipati, Global Supply Chain Leader at Estée Lauder

Across multiple sectors in manufacturing, the execution phase encompasses a diverse array of processes tailored to the specific requirements of each industry. For example, in the automotive sector, manufacturing involves processes like stamping, welding, painting, and assembling vehicle components. According to data from the International Organization of Motor Vehicle Manufacturers, global automobile production reached over 93 million units in 2023, underscoring the scale and complexity of automotive manufacturing operations.[1] Similarly, in the food and beverage industry, production processes encompass ingredient mixing, cooking, packaging, and labeling, with stringent QC measures to ensure product safety and compliance with regulatory standards.

Despite advancements in technology and automation, manufacturing operations are fraught with challenges and complexities that pose significant hurdles to efficiency and productivity. One such challenge is demand variability, where fluctuating customer demand and market dynamics necessitate agile and responsive manufacturing processes. Demand drives all activity within manufacturing, from acquiring raw material to planning production and shipping out finished products. According to a survey of chief procurement officers by McKinsey & Company, 73% of apparel manufacturing executives cite demand volatility as a major challenge in supplier relationships, impacting production planning and scheduling.[2]

The increasing complexity of products and processes, along with constantly changing technology, adds other layers of complexity to manufacturing operations. With the rise of customization and product diversification, manufacturers are tasked with managing a wide range of product variants and configurations while maintaining efficiency and cost-effectiveness. For instance, in

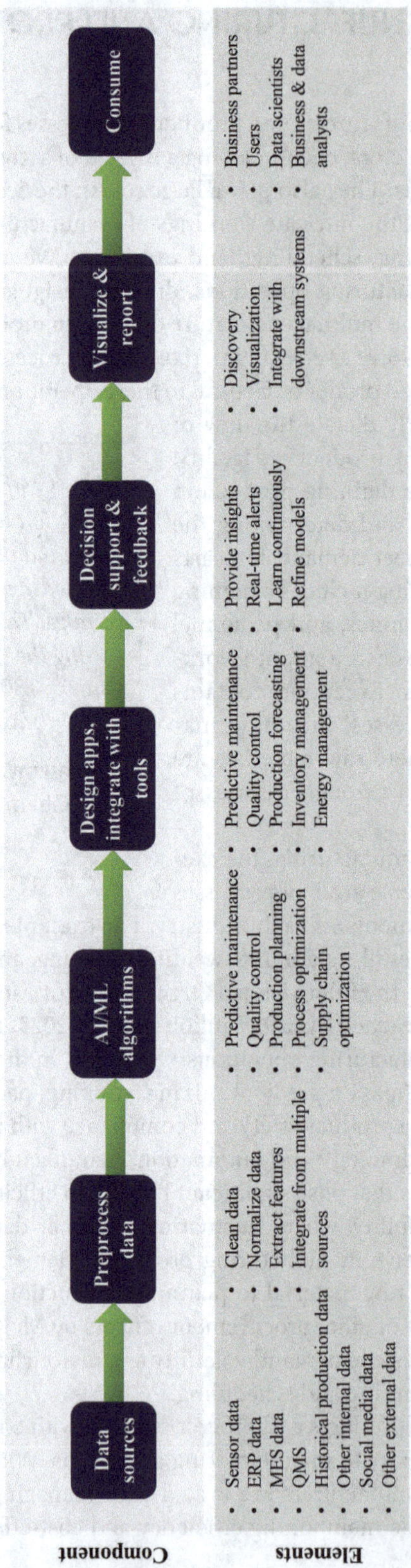

Figure 7.1 Sample of steps involved in manufacturing and production.

the aerospace industry, where aircraft components are highly complex and engineered to exact specifications (also known as make-to-order), production processes must adhere to stringent quality standards and regulatory requirements. Ever-lengthening supply chains have increased the risk of supply chain disruptions and equipment downtime as workforce shortages pose additional challenges to manufacturing operations. According to a report by Deloitte, unplanned downtime costs manufacturers an estimated $50 billion annually, highlighting the impact of equipment failures on production efficiency and profitability.[3] Similarly, workforce shortages, exacerbated by pandemics like COVID-19, demographic shifts, and skill gaps, hinder manufacturers' ability to meet production demands and maintain operational continuity.

Understanding manufacturing and production processes is essential for navigating the complexities of modern industrial operations. Key components, such as planning, scheduling, and execution, form the foundation of manufacturing operations, enabling the efficient conversion of raw materials into finished products. However, challenges such as demand variability, product complexity, and supply chain disruptions pose significant hurdles to manufacturing efficiency and productivity. By addressing these challenges through innovation, technology adoption, and agile operational strategies, manufacturers can enhance their competitiveness and adaptability in an ever-evolving marketplace.

Figure 7.2 illustrates the processes within traditional manufacturing and production. Visually, the size of the icons for computing versus human resources depicts a significant number of manual versus automated processes—most process steps are either fully or partially manual. The components in this figure describe the flow of materials, information, and processes in manufacturing. Note that this diagram is generic and may need to be customized for the target environment. Most of the items have a high level of human engagement—computing tools are used as necessary.

UNDERSTANDING DATA SOURCES AND INTEGRATION

In the dynamic realm of manufacturing and production, the convergence of advanced analytics, BI, AI, and ML has revolutionized data integration, offering unprecedented opportunities to optimize processes, enhance quality, and drive innovation. Here, we explore the pivotal role of advanced analytics, BI, AI, and ML algorithms in harnessing data from diverse sources such as IoT sensors, production machines, supply chain systems, and QC tools. By digging into key components of manufacturing operations and the complexities therein, we uncover how these technologies facilitate seamless data integration, analysis, and decision making, propelling manufacturing into a new era of efficiency and competitiveness.

Manufacturing operations generate vast amounts of data at every stage of the production process, from raw material procurement to finished product delivery. Advanced analytics, BI, AI, and ML algorithms are crucial in extracting actionable insights from this data deluge, enabling manufacturers to optimize key components of manufacturing operations such as planning, scheduling, and execution. By applying predictive analytics algorithms to historical production data, manufacturers can forecast demand, anticipate equipment failures, and optimize production schedules to maximize throughput and minimize downtime. In addition, AI and ML algorithms empower manufacturers to automate decision-making processes and enable autonomous operations throughout the manufacturing value chain. For instance, AI-powered PdM systems analyze sensor and IoT data from production machines in real time, detecting anomalies and predicting equipment failures before they occur. By proactively addressing maintenance issues,

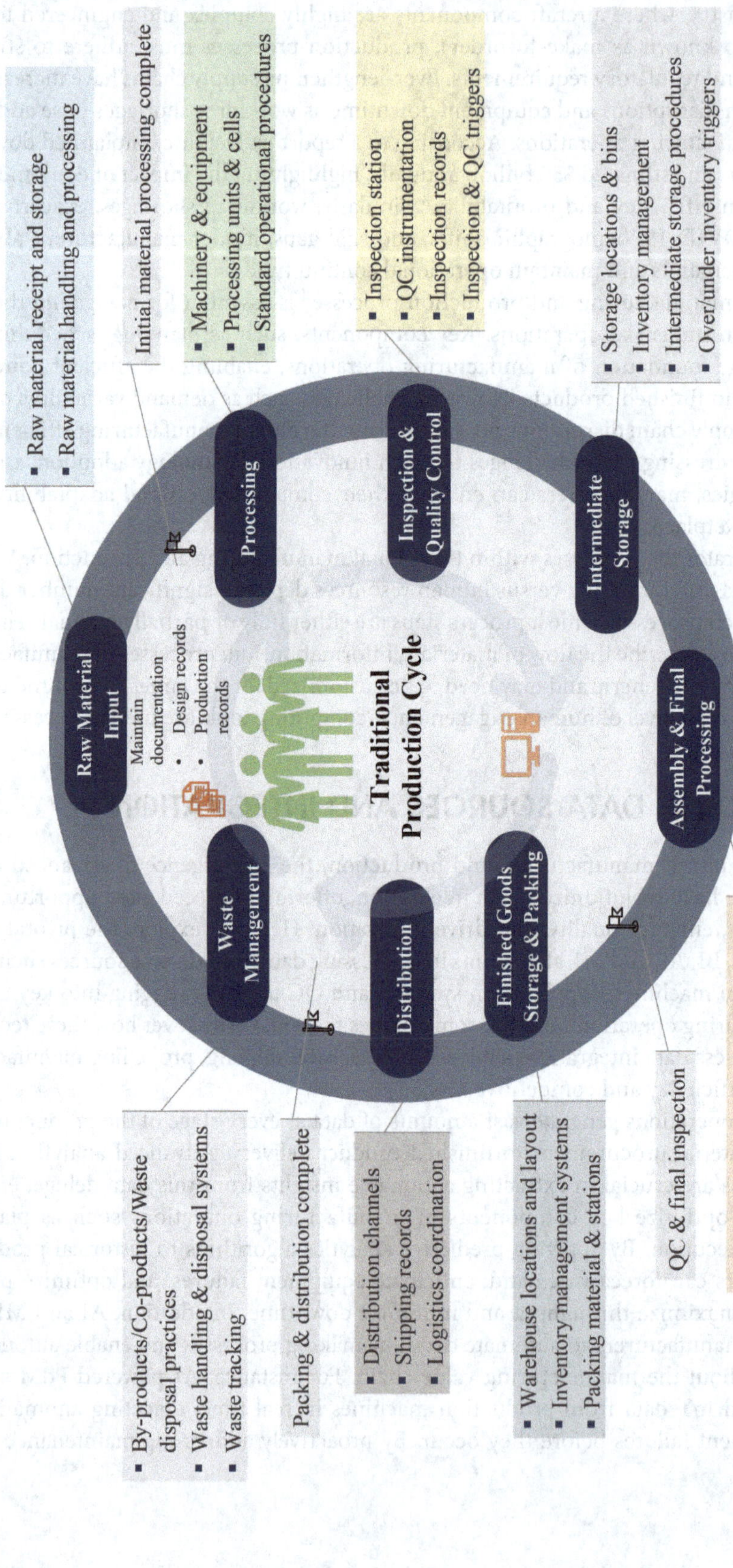

Figure 7.2 Processes within traditional manufacturing and production.

manufacturers can minimize unplanned downtime, reduce maintenance costs, and optimize asset utilization.

AI-driven quality control systems leverage computer vision and pattern recognition algorithms to inspect products for defects and deviations from quality standards. By analyzing images captured by automated inspection systems, these algorithms can detect subtle defects that may be imperceptible to the human eye, ensuring that only high-quality products reach the market. Also, AI-enabled anomaly detection algorithms analyze production data to identify deviations from expected process behavior, allowing manufacturers to detect and mitigate quality issues in real time. Figure 7.3 shows an example of an architecture for manufacturing and production that can be used to guide the development of your solution.

As mentioned, integrating and harmonizing data from diverse sources pose significant challenges for manufacturers. Data silos, disparate systems, and incompatible timelines and formats hinder seamless data exchange and interoperability, undermining the effectiveness of advanced analytics, BI, AI, and ML algorithms. Manufacturers must invest in robust data integration platforms, interoperability standards, and data governance frameworks to break down silos, harmonize data across disparate systems, and unleash the full potential of these technologies. Manufacturers must also invest in talent development to effectively work with the new tools, technologies, and processes coming their way.

KEY ALGORITHM CLASSES USED IN MANUFACTURING AND PRODUCTION

In manufacturing, various algorithms are employed to tackle different challenges and optimize processes. The following sections describe some examples of classes of algorithms used in manufacturing. Each class embeds algorithms that enable the necessary capability and serves a specific purpose in optimizing processes, improving efficiency, and driving innovation in the industry. Depending on the specific manufacturing application, different algorithms may be employed to address unique challenges and objectives. Often, these may be an ensemble of algorithms in a pipeline to achieve the orchestration for a desired outcome.

Predictive Maintenance Algorithms

PdM algorithms utilize ML techniques such as regression analysis, decision trees, and neural networks (NNs), such as long short-term memory (LSTM), to predict equipment failures before they occur. By analyzing historical data from sensors and equipment, these algorithms can identify patterns indicative of impending failures, enabling proactive maintenance to prevent unplanned downtime and optimize asset utilization.

Demand Forecasting Algorithms

Demand forecasting algorithms employ time-series analysis, moving averages, and exponential smoothing techniques to predict future demand for products based on historical sales data, market trends, and other factors. More recent AI algorithms, such as LSTM, capture the span and temporal information much better than traditional algorithms. These algorithms help manufacturers anticipate customer demand, optimize inventory levels, and adjust production schedules to meet demand fluctuations efficiently.

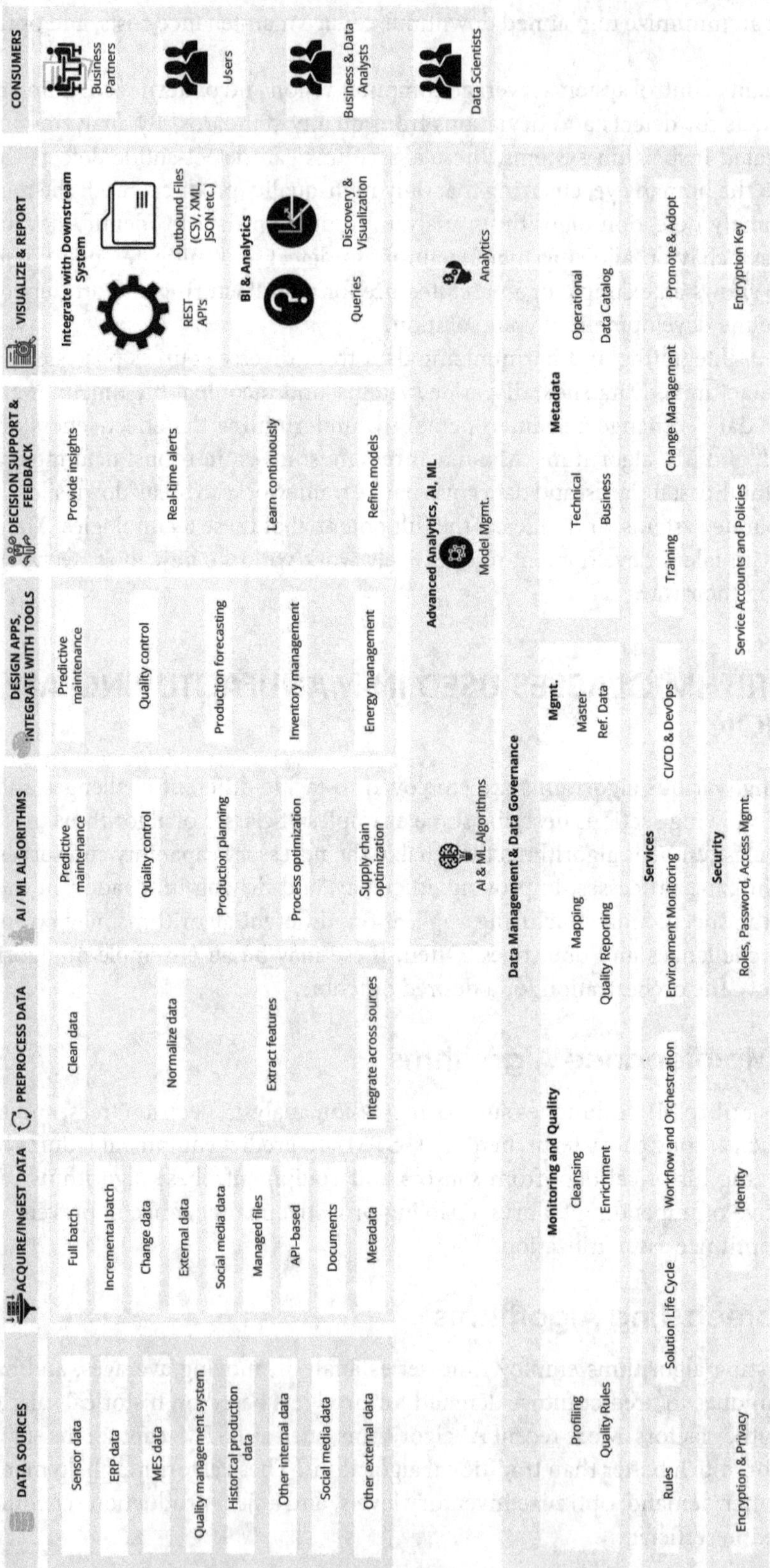

Figure 7.3 Example architecture for manufacturing and production.

Quality Control Algorithms

QC algorithms utilize statistical process control, Six Sigma methodologies, and ML algorithms, such as support vector machines and deep learning in image recognition, to inspect products for defects and deviations from quality standards. These algorithms enable manufacturers to ensure product quality, reduce defects, and maintain compliance with regulatory requirements.

Supply Chain Optimization Algorithms

Supply chain optimization algorithms leverage optimization techniques, such as linear programming, integer programming, deep learning-based AI algorithms, and genetic algorithms, to optimize supply chain networks, inventory management, and logistics operations. These algorithms help manufacturers optimize transportation costs, reduce lead times, and improve overall supply chain efficiency.

Production Scheduling Algorithms

Production scheduling algorithms use mathematical optimization models, including job shop scheduling, flow shop scheduling, and scheduling algorithms such as genetic algorithms and simulated annealing, to optimize production schedules and resource allocation. These algorithms help manufacturers maximize production throughput, minimize switchover and idle time, and meet delivery deadlines.

Anomaly Detection Algorithms

Anomaly detection algorithms employ statistical methods, clustering techniques, deep learning-based AI algorithms (e.g., computer vision algorithms), and ML algorithms, such as k-means clustering and isolation forests, to detect anomalies or outliers in production data. These algorithms help manufacturers identify abnormal patterns or deviations from expected behavior, indicating potential equipment malfunctions, process deviations, or quality issues.

Energy Optimization Algorithms

Energy optimization algorithms utilize optimization techniques and ML algorithms to minimize energy consumption and optimize energy usage in manufacturing processes. These algorithms analyze data from sensors and energy meters in the manufacturing environment to identify opportunities for energy savings, optimize equipment settings, and reduce environmental impact.

The following sections will provide further explanations of how these algorithms are used.

APPLICATION OF ADVANCED ANALYTICS AND BI IN MANUFACTURING AND PRODUCTION—ENHANCING OPERATIONAL VISIBILITY AND PERFORMANCE

The application of advanced analytics and BI has become instrumental in driving operational excellence and enhancing competitiveness. Here, we will investigate the numerous ways in which advanced analytics and BI tools empower manufacturers to gain real-time visibility into production metrics, equipment performance, and supply chain dynamics. By leveraging cutting-edge algorithms and tools, manufacturers can optimize processes, improve decision making, and achieve superior performance across the production value chain.

Advanced analytics and BI tools enable manufacturers to capture, analyze, and visualize vast amounts of data generated throughout the production process, providing real-time insights into key performance metrics. For example, sophisticated predictive analytics algorithms can forecast production output, identify potential bottlenecks, and optimize resource allocation to maximize efficiency. By leveraging historical production data and machine sensor readings, manufacturers can predict equipment failures, schedule preventive maintenance, and minimize unplanned downtime, thereby improving overall equipment efficiency (OEE). BI dashboards and visualization tools support advanced analytics through intuitive interfaces for monitoring production key performance indicators (KPIs), such as cycle time, yield, and scrap rates across multiple production lines or facilities. By aggregating and analyzing data from disparate sources, manufacturers can identify trends, anomalies, and areas for improvement. According to a study by McKinsey, companies across global manufacturers in multiple sectors that used real-time production monitoring systems reduced unit cost by 3.5%, unplanned downtime by 25%, and rework by 5%; increased production output by 6%; and improved OEE by 11% and perfect first pass throughput to 40%. These are impressive numbers—not across manufacturing, but across differing sectors.[4]

Several key algorithms are instrumental in driving operational insights and performance improvements in manufacturing and production. One such algorithm is PdM, which leverages ML techniques to analyze equipment sensor data and predict potential failures before they occur. By identifying early warning signs of equipment degradation or malfunctions, manufacturers can schedule maintenance proactively, minimize downtime, and extend asset life cycles. Another key algorithm is demand forecasting, which utilizes algorithms, such as time series analysis, regression models, and ML algorithms, to predict future demand for products or components. By accurately forecasting demand patterns, manufacturers can optimize inventory levels, reduce stockouts, and improve customer service levels.

> Demand forecasting plays an important role for Amazon, which uses it to leverage demand forecasting algorithms to anticipate customer demand for various products, enabling efficient inventory management and fulfillment operations. According to Amazon's report on its demand forecasting algorithm, Amazon Forecast achieved 50% better forecast accuracy, helped one of its clients, MRL, improve forecast accuracy from 24% to 76%, reduced waste by up to 30% on fresh produce, and increased gross profits by 25%.[5]

An example of PdM is General Electric's implementation of reliability analysis software, called APM Reliability, to assist in analysis and provide visibility to control costs and improve uptime. This resulted in a 5–10% reduction in inventory costs, a 2–6% increase in availability, and a 10–40% reduction in unplanned maintenance.[6]

Optimization algorithms play a crucial role in optimizing production schedules, resource allocation, logistics, and supply chain management. These algorithms, which include linear programming, integer programming, and genetic algorithms, enable manufacturers to maximize efficiency, minimize costs, and meet production deadlines.

For example, Toyota utilizes optimization algorithms in its production planning processes to minimize production lead times, reduce inventory levels, and synchronize production with customer demand fluctuations. Another large global automotive manufacturer used optimization algorithms in production planning. This helped them reduce production costs by over 10% and improve production cycle times by almost 20%.

As mentioned earlier, for food and beverage processing—an over $7.2 trillion industry in 2023—production processes involve mixing ingredients, cooking, packaging, and labeling, with stringent quality control measures to ensure product safety and compliance with regulatory standards. In the $5.72 trillion chemical manufacturing industry, execution involves batch preparation, mixing in reaction vessels, distilling and separating, filtering and purifying, drying and solidifying, and packaging and storing.[7] Each step can get pretty complex, so planning is critical. Similarly, in the $1.6 trillion pharmaceutical industry, comparable steps are involved but with many more regulations.[8] Further, the pharmaceutical industry involves two additional steps: quality testing and batch releasing. These are integral parts of the regulations, documentation, and provenance (ability to trace origin) of the drugs being sold.

HARNESSING ML AND AI IN MANUFACTURING AND PRODUCTION

In the realm of manufacturing and production, the integration of ML and AI has revolutionized traditional practices, offering unprecedented opportunities to enhance efficiency, productivity, and competitiveness. Here, we explore the many ways in which ML and AI are harnessed to gain real-time visibility into production metrics, equipment performance, and supply chain dynamics. By leveraging cutting-edge algorithms and tools, manufacturers can optimize processes, improve decision making, and achieve superior performance across the production value chain.

ML and AI algorithms enable manufacturers to capture, analyze, and interpret vast streams of data generated throughout the production process, providing real-time insights into key performance metrics. For instance, PdM algorithms leverage historical and real-time equipment data, sensor readings, and maintenance records to predict equipment failures before they occur. By identifying patterns indicative of impending failures, manufacturers can schedule maintenance proactively, minimize downtime, and optimize asset utilization. According to a study by McKinsey, companies that used PdM algorithms experienced an 18% to 25% reduction in maintenance costs and a 5% to 15% increase in asset availability.[9] AI-powered anomaly detection algorithms monitor production metrics in real time, identifying deviations from expected norms and triggering alerts for potential issues. By flagging anomalies such as unexpected machine downtime, quality defects, or production bottlenecks, manufacturers can take corrective actions promptly, minimizing disruptions and optimizing production efficiency. Earlier case studies and a report by McKinsey have indicated that using advanced analytics, including anomaly detection, can reduce downtime by 30-50% and increase asset life by close to 40%.[10]

Several key algorithms play pivotal roles in driving operational insights and performance improvements in manufacturing and production. One such algorithm is demand forecasting, which utilizes techniques such as time series analysis, regression models, and deep NNs (e.g., LSTM) to predict future demand for products or components. By accurately forecasting demand patterns, manufacturers can optimize inventory levels, reduce stockouts, and improve customer service levels. For instance, companies like Amazon and Walmart leverage demand forecasting algorithms to anticipate customer demand for various products, enabling efficient inventory management and replenishment operations.

Optimization algorithms are vital for optimizing production schedules, resource allocation, and supply chain logistics. These algorithms, including linear programming, genetic algorithms, and simulated annealing, enable manufacturers to maximize efficiency, minimize costs, and meet production deadlines.

When incorporating advanced analytics, BI, AI, and ML into the traditional manufacturing process, the components of each step are enhanced with technology. Figure 7.4 depicts the processes involved in modern manufacturing and development. Incorporating advanced technologies transforms each step with components that leverage data and automation to improve efficiency, quality, and decision making throughout the manufacturing process. Unlike Figure 7.2, all of the processes are semi-automated with a much higher contribution from computing resources instead of manual resources (depicted by the icon size for computing and human resources). Although semi-automated, most of the *hard* work is done by the models. These models provide guidance to human resources on actions to incorporate in operations. The manual intervention ensures checks and balances are in place by incorporating the *human-in-the-loop* framework. This allows people to quickly react and manually override poor outcomes from the modern tools, preventing them from getting pushed downstream.

ADVANCED ANALYTICS, BI, ML, AND AI IN SUPPLY CHAIN MANAGEMENT: DRIVING EFFICIENCY AND OPTIMIZATION

Efficient supply chain management is pivotal for organizations striving to maintain operational excellence and meet evolving customer expectations. Traditionally, supply chain management involved a complex coordination of various processes, including procurement, production, logistics, and distribution. However, with the advent of advanced analytics, BI, ML, and AI technologies, manufacturers now have powerful tools at their disposal to revolutionize their supply chain operations. The integration of advanced analytics, BI, ML, and AI in manufacturing supply chain management represents a paradigm shift, offering manufacturers unprecedented capabilities to optimize processes, enhance visibility, improve velocity, and drive competitive advantage. In this section, we examine the role of advanced analytics, BI, ML, and AI in manufacturing supply chain management, focusing on their applications, benefits, and real-world examples. By probing into the transformative potential of these technologies, we aim to provide manufacturers with a comprehensive understanding of how they can leverage data-driven insights to streamline operations, reduce costs, and deliver superior value to customers.

As organizations continue to navigate the complexities of the modern manufacturing landscape, the role of advanced analytics, BI, ML, and AI in supply chain management will become increasingly vital. By embracing these technologies and harnessing the power of data-driven insights, manufacturers can unlock new opportunities for innovation, agility, and growth in the digital era. One significant challenge in manufacturing supply chain management is achieving real-time visibility into operations and gaining actionable insights from vast volumes of data generated across the supply chain. Advanced analytics, BI, ML, and AI technologies play a crucial role in addressing this challenge by providing manufacturers with the tools and capabilities to monitor, analyze, and optimize their supply chain operations in real time.

Real-time visibility enables manufacturers to track KPIs, monitor production metrics, and identify disruptions as they occur. It also helps them recognize potential bottlenecks that can be

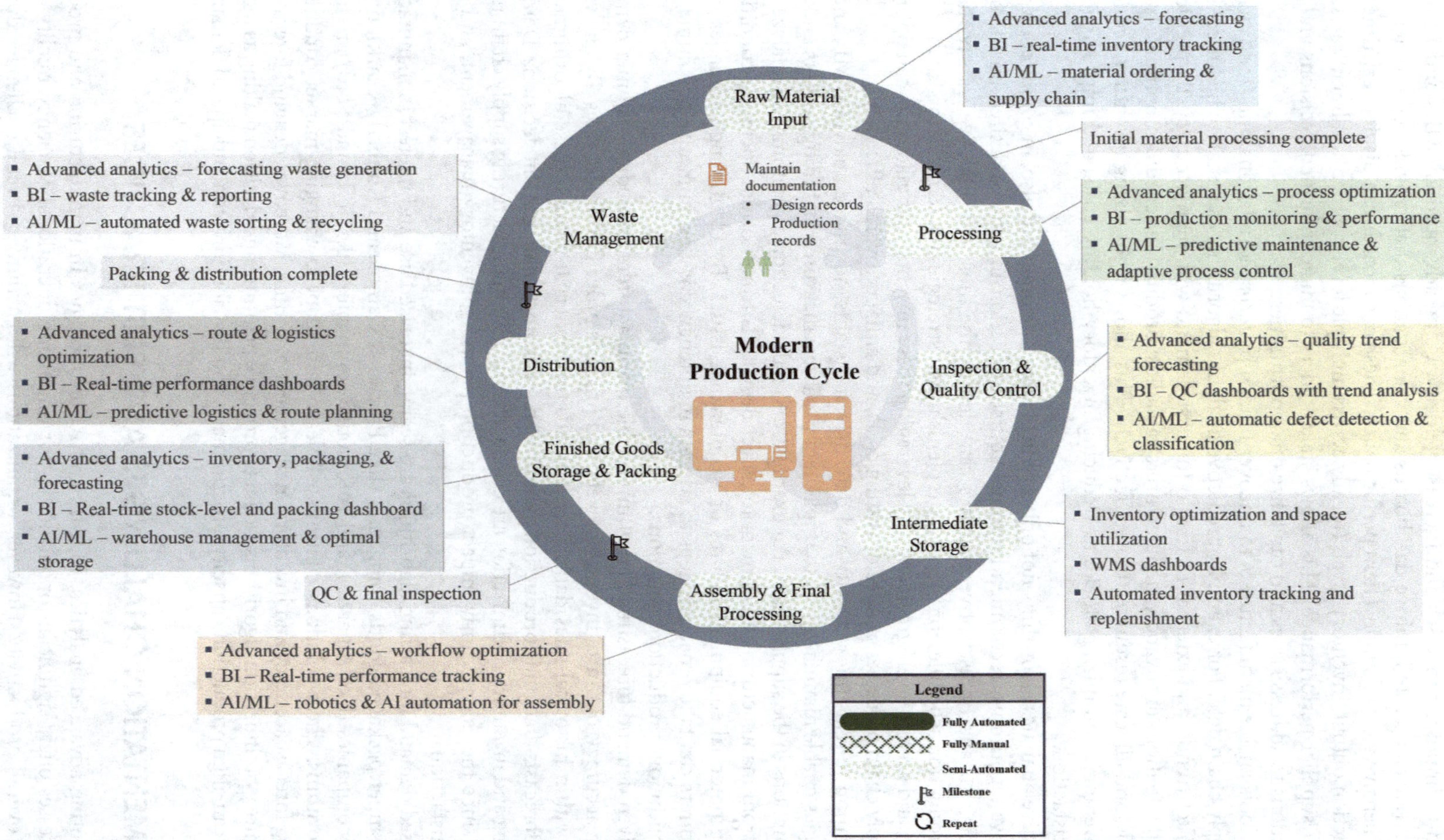

Figure 7.4 Processes within modern manufacturing and production.

leveraged to improve processes and increase velocity. Using advanced analytics and BI tools, manufacturers can create intuitive dashboards and visualizations that provide stakeholders with a comprehensive view of supply chain performance. These dashboards can display real-time data on production output, inventory levels (ideally across multiple levels), order fulfillment rates, and multi-tier supplier performance. This allows stakeholders to make informed decisions and take timely actions to address issues and capitalize on opportunities.

These tools enable manufacturers to uncover hidden patterns, correlations, and trends within their supply chain data. For example, predictive analytics algorithms can analyze historical sales data, market trends, and external factors to forecast demand with high accuracy. By predicting demand fluctuations in advance, manufacturers can adjust production schedules, optimize inventory levels, and allocate resources more effectively. This allows right-sizing inventory, reducing excess inventory, minimizing stockouts, reducing inventory holding costs, and improving customer satisfaction.

Predictive analytics algorithms not only predict future outcomes but also, in conjunction with prescriptive algorithms, recommend optimal courses of action to achieve desired business objectives. For instance, prescriptive analytics can identify the most cost-effective transportation routes, production schedules, or inventory replenishment strategies based on factors such as cost, lead time, and service level agreements. By leveraging these insights, manufacturers can optimize their supply chain processes, reduce operating costs, and enhance overall efficiency.

In addition to real-time visibility and data-driven insights, advanced analytics, ML, and AI technologies enable manufacturers to implement proactive measures that mitigate risks and prevent disruptions in the supply chain. For example, anomaly detection algorithms can analyze sensor data from manufacturing equipment to detect deviations from normal operating conditions, indicating potential equipment failures or maintenance issues. By identifying anomalies early, manufacturers can take proactive steps to address maintenance issues, minimize downtime, and ensure continuous production operations.

AI-driven demand forecasting algorithms enable manufacturers to predict customer demand with high accuracy, optimize inventory levels, and improve order fulfillment rates. By analyzing patterns within historical sales data, market trends, and external factors, these algorithms can generate forecasts that guide production planning and inventory management decisions. Demand forecasting and inventory optimization are crucial for effective manufacturing supply chain management since they significantly impact production planning, inventory management, and customer satisfaction.

AI-based demand forecasting algorithms are seeing increased use and interest in manufacturing. Consumer packaged goods companies (like Procter & Gamble and Kimberly-Clark), multi-industry companies (like Amazon), and automotive companies (like Tesla and Ford) are using these algorithms. These solutions continuously learn and adapt to changing market conditions, improving forecast accuracy and inventory management over time. Moreover, demand forecasting and inventory optimization algorithms facilitate proactive planning and decision making as manufacturers can align production capacity, procurement, and distribution with anticipated demand.

IMPLEMENTATION CHALLENGES AND BEST PRACTICES

Implementing advanced analytics, BI, ML, and AI technologies in manufacturing and production processes offers significant opportunities for improving operational efficiency, quality control, and overall productivity. However, organizations will encounter several challenges during the implementation phase, ranging from data quality issues to cultural resistance and lack of talent. Each obstacle represents a unique hurdle that requires careful consideration and strategic approaches to overcome. This section explores some of the key implementation challenges and discusses how algorithms play critical roles in addressing these challenges.

One of the primary challenges in implementing advanced analytics and AI in manufacturing is ensuring data quality and integrating disparate data sources. Manufacturing organizations generate vast amounts of data from various sources, including production machines, sensors, enterprise resource planning systems, and supply chain partners. However, this data may be incomplete, inconsistent, or siloed, making it challenging to derive actionable insights. ML algorithms, such as data cleansing algorithms and data integration techniques, play a crucial role in addressing data quality issues and harmonizing diverse datasets. Many global companies have implemented data cleansing algorithms to clean and integrate data from multiple sources, resulting in improved data accuracy and reliability. These algorithms enhance the accuracy and reliability of data analysis, enabling organizations to derive actionable insights and drive operational improvements.

A significant challenge in implementing AI and ML in manufacturing is selecting the appropriate algorithms along with proper hyperparameters, ensuring model interpretability (explainable), and making the algorithms trustworthy and responsible AI. Manufacturing is inherently complex, involving numerous variables and interactions, so identifying the most suitable algorithms for specific use cases can be daunting. Additionally, the opacity of some ML models may hinder stakeholders' ability to interpret model outputs and make informed decisions. To mitigate this challenge, organizations must leverage explainable AI techniques and model interpretability tools to enhance transparency and trust in AI-driven insights. For instance, large global conglomerates like General Electric have developed or are using explainable AI frameworks.[11] These provide clear explanations of model predictions, enabling stakeholders to understand the underlying factors driving production outcomes and facilitating informed decision making. In terms of trustworthy AI, Beena Ammanath provides a framework based on nine dimensions of Trustworthy AI.[12]

Maintaining data privacy and security is a critical concern for manufacturing organizations, especially when implementing advanced analytics and AI technologies. The proliferation of connected devices and IoT sensors in manufacturing environments heightens the risk of cybersecurity threats and data breaches. ML algorithms, including anomaly detection algorithms and predictive analytics models, play a crucial role. They will detect and mitigate security risks by monitoring data patterns and identifying suspicious activities. For instance, companies like Bosch utilize anomaly detection algorithms in their industrial automation and control systems to identify unusual behavior, thereby preventing potential cyberattacks within their manufacturing environment.[13] Implementing robust security protocols and staying vigilant against emerging threats are key steps in maintaining data privacy and security in manufacturing environments.

Bridging the skill gap and fostering a data-driven culture are essential for successfully implementing advanced analytics and AI in manufacturing. Many organizations lack the necessary expertise and talent to develop, deploy, and maintain AI-driven solutions. Additionally, cultural resistance toward change can hinder adoption efforts. To overcome these challenges, organizations must invest in employee training programs, recruit data science and AI/ML talent, and cultivate a culture of innovation and continuous learning. Many companies across different manufacturing sectors have implemented comprehensive training programs to upskill employees in data analytics and AI, enabling them to drive process improvements and innovation. Empowering employees with the skills and knowledge required to leverage advanced analytics and AI technologies is crucial for driving meaningful change and achieving lasting success.

FUTURE TRENDS AND OUTLOOK

As we look to the future of the manufacturing industry, the use of advanced analytics, BI, ML, and AI is poised to revolutionize operational practices and drive unprecedented levels of efficiency and innovation. One of the key trends shaping the future of manufacturing is the adoption of transformer models, a type of deep learning architecture known for its ability to process sequential data with remarkable accuracy and efficiency. Transformer models, such as the Transformer-XL and the Generative Pre-trained Transformer series, are increasingly being leveraged to analyze vast amounts of manufacturing data, including sensor readings, production logs, and quality control records. These models excel at capturing complex patterns and dependencies, including temporal dependencies within sequential data, enabling more accurate predictive modeling and proactive maintenance strategies.

Further, the emergence of generative AI algorithms is transforming traditional manufacturing workflows by enabling the generation of synthetic data and simulations for testing and optimization purposes. Generative adversarial networks and variational autoencoders are being used to create realistic virtual environments that mimic real-world manufacturing processes. By generating synthetic data, manufacturers can conduct extensive simulations and scenario analyses to identify potential bottlenecks, optimize production schedules, and mitigate risks without needing costly physical prototypes.

In addition to these algorithms, the future of manufacturing will also see an increased focus on autonomous decision-making systems powered by reinforcement learning (RL) algorithms. RL algorithms enable machines to learn optimal decision-making policies through trial and error, gradually improving their performance over time. In manufacturing, RL algorithms are being applied to autonomous robotic systems for tasks such as material handling, assembly, and quality inspection. By continuously learning from their interactions with the environment, autonomous systems can adapt to changing production conditions and optimize processes in real time, leading to significant improvements in productivity and resource utilization.

SUMMARY AND KEY INSIGHTS

As we conclude this chapter on the transformative impact of advanced analytics, BI, ML, and AI in manufacturing and production processes, it becomes evident that these technologies have reshaped the industry landscape, ushering in a new era of efficiency, innovation, and competitiveness. Throughout this chapter, we explored numerous ways in which manufacturers are leveraging data-driven insights to optimize operations, improve product quality, and drive strategic decision making.

From PdM and demand forecasting to supply chain optimization and autonomous manufacturing, the applications of advanced analytics and AI in manufacturing are vast and diverse. Organizations that have embraced these technologies have witnessed tangible benefits, including increased productivity, reduced costs, and enhanced customer satisfaction. However, it is crucial to acknowledge that the journey toward digital transformation has challenges, ranging from data quality issues and algorithm selection to cultural resistance and skill gaps. Overcoming these hurdles requires a concerted effort involving investment in technology, infrastructure, talent development, and organizational change management. By addressing these challenges head-on and adopting a data-driven mindset, manufacturers can unlock the full potential of advanced analytics and AI.

Looking ahead, the next chapter will delve into the critical role of distribution and logistics in the manufacturing value chain. Just as advanced analytics and AI have revolutionized design, procurement, and production processes, they are also reshaping the way manufacturers manage their distribution networks, optimize inventory levels across multiple echelons, and fulfill customer orders. It is essential to recognize the interconnected nature of distribution and logistics processes and their impact on the entire manufacturing ecosystem.

Some key takeaways from this chapter include the following:

1. Advanced analytics, BI, ML, and AI are reshaping the manufacturing landscape, driving improved operational efficiency, QC, and productivity gains. They enable manufacturers to optimize processes, enhance quality, and drive innovation, allowing them to gain a competitive advantage and achieve sustainable growth.
2. Data-driven insights, real-time visibility, and efficient production operations are essential for maintaining competitive advantage in the dynamic manufacturing world.
3. To overcome challenges, organizations must invest in employee training programs, recruit data science and AI/ML talent, and cultivate a culture of innovation and continuous learning.

8

LEVERAGING ADVANCED ANALYTICS, BI, ML, AND AI IN DISTRIBUTION AND LOGISTICS

The AI-Driven Transformation at Stellar Electronics: Optimizing Distribution and Logistics

At Stellar Electronics, the integration of artificial intelligence (AI) and machine learning (ML) has revolutionized the company's logistics and distribution networks. Leading the initiative, John Davies, the chief operations officer, saw an urgent need to address the inefficiencies in inventory management, shipment delays, and route planning, which were causing rising costs and missed customer deadlines.

In collaboration with Emma Patel, Stellar's head of data science, the company implemented an ML-based demand forecasting system. This system not only analyzed historical sales data but also incorporated external factors such as weather patterns, market trends, and consumer behavior to provide accurate predictions. Within the first year, the AI-driven demand forecasting reduced stockouts by 15% and lowered excess inventory levels by 20%, ensuring that product availability matched customer demand in real time.

To enhance distribution efficiency, Stellar Electronics also adopted dynamic routing algorithms powered by AI. Michael Zhang, the logistics manager, integrated this system across the company's fleet. The AI system dynamically adjusted routes based on traffic conditions, fuel costs, and delivery schedules. This led to a 10% reduction in fuel costs and improved delivery times by 12%, significantly enhancing the overall customer experience.

In addition to route optimization, the company employed predictive analytics for warehouse operations. Emma Patel's team introduced AI-powered automation to optimize storage layouts and picking paths in the warehouse. This reduced the time needed to fulfill orders by 25%, further accelerating the company's logistics operations. Additionally, predictive maintenance (PdM) algorithms were deployed to monitor the performance of warehouse machinery, resulting in a 30% reduction in equipment downtime.

With AI and ML at the heart of Stellar Electronics' operations, the company is not only meeting its logistics challenges but also creating new value streams through data-driven insights. The seamless collaboration between these leaders has positioned Stellar as a cutting-edge player in the electronics industry, with a supply chain that is agile, efficient, and future-proof.

"By offloading process-oriented tasks to AI, freight forwarders can focus on real problem solving."

—Will Urban, chief revenue officer at Flexport

The importance of distribution and logistics in the manufacturing sector cannot be overstated. These processes serve as the vital link between goods production and their delivery to the end customer, impacting the entire manufacturing value chain. Effective management of distribution and logistics ensures smooth operations, reduces package damages and lead times, and enhances overall customer satisfaction, all of which are crucial for manufacturers to remain competitive globally. As the manufacturing industry continues to evolve, the integration of advanced analytics, business intelligence (BI), ML, and AI is reshaping the way distribution and logistics are managed and optimized.

Advanced technologies enable manufacturers to gain real-time visibility into their supply chains, allowing them to make data-driven decisions that improve efficiency and reduce costs. For instance, ML algorithms can be used to predict demand patterns, helping manufacturers adjust production schedules and inventory levels to meet customer needs more effectively. Additionally, AI-powered route optimization tools streamline transportation planning, identifying the most efficient delivery paths and reducing fuel consumption. This speeds up delivery times, minimizes greenhouse gas (GHG) emissions, and reduces the carbon footprint of distribution activities. Advanced analytics and BI also provide manufacturers with actionable insights into their logistics networks, enabling them to identify bottlenecks and areas for improvement. Predictive analytics can forecast potential disruptions, such as delays or shortages, allowing manufacturers to take proactive measures to mitigate risks and maintain seamless operations. These technologies also facilitate better collaboration and communication across the supply chain.

The impact of integrating advanced analytics, BI, ML, and AI into distribution and logistics extends beyond operational efficiency. By optimizing these processes, manufacturers can improve customer satisfaction through faster and more reliable deliveries. This, in turn, enhances brand reputation and fosters customer loyalty. Moreover, these innovations support the demands of modern manufacturing, such as just-in-time production and mass customization, enabling manufacturers to adapt quickly to changing market conditions and customer preferences.

AI-driven personalization can significantly improve efficiency and customer satisfaction within the distribution and logistics industry. By leveraging customer data, AI can enable more tailored experiences throughout the supply chain, such as optimizing delivery schedules and methods based on individual customer preferences. This approach can enhance customer relationships, increase brand loyalty, and lead to more efficient operations. AI-driven personalization can also help businesses identify customer personas, patterns, and trends in customer behavior, enabling them to forecast demand and adjust inventory and distribution strategies accordingly.

AI and ML applications have a significant impact on various manufacturing sectors, like pharmaceuticals, food and beverages, and automotive, in the context of distribution and logistics. Within the pharmaceutical industry, AI and ML can optimize the supply chain by predicting demand for specific drugs and adjusting production and distribution accordingly. These technologies can also help ensure compliance with regulatory requirements, such as tracking and tracing drugs throughout the supply chain to prevent counterfeiting and excursions in ambient conditions (e.g., temperature rising above or falling below the appropriate threshold). AI-driven demand forecasting and inventory management enable companies in the food and beverage industry to minimize waste and ensure product freshness. ML algorithms can also optimize delivery routes and schedules to maintain cold chain integrity and ensure timely deliveries. In the automotive industry, AI and ML can enhance supply chain resilience by predicting potential disruptions and suggesting alternatives like sourcing materials from different suppliers. These technologies also enable efficient inventory management and just-in-time delivery to support manufacturing processes.

By leveraging these technologies, manufacturers can achieve greater efficiency, responsiveness, and sustainability in their operations, positioning themselves for success in an increasingly competitive market. Figure 8.1 shows sample steps involved in distribution and logistics. In this chapter, the data used in distributions and logistics will be discussed, along with how to integrate it within the manufacturing value chain ecosystem. Then, the use of advanced analytics, BI, ML, and AI within distribution and logistics will be reviewed. Sustainability and risks within distributions and logistics are covered next. Finally, before concluding this chapter, implementation challenges will be covered, hopefully providing a crystal ball perspective into future trends.

UNDERSTANDING DATA SOURCES AND INTEGRATION

In the distribution and logistics sector, managing diverse data from diverse players is crucial to optimizing operational efficiency, customer satisfaction, and overall business performance. These data sources are derived from multiple aspects within distribution, including inventory databases, transportation management systems (TMS), route conditions, and customer order records. Integrating this data from disparate sources to create a unified and comprehensive view of the distribution network can be challenging, yet it is essential for success.

One of the primary data sources in the distribution and logistics sector is inventory databases. These databases contain critical information about product quantities and locations, including stock levels at various warehouses and distribution centers. Most companies forget to account for inventory in motion (i.e., inventory that is in transit). Managing inventory data effectively is essential for making informed decisions about stock replenishment, demand forecasting, and warehouse optimization. TMS are key data sources, as they provide data on the movement of goods throughout the supply chain. This includes information on routing, delivery schedules, and carrier performance metrics. The efficient management of transportation data helps organizations optimize routes, reduce delivery times, and minimize transportation costs. Customer order records are also a vital data source in distribution and logistics. These records provide data about orders placed by customers, including order details, delivery and labeling preferences, and historical purchasing patterns. Leveraging this data allows organizations to better understand customer needs, tailor delivery experiences, and enhance overall satisfaction.

Figure 8.2 provides a high-level process for distribution and logistics. Components in Figure 8.2 help visualize and manage the flow of materials, products, and information through the distribution and logistics processes. These steps and sub-steps focus on managing the flow of goods from manufacturing to customers, ensuring timely and accurate delivery while handling returns and continuously improving the process.

Integrating data from these diverse sources presents several challenges. One major hurdle is the presence of data silos, which can lead to inconsistencies and fragmentation. Data silos occur when

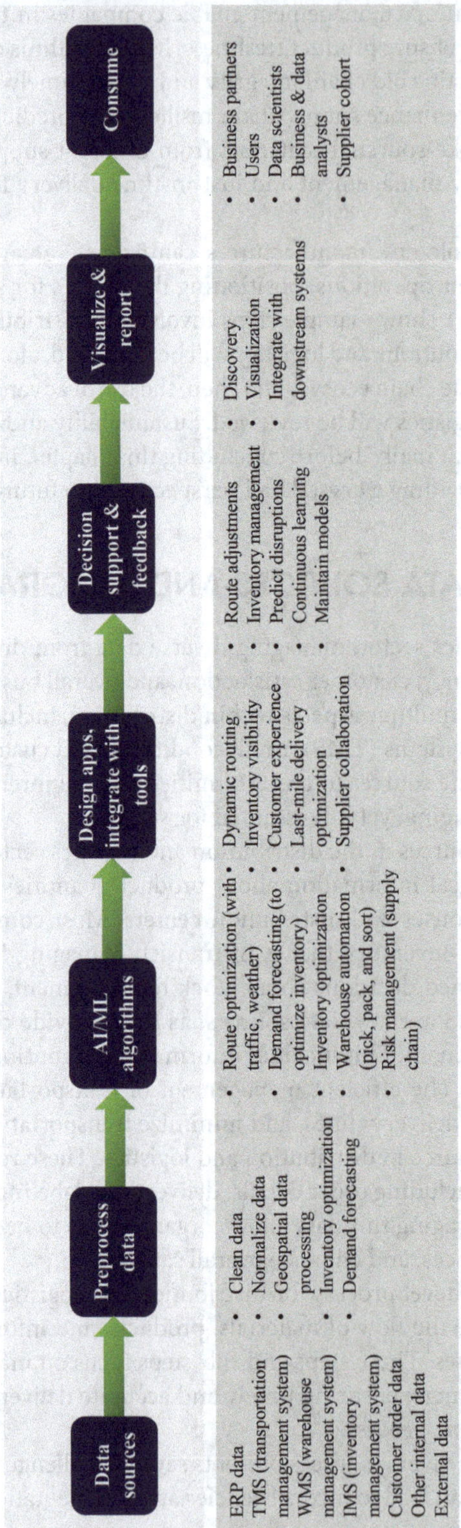

Figure 8.1 Sample of steps involved in distribution and logistics.

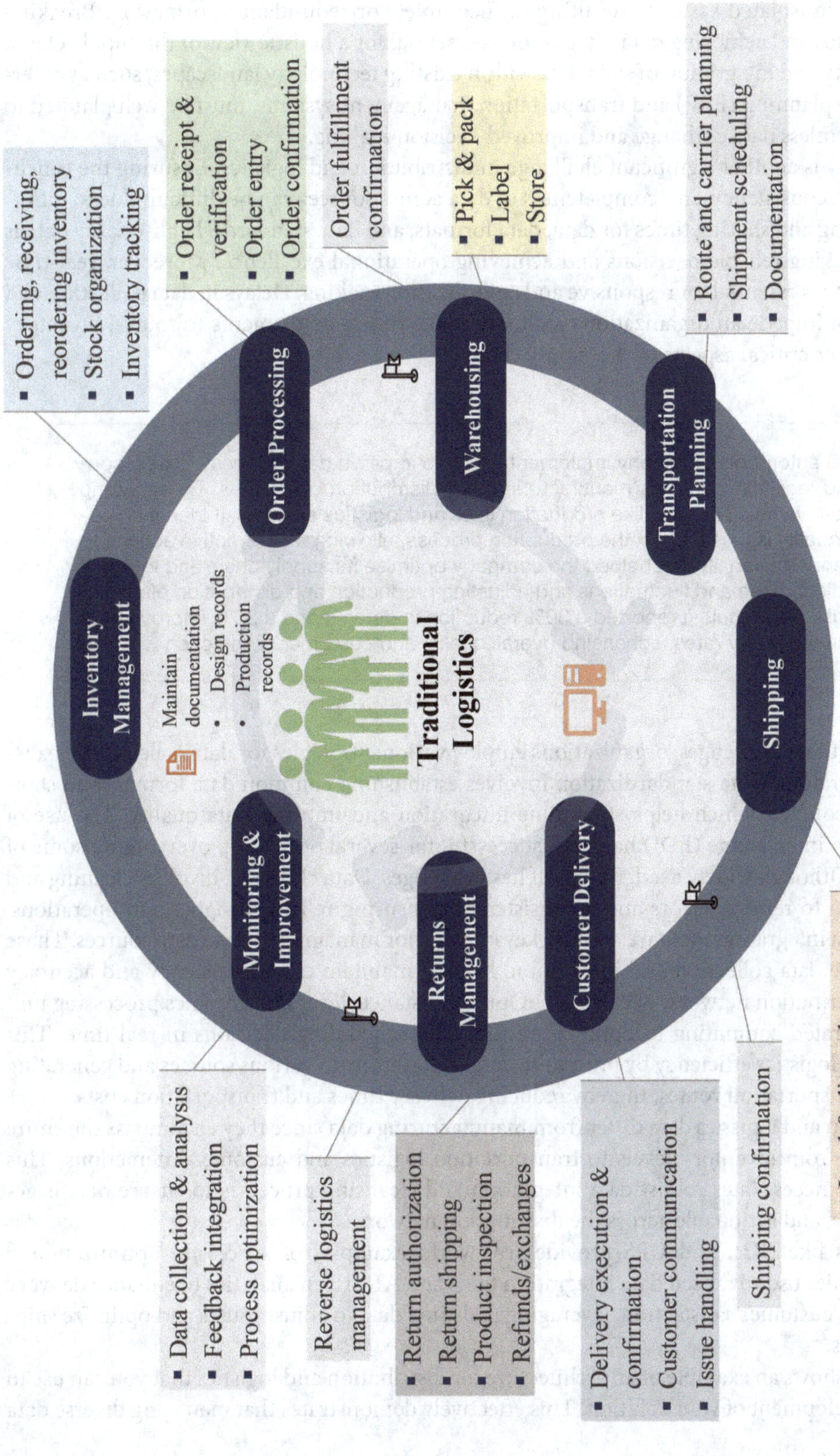

Figure 8.2 Processes within traditional distribution and logistics.

data is stored in isolated systems, resulting in incomplete or redundant information. Breaking down these silos and achieving data integration is essential for a holistic view of the supply chain. Interoperability and integration of such data within existing technology landscapes, such as enterprise resource planning (ERP) and transportation management systems, must be well planned to allow for a seamless data exchange and improved decision making.

Data quality is another significant challenge in distribution and logistics. Ensuring the timeliness, accuracy, consistency, and completeness of data across sources can be difficult due to differences in posting and sharing times for data, data formats, and data standards. High-quality data is crucial for making reliable decisions and achieving operational excellence. Moreover, real-time data availability is essential for responsive and agile decision making. Delays in data collection and processing can impede an organization's ability to make timely adjustments to routes, inventory levels, and other critical aspects of the supply chain.

> A global automobile company implemented ML to improve data, process quality control, and logistics within its manufacturing and distribution operations. By analyzing data from various sources, like production lines and logistics systems, it identified potential quality issues early in the production process, allowing for corrective actions to be taken. Additionally, ML helped the company optimize its supply chain and logistics by predicting demand fluctuations and adjusting production and distribution plans accordingly. As a result, it reported a 30% reduction in defects and a 25% improvement in on-time delivery rates, enhancing overall quality and customer satisfaction.

To address these challenges, organizations employ various strategies for data collection, cleansing, and integration. Data standardization involves establishing common data formats and standards across sources, which helps streamline integration and improve data quality. The use of electronic data interchange (EDI) has been successful for several decades in overcoming some of these issues. Although widely used, EDI itself has challenges. Data cleansing involves cleaning and validating data to remove errors and inconsistencies, ensuring reliable insights into operations. Advanced data integration tools are another key strategy for managing diverse data sources. These tools automate data collection and integration, helping maintain data consistency and accuracy across the distribution network. NVIDIA's cuOpt, for instance, leverages graphics processing unit (GPU)-accelerated computing to optimize routing and scheduling decisions in real time. This tool enhances logistics efficiency by processing large datasets from various sources and generating optimized transportation routes, thereby reducing delivery times and transportation costs.

Distribution and logistics data differ from manufacturing data since they encompass the entire supply chain, from inventory levels to transportation logistics and customer interactions. This broader scope necessitates robust data integration and cleansing processes to ensure data is accurate, reliable, and actionable across the distribution network.

Companies like DHL and UPS provide real-world examples of successful optimization.[1,2] These companies use advanced data integration tools and AI-driven analytics to enhance delivery efficiency and customer satisfaction, leveraging real-time data to adjust routes and optimize shipping schedules.

Figure 8.3 shows an example of an architecture for distribution and logistics that you can use to guide the development of your solution. This effectively demonstrates that managing diverse data

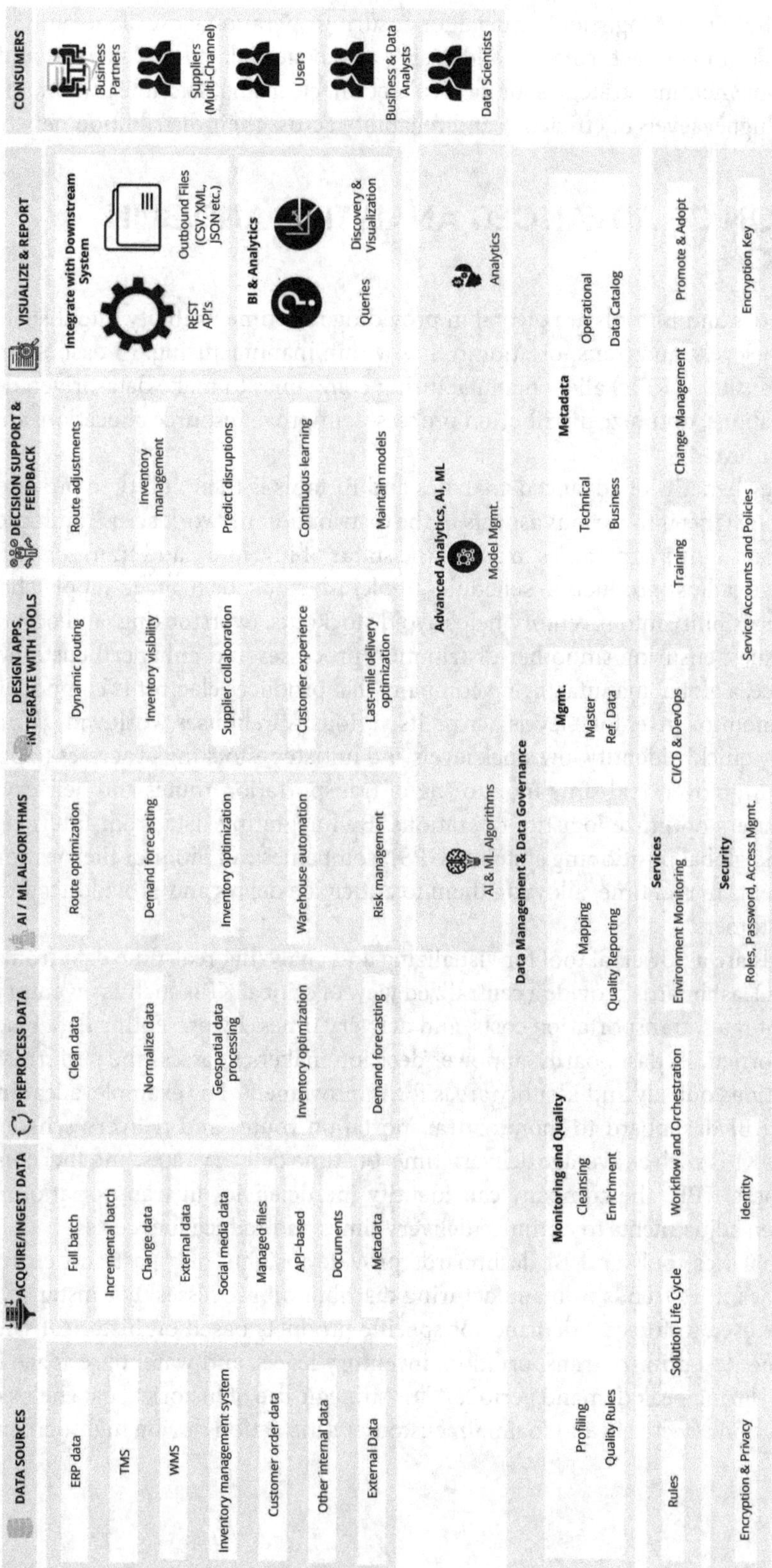

Figure 8.3 Example architecture for distributions and logistics.

sources in distribution and logistics is crucial for ensuring operational excellence and meeting the demands of modern manufacturing. By addressing challenges related to data silos and inconsistencies while implementing strategies for data collection, cleansing, and integration, manufacturers can achieve higher levels of efficiency and reliability across their distribution networks.

APPLICATION OF ADVANCED ANALYTICS AND BI IN DISTRIBUTION

Advanced analytics and BI tools are pivotal in providing real-time visibility into distribution networks, inventory levels, and transportation routes within manufacturing. BI dashboards and key performance indicators (KPIs) allow manufacturing companies to make data-driven decisions to streamline operations, optimize distribution processes, improve resource allocation, and enhance customer service levels.

One of the key benefits of advanced analytics and BI tools is their ability to provide real-time monitoring and insights into various aspects of the distribution network across multiple echelons. This real-time data allows companies to assess the status of inventory and transportation quickly, making it easier to adjust production schedules, replenish stock, or manage supply chain disruptions. Real-time visibility into inventory helps avoid stockouts, overstocking, and other inventory management issues, ensuring smoother distribution processes and enhanced customer satisfaction. For instance, a global manufacturing company that produces electronic components uses BI dashboards to monitor inventory levels across its various warehouses worldwide. This visibility lets the company quickly identify low stock levels and initiate restocking processes to prevent production delays. Moreover, real-time monitoring of transportation routes and delivery schedules helps manufacturers optimize logistics operations. By integrating data from Internet of Things (IoT) devices and global positioning systems (GPS), companies can monitor the location and condition of shipments in real time, allowing them to anticipate delays and provide accurate delivery estimates to customers.

BI dashboards are a powerful tool for visualizing and analyzing real-time data from the distribution network. Dashboards provide a centralized view of critical KPIs such as inventory turnover, order fulfillment rates, transportation costs, and delivery times. By presenting data in an intuitive and accessible format, BI dashboards empower decision makers to assess the performance of distribution operations quickly and identify areas for improvement. For example, a leading logistics company uses a BI dashboard to monitor transportation routes and delivery schedules in real time, displaying KPIs such as average delivery time, on-time delivery rates, and fuel consumption. By analyzing these KPIs, the company can identify inefficiencies in transportation routes and make data-driven adjustments to optimize delivery times and reduce fuel costs.

Advanced analytics tools and BI dashboards provide insights into past and current performance, as well as future trends in manufacturing distribution processes. For instance, predictive analytics can be used to forecast demand for specific products based on historical sales data, allowing companies to optimize transportation, inventory levels, and warehouse space in order to avoid stockouts during peak demand periods. This strategic use of historical data helps companies manage distribution effectively and maintain customer satisfaction during high-demand seasons.

Chapter 7 presented a notable example of the use of advanced analytics in Amazon distribution. Amazon employs predictive analytics to anticipate customer demand and strategically position products in warehouses close to customer locations. This approach reduces delivery times and enhances customer satisfaction. By optimizing inventory placement and transportation routes using advanced analytics, Amazon has been able to set industry benchmarks for efficiency and speed.

FedEx has also implemented predictive analytics to optimize its distribution and logistics operations. By using advanced analytics and BI to analyze historical and real-time data, FedEx can anticipate potential delays and disruptions in its delivery network. The company's system continuously adjusts delivery routes based on traffic patterns, weather forecasts, and other factors to minimize delays. As a result, FedEx has improved its on-time delivery rate and reduced fuel consumption. The company reports that these initiatives have contributed significantly to savings and have enhanced customer satisfaction by providing more accurate delivery estimates.

Another aspect of distribution optimization involves improving resource allocation. A proof of concept of cuOpt has demonstrated significant improvement in time and quality of optimal routes using GPUs. By analyzing data on transportation routes and inventory levels, companies can allocate resources such as delivery vehicles and warehouse space more efficiently. This can lead to cost savings and improved operational performance. To further enhance the optimization of distribution processes, companies can integrate data from IoT devices and sensors placed on vehicles, warehouses, and products. These sensors provide real-time data on the location and condition of goods in transit, allowing companies to monitor shipments and respond quickly to any issues that arise. For example, IoT-enabled sensors on delivery trucks can provide real-time updates on the truck's location, speed, and fuel consumption. For temperature-dependent products like frozen foods, fresh produce, and pharmaceutical products, these sensors play a critical role in quickly detecting unexpected excursions. By analyzing this data, companies can optimize fuel efficiency, minimize delivery times, promptly react to conditions, and take appropriate action.

Figure 8.4 provides a high-level process for distribution and logistics with advanced analytics, BI, AI, and ML incorporated. The components in Figure 8.4 help visualize and manage the flow of materials, products, and information through distribution and logistics processes. These steps and sub-steps focus on managing the flow of goods from manufacturing to customers, ensuring timely and accurate delivery while handling returns and continuously improving the process.

> Numerous manufacturing companies have leveraged advanced analytics and BI tools to optimize distribution networks and improve operational efficiency. One success story comes from a leading consumer electronics manufacturer that implemented advanced analytics to optimize its distribution processes. This manufacturer used real-time data from IoT sensors on vehicles and in warehouses, enabling them to monitor inventory levels and transportation routes in real time. This allowed it to reduce transportation costs by nearly 15% and achieve a 98% on-time delivery rate.
>
> Another firm used NVIDIA's cuOpt instead of traditional optimization solutions to optimize its last-mile deliveries. This reduced costs by over 15% per delivery, reduced failure-to-deliver to under 5% (a significant improvement), and improved asset utilization by 27%. Dynamic route generation, a critical element in the current business environment, took under one hour as opposed to a non-dynamic solution taking several hours during a daily run.

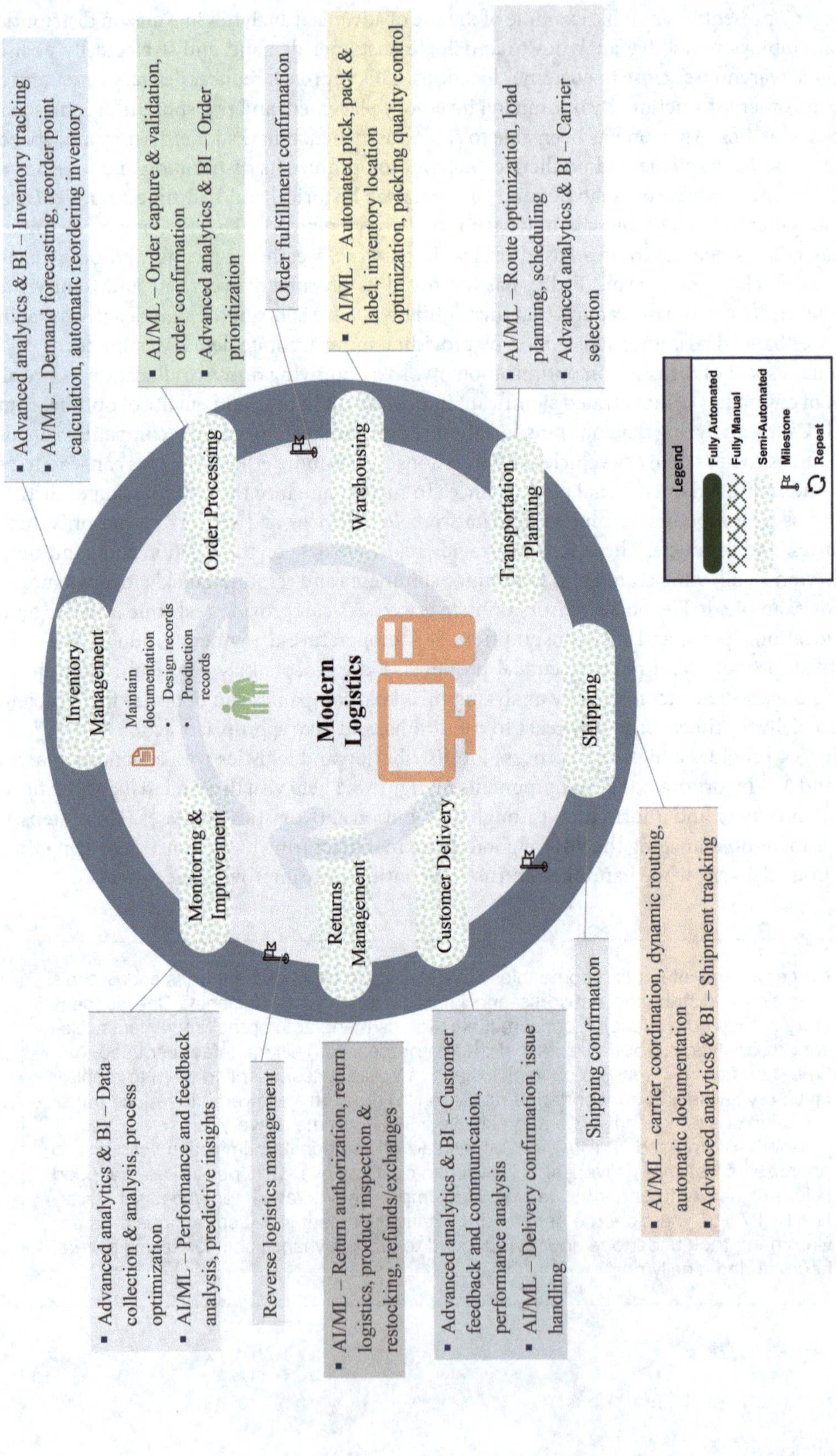

Figure 8.4 Processes within modern distribution and logistics.

HARNESSING ML AND AI FOR ROUTE OPTIMIZATION

AI and ML algorithms have transformed the manufacturing industry by optimizing transportation routes and logistics operations, reducing delivery times, and minimizing transportation costs through predictive analytics and dynamic route planning. By leveraging these advanced technologies, manufacturers can achieve higher levels of efficiency, improve customer satisfaction, and streamline their distribution processes.

Predictive analytics utilizes ML algorithms to analyze historical and real-time data, including sales trends, production schedules, and customer demand, to forecast potential disruptions and optimize logistics operations. By leveraging this data, manufacturers can anticipate future requirements and plan transportation routes accordingly, adjusting inventory levels and managing lead times effectively. A major global manufacturer of consumer electronics implemented an ML-based predictive analytics solution to forecast demand for its products. Through the analysis of sales data, the company optimized its transportation routes and delivery schedules, resulting in a reduction in delivery times of almost 25% and an 18% reduction in transportation costs. This improved customer satisfaction and competitive position in the market.

> *"Looking ahead, we believe AI has the opportunity to significantly augment current activities in logistics from end to end. AI will fundamentally extend human expertise in terms of reach, quality, and velocity by eliminating mundane and routine work, allowing logistics workforces around the world to focus on more meaningful and impactful work."*
>
> —Keith W. Dierkx, IBM Global Industry Leader for Rail, Freight, and Logistics; Matthias Heutger, DHL Senior Vice President for Strategy, Marketing, and Innovation

Dynamic route planning leverages real-time data from sources such as GPS, traffic conditions, and weather forecasts to determine the most efficient routes for delivery vehicles. Think about the use of your mobile phone with map software (Apple maps, Google maps, etc.) providing the best route considering the current road conditions. This approach minimizes travel times and fuel consumption, enhancing operational efficiency and reducing emissions. By continuously adapting to changing conditions, dynamic route planning ensures timely and reliable deliveries. A leading logistics provider in the automotive industry implemented dynamic route planning using AI-powered software. This reduced delivery times by over 20% and fuel costs by almost 25%. The software considered factors such as traffic congestion, road closures, and weather conditions to optimize delivery routes, resulting in faster and more reliable service for customers.

When incorporating advanced analytics, BI, AI, and ML into traditional distribution and logistics processes, the components of each step are enhanced with technology. Figure 8.4 depicts the processes involved in modern manufacturing and development. Incorporating advanced technologies transforms each step with components that leverage data and automation to improve efficiency, quality, and decision making throughout the manufacturing process. Unlike Figure 8.2, all processes are semi-automated (depicted by the icon size for computing and human resources), with a much higher contribution from computing resources instead of manual resources. Although semi-automated, most of the *hard* work is done by the models. These models provide guidance to human resources on actions to incorporate in operations. The manual intervention ensures checks and balances are in place by incorporating the *human-in-the-loop* framework. This allows people to quickly react and manually override poor outcomes from the modern tools, preventing them from getting pushed downstream.

The following paragraphs include some real-world case studies. These companies in the distribution and logistics industry, many supporting manufacturing, have used AI and ML tools for optimizing transportation routes, logistics operations, and overall distribution processes. They have significantly improved efficiency, reduced costs, and increased customer satisfaction.

DHL, mentioned earlier, implemented ML algorithms to analyze real-time data from its fleet. DHL has improved its sorting capacity by 40% and its sorting accuracy to 99%.[3] DHL believes that robots-only or humans-only are inferior to *cobots*—the best of both humans and robots—which is what it uses in its distribution network. Similarly, UPS uses its proprietary ORION (On-Road Integrated Optimization and Navigation) system, which incorporates AI and ML algorithms to optimize delivery routes for its drivers. By using ORION, UPS can save an estimated 100 million miles driven annually, leading to significant reductions in fuel consumption and emissions.[4]

Procter & Gamble (P&G) utilizes AI and ML algorithms to optimize its supply chain and distribution networks. In collaboration with its retailers and wholesalers, the company implemented AI-based demand forecasting and reduced its delivery truck fleet by 30% in Japan.[5] P&G has also used AI to optimize its transportation routes, resulting in significant reductions in transportation costs.

The Apple and Google Maps platforms offer application programming interfaces (APIs) for real-time mapping and routing services, enabling manufacturers to integrate real-time data on traffic, road conditions, and estimated arrival times into their logistics operations. This helps optimize delivery routes and improve customer service.

Tableau is a business intelligence tool that provides data visualization capabilities and real-time analytics. Manufacturers can leverage Tableau's dashboards and visualization tools to monitor KPIs and track performance metrics related to logistics and distribution processes.

IBM Watson Studio offers a collaborative environment for data scientists, developers, and business analysts to build, train, and deploy ML models. Manufacturers can use Watson Studio to develop AI-powered solutions for route optimization and logistics management.

HERE offers a range of location-based services, including real-time routing, geocoding, and traffic information. Manufacturers can leverage HERE's APIs to optimize transportation routes and improve delivery efficiency.

Manhattan Active Transportation Management offers AI-powered tools for real-time transportation planning and execution. Manufacturers can use the platform to optimize delivery routes, reduce transportation costs, and improve on-time delivery rates.

Oracle's cloud-based solution provides advanced logistics planning and execution tools, including AI-driven dynamic route optimization. This enables manufacturers to streamline transportation operations and improve overall efficiency.

As AI and ML technologies continue to advance, manufacturers can expect further enhancements in route optimization and logistics management practices. The integration of IoT devices and real-time data analytics will enable more sophisticated dynamic route planning, allowing for greater agility and adaptability in transportation operations. In the future, the use of AI-powered autonomous vehicles may become more widespread, further revolutionizing the logistics landscape. By embracing these innovations, manufacturers can stay ahead of the competition, achieve sustainable growth, and deliver exceptional value to their customers.

While AI and ML technologies offer numerous benefits, manufacturers encounter challenges such as data quality issues, integration complexities, and the need for specialized talent. To overcome these obstacles, manufacturers should adopt best practices, such as investing in robust data management solutions, fostering cross-functional collaboration, and providing ongoing employee training. Manufacturers should establish clear KPIs to measure the performance of their logistics operations and drive continuous improvement.

Other challenges include constantly evolving local, state, and/or federal regulations. In addition, the availability of drivers, containers, warehouse capacity, and material handling capabilities are also areas of concern, as is the need for specialized containers such as oil tankers, temperature-controlled containers, etc. Another challenge is backhaul, where the supply chain incurs the unrecoverable expense of an empty load backhaul. Risk management, environmental issues, and sustainability concerns are additional risks that must be addressed. By addressing these challenges effectively, manufacturers and logistics providers can enhance operational performance and customer satisfaction.

OPTIMIZATION OF WAREHOUSE OPERATIONS WITH AI

AI has become an indispensable tool for optimizing warehouse operations in the manufacturing industry. By leveraging AI-driven technologies, manufacturers can enhance various aspects of warehouse management, including inventory control, order pick and pack, and warehouse layout optimization. These advancements lead to increased efficiency and reduced operational costs, making AI a critical component of modern manufacturing and logistics operations. In this section, we will examine how AI is transforming warehouse operations, consider the benefits it brings, and explore some real-world examples illustrating its impact on logistics management practices.

Inventory Management with AI

AI-driven inventory management solutions use predictive analytics to anticipate demand and optimize stock levels. These tools analyze historical sales data, market trends, and external factors to forecast future demand accurately. By maintaining optimal inventory levels, manufacturers can reduce carrying costs, minimize stockouts, and avoid overstocking. Earlier examples of Amazon, P&G, etc., have successfully demonstrated the use of AI in inventory management.

Order Picking and Packing with AI

AI-powered order picking and packing systems can significantly improve efficiency and accuracy in warehouse operations. These systems use computer vision (CV), robotics, and natural language processing to identify and locate items quickly. By automating order picking, manufacturers can reduce labor costs and minimize human error. DHL has used AI-powered sorting with the DHL-Bot (an AI-powered robotic arm), resulting in the ability to sort over 1,000 small parcels in an hour with a 99% accuracy rate.[6] Ocado, an online grocery retailer, is another notable example of a company that uses AI-driven robotic systems to manage its order-picking process. These robots navigate the warehouse, identify products, and transport them to the packing area. As a result, Ocado can fulfill orders quickly and accurately. The company reported that its AI-driven automated systems enable it to reduce fresh produce waste to 0.5%, compared to the industry average of 3% to 5%, and can accurately fulfill customer orders within minutes of receiving them.[7]

Warehouse Layout Optimization

AI can optimize warehouse layout by analyzing data on product demand, storage utilization, and order fulfillment patterns. This allows manufacturers to design efficient warehouse layouts that shave off seconds to minutes of time for workers, robots, and material handling equipment to identify and reduce bottlenecks. In addition to warehouse layout, AI can help optimize rack loading to ensure a balanced load throughout the warehouse and ease in order fulfillment.

Amazon is an example of a company that employs AI to optimize its warehouse layouts. By analyzing data on product demand and customer preferences, Amazon can strategically position products within its warehouses, improving order fulfillment speed and reducing operational costs. AI-driven warehouse optimization strategies have contributed to a 15% reduction in logistics costs, a 35% decrease in inventory, and a 65% improvement in service levels.[8]

Real-Time Insights and Automation

AI solutions provide real-time insights into warehouse operations, enabling manufacturers to make data-driven decisions. Operational performance can be tracked through BI dashboards connected to AI solutions by using appropriate KPIs. By automating repetitive tasks, such as inventory tracking and replenishment, manufacturers can focus on more strategic initiatives like improving order fulfillment cycle time. UPS uses AI-driven route optimization and logistics management practices to enhance operational efficiency. By implementing AI and ML algorithms for dynamic route planning and package sorting, UPS has reduced delivery times and transportation costs. The company's ORION system uses AI to determine the most efficient delivery routes for drivers, resulting in a reduction of 10 million gallons in fuel consumption per year.[9]

SUSTAINABILITY AND RISK MITIGATION

AI-driven technologies are making a significant impact on sustainability and risk mitigation across various aspects of distribution and logistics operations. By leveraging AI, companies can enhance environmental sustainability, manage social and ethical considerations, ensure governance within this important set of processes, and mitigate risks associated with disruptions and security threats.

Environmental Sustainability

Rivers are one of the age-old transportation channels, and climate change has made them highly unreliable. For example, low water levels in the Rhine River disrupted a traditional route to transport goods.[10] Similarly, water levels in the drought-stricken Mississippi River impacted goods transport on barges.[11] Although water levels eventually rose, both these examples show that transporting goods through these channels can be unreliable.

Effective energy management is a critical sustainability goal. AI-driven energy management systems optimize energy usage in manufacturing and warehouse operations. These systems can predict peak energy usage times and adjust operations accordingly, resulting in reduced carbon emissions and lower energy costs. For example, Google uses AI to optimize data center cooling, resulting in a 40% reduction in energy consumption.[12] Another essential sustainability goal is to minimize waste. AI can help by providing automated sorting systems, route optimization, and data-driven decision making.[13] In manufacturing, industrial waste constitutes over 50% of waste globally, mainly from poor-quality products. AI algorithms can monitor production processes to minimize waste and identify opportunities for recycling and reusing materials. AI-enabled waste management systems can analyze waste streams and provide actionable insights on waste reduction strategies.[14] Finally, emission control significantly improves when using AI technologies that track emissions from production and logistics operations. This provides insights that enable companies to reduce their carbon footprint. For instance, ABB uses AI to reduce emissions through its Predictive Emission Monitoring System.[15] This system, based on predictive technologies for gas providers in the Gulf of Mexico, was the first of its kind to be validated by the Environmental Protection Agency.

Social and Ethical Considerations

Worker safety is positively impacted using AI-powered safety monitoring systems. These systems detect unsafe conditions and alert employees or supervisors in real time, leading to a safer working environment and fewer accidents. AI-powered CV systems can also monitor worker behavior to flag potential safety violations. Such CV systems aid in recognizing workers and verifying their qualifications to operate specific equipment, resulting in safer operations.

AI helps companies monitor labor practices across their supply chains and ensure compliance with ethical standards. By analyzing employee work patterns and schedules, companies can identify potential risks of overwork or unfair treatment and make necessary adjustments. AI provides insights into the social impact of a company's operations, helping guide corporate social responsibility initiatives. For example, AI tools can analyze customer feedback and community sentiments to gauge the success of initiatives and guide future actions. Companies like Dow Chemical and Komatsu are using CV to enhance personal safety and security.[16]

Disruption and Security Risk Mitigation

AI can forecast potential disruptions in the supply chain, such as weather events, political instability, or transportation delays. For example, IBM has used AI-powered predictive analytics and weather data to anticipate and mitigate risks in its supply chain. AI-powered cybersecurity solutions safeguard data and systems from cyberattacks by monitoring network traffic and identifying anomalies that may indicate an attack. Companies must plan to incorporate deep cybersecurity measures and risk management strategies. These become more relevant when implementing AI and ML in distribution and logistics. Techniques such as data encryption, network monitoring, and incident response play a role here. For instance, AccuKnox's Zero Trust Cloud Security uses AI to detect and prevent cyber threats in real time, providing advanced security for organizations. AI-powered solutions also optimize transportation routes based on real-time data, helping companies avoid delays and disruptions.

DHL uses AI-driven technologies to improve sustainability and risk mitigation in its operations. By leveraging CV systems that track personal protective equipment compliance, DHL can monitor safety and compliance in its warehouses, reducing workplace accidents and improving overall operational efficiency. Another example of using AI at DHL, specifically CV, is to detect driver fatigue, an important consideration for logistics and transportation providers.[17]

IMPLEMENTATION CHALLENGES AND BEST PRACTICES

Implementing AI and ML in distribution and logistics can produce numerous benefits, including optimized transportation routes, reduced delivery times, and improved overall efficiency. However, there are several common and esoteric challenges that organizations face during the implementation process. By understanding these challenges and employing best practices, companies can successfully harness the power of AI and ML technologies in their distribution and logistics operations.

As mentioned, AI and ML rely on high-quality, clean, and integrated data to provide accurate insights and predictions. In distribution and logistics, data comes from various sources, such as transportation management systems, inventory databases, and customer order records. Ensuring data consistency and accuracy across these sources is essential. Further, complex supply chains are not beneficial, and modern supply chains can be long, highly complex, and interconnected. Integrating AI into such supply chains requires a deep understanding of data flows, dependencies, and relationships between different supply chain nodes. AI and ML technologies require robust computational resources and data storage capabilities. Companies need to invest in the necessary infrastructure to support AI and ML implementations, including cloud computing and data analytics platforms. As the regulatory landscape and compliance requirements continue to shift, companies must navigate these when implementing AI and ML in distribution and logistics. This includes discussing data protection laws, safety regulations, and other government- and industry-specific standards.

Adopting AI and ML in distribution and logistics requires significant changes in processes and workflows across multiple parties and trading partners. This demands robust change management systems across people, processes, and technology dimensions. Employees must be trained to work with new technologies and adapt to changing roles and responsibilities. With data coming from various sources, privacy and security concerns become paramount to ensure the manufacturing value chain data is secure. AI-driven technologies often involve handling sensitive data, such as supplier and customer information, as well as proprietary business data like demand, production rates, etc. Companies must address privacy and security concerns to ensure compliance with regulations and protect customer trust.

From an AI and ML algorithms perspective, algorithmic bias can inadvertently perpetuate biases in the training data. Companies must identify and mitigate bias to ensure fair and equitable decision making. Furthermore, understanding how AI and ML models make decisions is crucial for building trust and accountability. Companies should prioritize explainability and transparency in their AI implementations. Scaling AI and ML technologies across large distribution networks can be challenging. Companies must ensure their AI architectures and solutions are highly scalable and can handle increasing data volumes and complexity as business grows.

Although there are challenges, they are not insurmountable. Some ways to overcome these issues include investing in and establishing data management practices, such as data governance, data cleansing, integration, and standardization, to ensure data quality and reliability across the distribution network. It is important to foster collaboration between various departments, such as information technology (IT), operations, logistics, and business partners, to guarantee the smooth integration and alignment of AI initiatives. A key practice is not to take a *big-bang* approach; instead, use an incremental deployment approach. Start with pilot projects and gradually scale up AI and ML implementations as the organization becomes more comfortable with the technologies. When using an incremental approach, companies need to continuously monitor AI and ML models to ensure optimal performance and make adjustments as needed. Keep track of KPIs to measure success and follow them throughout the life of the initiatives. Companies must invest in comprehensive employee training to understand AI and ML technologies and adapt to new roles and responsibilities. Effective change management can help ease the transition. Companies should ensure that AI and ML implementations are fair, unbiased, and ethical, starting with the data. They must establish guidelines and oversight mechanisms to address ethical concerns.

FUTURE TRENDS AND OUTLOOK

Future trends point toward more efficient, agile, and customer-centric operations while companies leverage data-driven decision making, automation, and optimization across distribution and logistics. The current application of predictive analytics will also continue to grow through enhanced algorithms. The following sections contain examples of different technologies that will work in conjunction with people and processes to improve distribution and logistics metrics.

Advanced Autonomous Systems

The future of logistics will involve deploying autonomous delivery vehicles, including drones (unmanned aerial vehicles) and self-driving trucks. Companies like Amazon and small regional players have experimented with last-mile drone delivery, and many new companies are jumping in the fray. These technologies will lead to faster, safer, and more cost-effective delivery methods. In addition to autonomous delivery, robotics is getting increased visibility, especially in warehouses. The integration of advanced robotics with AI will create highly efficient and flexible warehouse environments. Automated picking and sorting systems will improve productivity and accuracy in order fulfillment.

Intelligent Inventory Management

AI will enable *smart warehousing* that can optimize storage space, improve inventory turnover, and reduce holding costs. Warehouses have many moving parts—think of conveyors, robots, and other material-handling equipment. IoT-connected sensors paired with AI will allow for PdM of warehouse equipment and transportation assets. This can minimize downtime and extend asset lifespan.

Sustainability and Ethical Considerations

AI will play a key role in helping companies meet sustainability goals by optimizing routes and modes of transportation to reduce GHG emissions. AI can also help identify sustainable sourcing options within the supply chain. In addition, future AI deployments in logistics will focus on ethical considerations, such as data privacy and the fair treatment of workers, ensuring that AI applications do not inadvertently perpetuate biases. Other dimensions, like transparency and accountability within ethical AI principles, should also be considered. AI and ML technologies impact various aspects of international logistics. This becomes more important when transit goes through highly regulated countries (e.g., those in the EU).

Human-AI Collaboration

AI will serve as supplemental intelligence to augment human decision making, providing insights and recommendations to improve logistical efficiency while maintaining a human-centered approach. As AI takes on more repetitive and routine tasks, companies will invest in upskilling their workforce to work alongside AI technologies and focus on higher level strategic activities. Amazon has demonstrated the potential of AI and robotics in logistics, using advanced warehouse automation and autonomous delivery technologies. The company continues to innovate with projects like Prime Air for drone deliveries, which may become mainstream soon. Amazon has also focused heavily on the use of robots, with over 750,000 currently in its warehouses.[18,19]

Additional Considerations

Other emerging trends in AI and ML, such as the rise of edge computing, blockchain technology, and advancements in quantum computing, will play a significant role in the future of distribution and logistics. Edge computing is transforming distribution and logistics by enabling real-time data processing and decision making closer to where the event occurs, improved connectivity and integration, and decentralized computing for reliable operations in remote areas. It enhances security and privacy by reducing data transmission over networks while also supporting scalability and flexibility to accommodate growing data volumes. Edge computing optimizes resource usage and facilitates remote monitoring and maintenance for predictive upkeep, improving equipment lifespan and efficiency. Additionally, it supports autonomous systems like self-driving vehicles and drones, allowing for real-time sensor data processing for safe and efficient operations. This trend boosts AI and ML effectiveness and leads to cost savings and enhanced logistics performance.

Blockchain technology is revolutionizing distribution and logistics by providing secure, transparent, and immutable records for transactions across the supply chain. This technology enhances traceability and accountability and includes provenance, allowing companies to track products from origin to destination and verify authenticity, reducing fraud and counterfeiting risks. Blockchain's decentralized nature ensures data integrity and fosters trust among partners and customers. By improving efficiency and collaboration, blockchain technology leads to cost savings, better inventory management, and enhanced customer satisfaction. Blockchain plays a crucial role in pharmaceuticals, food (fresh produce, meats, seafood, etc.), and regulated components (e.g., in aircraft, heavy machinery, high-volume public transportation, etc.).

Quantum computing holds the potential to revolutionize distribution and logistics by solving highly complex and time-consuming optimization problems much more efficiently than classical computing methods. In logistics, quantum computing could aid in optimizing route planning and delivery schedules, potentially reducing transportation costs and carbon emissions. It could also improve supply chain forecasting, enabling companies to make more accurate predictions about demand and supply. Quantum computing could enhance cybersecurity by providing stronger encryption methods for sensitive data within the supply chain. Although still in its early stages, using quantum computing in distribution and logistics could transform the logistics industry and lead to greater operational efficiency and cost savings.

SUMMARY AND KEY INSIGHTS

Distribution and logistics play a pivotal role in the manufacturing value chain by acting as the crucial bridge between production and the ultimate consumer. This segment has transformed due to the integration of advanced analytics, BI, ML, and AI, which have all contributed to revolutionizing the entire process. Real-time insights into distribution networks, inventory levels, and transportation routes have empowered manufacturers to streamline operations, enhance efficiency, and improve overall customer service levels.

Part of the transformative potential of ML and AI is their ability to optimize transportation routes and logistics operations, which in turn reduces delivery times and minimizes transportation costs through predictive analytics and dynamic route planning. Predictive analytics allow companies to anticipate changes in demand and adjust their operations accordingly. Tools such as AI-driven algorithms, dynamic route optimization, and autonomous vehicle technologies are also making significant strides in transforming distribution and logistics. AI and ML play a vital role in improving warehouse operations, from inventory management to order picking/packing and warehouse layout optimization. By automating repetitive tasks and providing real-time insights, AI solutions enable manufacturers to enhance productivity and reduce operational costs. This leads to greater efficiency in processes and overall cost savings. In the context of sustainability and risk mitigation, AI technologies help companies reduce carbon emissions and promote responsible sourcing. These technologies are also instrumental in mitigating potential risks, such as supply chain disruptions and security threats, by providing real-time data and adaptive strategies.

Looking ahead, the future of distribution and logistics in manufacturing will continue to evolve with trends such as autonomous delivery, smart warehousing, and intelligent inventory management reshaping the industry. As companies begin integrating AI and ML into their processes, they unlock new levels of efficiency, agility, and customer satisfaction, ensuring they remain competitive and responsive in a rapidly changing market. This ongoing evolution promises to redefine industry and set new standards for excellence in distribution and logistics within manufacturing. A few key insights from this chapter include the following:

1. Advanced analytics, BI, ML, and AI are reshaping how distribution and logistics are managed and optimized. These algorithms can predict demand, optimize inventory, minimize waste, ensure product freshness, determine shipping routes, and automate warehouses.
2. These technologies enable manufacturers to gain real-time visibility into their supply chains, allowing them to make data-driven decisions that improve efficiency and reduce costs.
3. Risk management and employee upskilling are important considerations for manufacturers when managing their distribution and logistics operations.

9

ADVANCED ANALYTICS, BI, ML, AND AI IN SALES & MARKETING AND AFTER-SALES SERVICE

AI-Enhanced Sales, Marketing, and After-Sales Service at Stellar Electronics: Driving Growth and Customer Satisfaction

Stellar Electronics has leveraged artificial intelligence (AI) and machine learning (ML) to elevate its sales, marketing, and after-sales service strategies, creating a seamless customer experience and maximizing operational efficiency. The initiative was led by Julia Sanders, VP of marketing, who collaborated with Ryan Cho, Stellar's head of customer insights, to implement an AI-driven strategy that could personalize customer interactions, optimize sales, and enhance post-sale support.

The AI-powered customer segmentation and personalization system developed by Ryan's team enabled the marketing department to analyze massive datasets from customer demographics, browsing behaviors, and purchase history. By identifying key customer segments and predicting preferences, the AI system created highly tailored marketing campaigns that resonated with different audience groups. This resulted in a 25% increase in customer engagement and improved conversion rates by 18%, allowing Stellar to meet diverse customer needs with precision-targeted campaigns.

Stellar implemented predictive analytics to support its sales teams in understanding demand patterns and sales cycles. By analyzing historical sales data, seasonality, and market trends, the AI algorithms provided the sales team with actionable insights, enabling them to forecast demand more accurately and make data-driven sales decisions. Carlos Varela, the director of sales, reported that these insights improved quarterly sales forecasts by 20% and led to a 15% boost in sales performance since sales reps were better equipped to prioritize high-potential leads and cross-sell relevant products.

In the realm of after-sales service, Stellar implemented a chatbot powered by natural language processing (NLP) and ML to handle customer queries, provide troubleshooting support, and escalate complex issues to human agents when necessary. This chatbot, overseen by Linda Nguyen, the customer service manager, resolved 70% of common inquiries autonomously, reducing response times and freeing up service agents to focus on more complex cases. The proactive use of AI-enabled predictive maintenance (PdM) insights allowed Stellar to inform customers of potential product issues before they arose, improving customer satisfaction and reducing product returns by 15%.

Stellar's AI-driven approach to sales, marketing, and customer service has not only enhanced customer experience but also increased revenue and operational efficiency. By embracing AI in these areas, Stellar Electronics is well-positioned to lead customer-centric innovation and maintain strong, long-term relationships with its clients.

"In the modern world of business, it is useless to be a creative, original thinker unless you can also sell what you create."

—David Ogilvy, father of advertising

Sales and marketing, along with after-sales service, play a critical role in the manufacturing sector because they connect manufacturers directly with their customers and ensure that products meet customer expectations. Effective sales and marketing strategies drive revenue growth, establish market competitiveness, and enhance brand reputation. Meanwhile, after-sales service is essential for nurturing long-term customer relationships, improving customer satisfaction, and fostering brand loyalty.

In recent years, the adoption of advanced analytics, business intelligence (BI), ML, and AI has revolutionized these functions in manufacturing. These technologies give manufacturers powerful tools to gain data-driven insights, optimize marketing and sales strategies, and deliver exceptional customer experiences. AI and ML, in particular, offer opportunities for personalized customer interactions, targeted marketing campaigns based on customer personas, and efficient after-sales support. One of the key impacts of advanced analytics, BI, ML, and AI is the ability to enhance customer satisfaction and loyalty. By leveraging these technologies, manufacturers can better understand customer preferences and behaviors, enabling them to create tailored products, services, and messaging that resonate with their target audiences. For instance, AI-driven personalization can lead to more relevant and engaging marketing content, improving chances of conversion, potential to upsell/cross-sell, and customer retention. Moreover, advanced analytics and BI enable manufacturers to monitor market trends, track sales performance, and identify potential opportunities for growth. By analyzing data on customer preferences, buying patterns, and competitive landscapes, manufacturers can make informed decisions about product development, pricing, and distribution strategies. This, in turn, can lead to improved market competitiveness and increased revenue.

When it comes to after-sales service, AI and ML can streamline processes such as customer support, maintenance, and warranty and returns management. By automating routine tasks and providing real-time assistance using technologies like multichannel conversational AI, firms can enhance customer experiences and reduce response times. Additionally, predictive analytics can anticipate customer needs and address potential issues before they arise, further improving customer satisfaction, net promoter score, and brand loyalty. Figure 9.1 shows sample steps involved in sales and marketing as well as after-sales service.

This chapter will explore various aspects of sales and marketing and after-sales service in detail. Readers can expect to learn about the role of AI-driven personalization in enhancing marketing strategies and customer experiences. In addition, the application of advanced analytics and BI in optimizing sales and marketing campaigns will be discussed. After that, the ways in which ML and AI are transforming after-sales service by enabling proactive customer support and PdM will be addressed. Data is crucial to any such initiatives, so the importance of data integration and real-time insights in driving decision-making processes is covered next. Finally, case studies highlighting successful implementations of AI and ML will showcase the use of these technologies.

> *"If you make customers unhappy in the physical world, they might each tell six friends. If you make customers unhappy on the internet, they can each tell 6,000."*
>
> —Jeff Bezos, Executive Chairman at Amazon

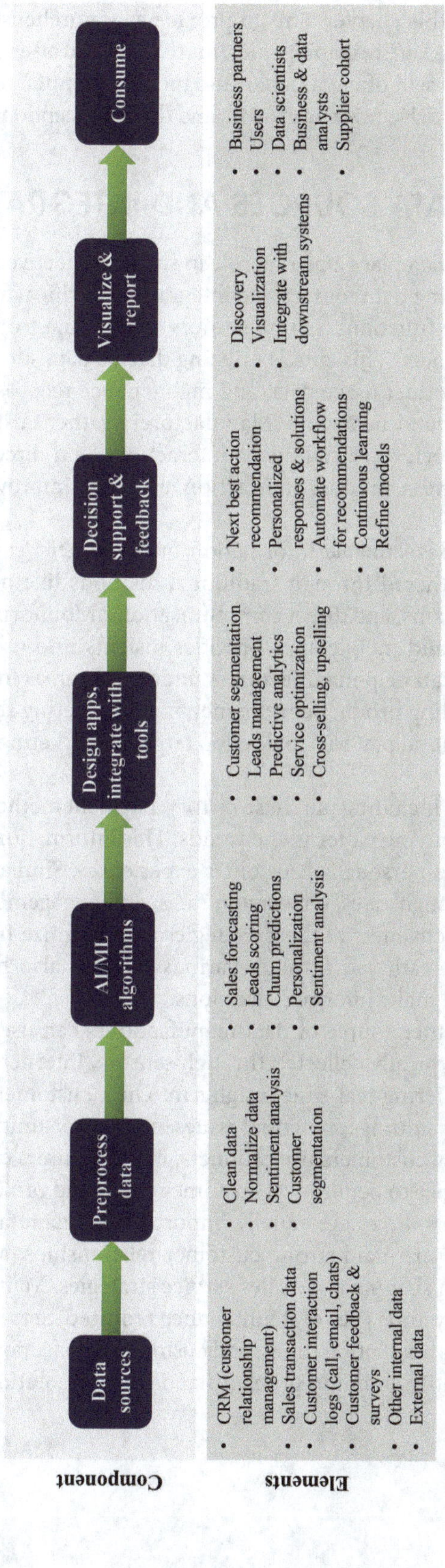

Figure 9.1 Sample steps involved in sales and marketing and after-sales service.

By exploring these topics, this chapter aims to provide a comprehensive overview of how advanced analytics, BI, ML, and AI are reshaping sales, marketing, and after-sales service in manufacturing. The insights presented will offer valuable guidance for manufacturers seeking to enhance their market competitiveness, drive revenue growth, and deliver exceptional customer experiences.

UNDERSTANDING DATA SOURCES AND INTEGRATION

In the manufacturing sector, data plays a pivotal role in shaping effective strategies and informing decision making. Understanding data sources and integration is critical for optimizing sales, marketing, and after-sales service functions. Manufacturers collect data from a variety of sources to gain insights into their customers. This entails utilizing diverse data sources, including customer feedback, service requests, product usage data, and maintenance records, to gain comprehensive insights into customer needs and behaviors. Manufacturers gather customer feedback through embedded sensors, surveys, reviews, social media interactions, and direct communications, providing invaluable information on product satisfaction, areas for improvement, and overall customer sentiment.

Customer feedback provides valuable information on customers' experiences with products and services. This data is gathered through traditional methods like market research, surveys, reviews, social media interactions, and direct communications. Monitoring customer interactions with social media, websites, and mobile apps provides insights into user behavior, preferences, and interests. This feedback can help manufacturers understand areas for improvement and customer satisfaction levels, guiding product development and marketing strategies. Using this data, manufacturers analyze behavioral patterns to develop targeted marketing campaigns and improve conversion rates.

Data on sales transactions, including purchase history, payment methods, and customer demographics, offers insights into buying patterns and trends. This information can help manufacturers tailor their sales strategies and personalize customer experiences. Similarly, data from marketing campaigns, such as click-through rates, conversion rates, and engagement metrics, helps manufacturers understand the effectiveness of their strategies and optimize future campaigns. Market and competitive intelligence, gathered through various sources, also helps manufacturers stay ahead of the competition and make informed decisions.

Product usage data is another source of data manufacturers can use in marketing and after-sales planning. This data is typically collected through sensors, Internet of Things (IoT) devices, and connected equipment, offering real-time insights into how customers interact with products. This data can reveal patterns in usage, potential issues, and opportunities for product enhancements. By understanding how customers use products, manufacturers can tailor their marketing and after-sales service strategies to better meet customer needs and prolong product life cycles.

In after-sales service, data sources are equally important for manufacturers to deliver exceptional customer support and maintain strong customer relationships. Maintenance records and service requests are crucial in shaping after-sales service strategies. Maintenance records provide information on the frequency and types of maintenance required. Service requests data offer insights into the nature of issues, resolution times, and customer satisfaction levels. By analyzing this data, manufacturers can identify common problems and develop solutions to proactively address them efficiently, including redesigning new products. This also helps manufacturers optimize maintenance schedules and improve product reliability. Monitoring how customers use products can also reveal patterns and potential issues, enabling manufacturers to improve product design. Additionally, data from spare parts inventory and replacement rates can provide insights into product longevity and possible areas for improvement. Another source of data is warranty claims. Tracking warranty claims offers insights into product quality and customer concerns, allowing manufacturers to use this additional data to improve product design and support.

Figure 9.2 provides a high-level process for sales, marketing, and after-sales service. Components in Figure 9.2 help visualize and manage the flow of materials, products, and information through the distribution and logistics processes. These steps and sub-steps focus on managing marketing strategies, generating and managing leads, order processing, and getting customer feedback, ensuring timely responses to customers and a continuously improving process.

Despite the wealth of information provided by these diverse data sources, integrating them can be challenging. Different organizations, departments, and systems generally collect data in isolation, leading to data silos that hinder cross-functional insights. Inconsistent data formats, structures, and standards across sources can make integration difficult and reduce data quality. The sheer volume of data generated across sales, marketing, and after-sales service can be overwhelming, requiring robust data management and storage solutions. Advanced data integration tools and platforms can help manufacturers unify data from disparate sources, providing a holistic view of customer interactions and operations. This comprehensive view enables manufacturers to create detailed customer profiles, enabling personalized marketing and targeted after-sales service strategies. Real-time data processing capabilities also allow manufacturers to access and analyze data as it is generated, enabling faster decision making and responsiveness to customer needs.

Advanced data integration tools and platforms can help manufacturers unify data from diverse sources, providing a single view of customer interactions and operations. Moreover, data integration enables manufacturers to monitor and evaluate the performance of their sales, marketing, and after-sales service strategies. By analyzing metrics such as sales conversions, campaign performance, customer returns, and customer satisfaction, manufacturers can continuously refine their approaches and achieve better outcomes. Standardizing data collection methods across departments and channels can improve data quality and facilitate integration. Data cleansing, which involves identifying and correcting errors, removing duplicates, and standardizing data formats, is essential for maintaining high-quality data.

Manufacturers should also invest in data analytics and AI tools to extract meaningful insights from integrated data sources. By leveraging these technologies in conjunction with BI, manufacturers can identify trends, customer preferences, and areas for improvement, leading to more targeted and effective strategies. For instance, AI-driven predictive analytics can help manufacturers anticipate customer needs and optimize inventory management and production processes. Establishing data governance frameworks ensures data accuracy, quality, and compliance with regulations while also defining roles and responsibilities for data management. Cross-functional collaboration between departments can help break down data silos and facilitate data sharing. Prioritizing data integration strategies focusing on customer needs and experiences can lead to more effective marketing and service initiatives. Figure 9.3 shows an example of an architecture for sales and marketing, as well as after-sales service, that can be used to guide the development of your solution.

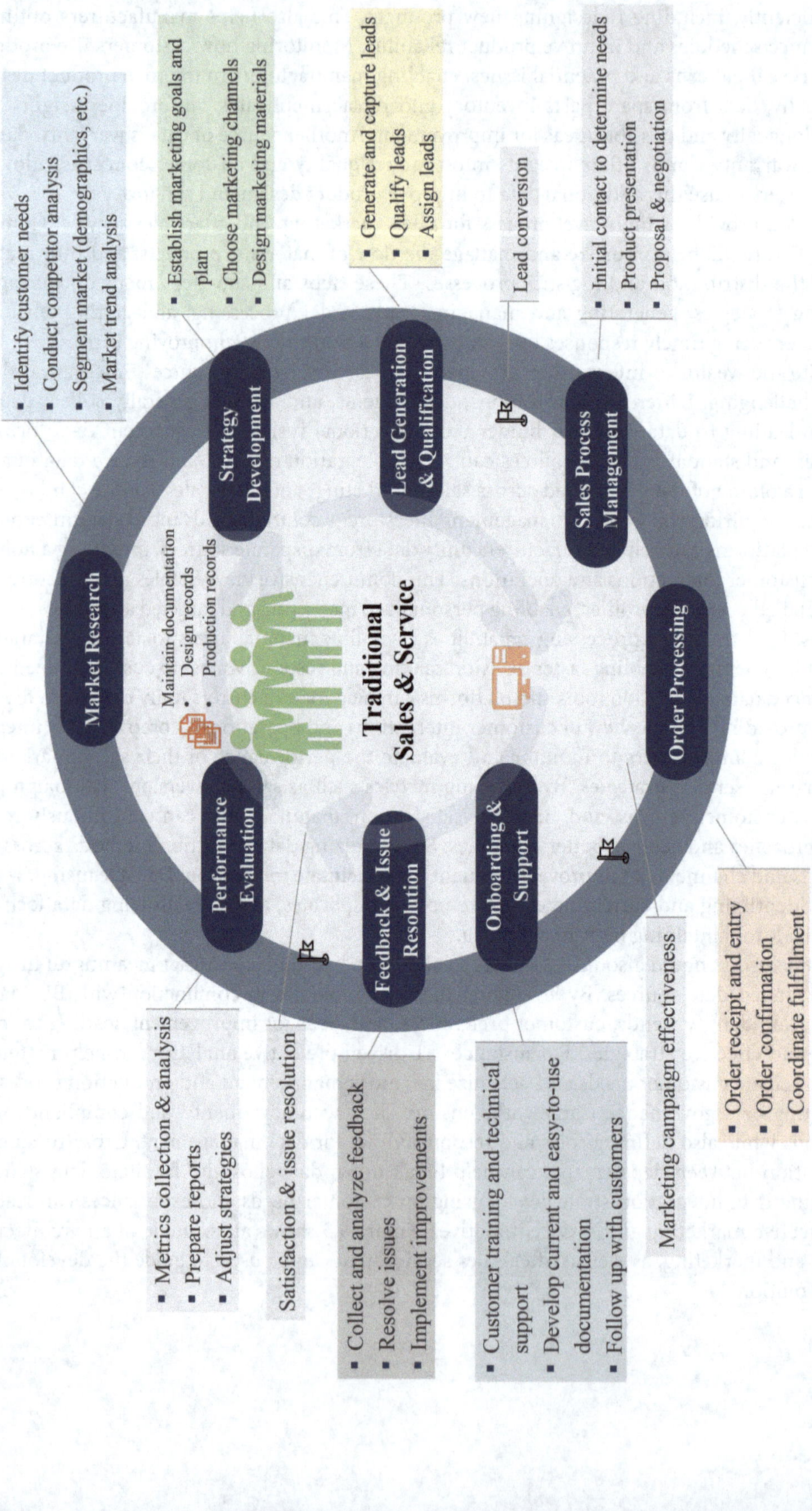

Figure 9.2 Processes within traditional sales, marketing, and after-sales service.

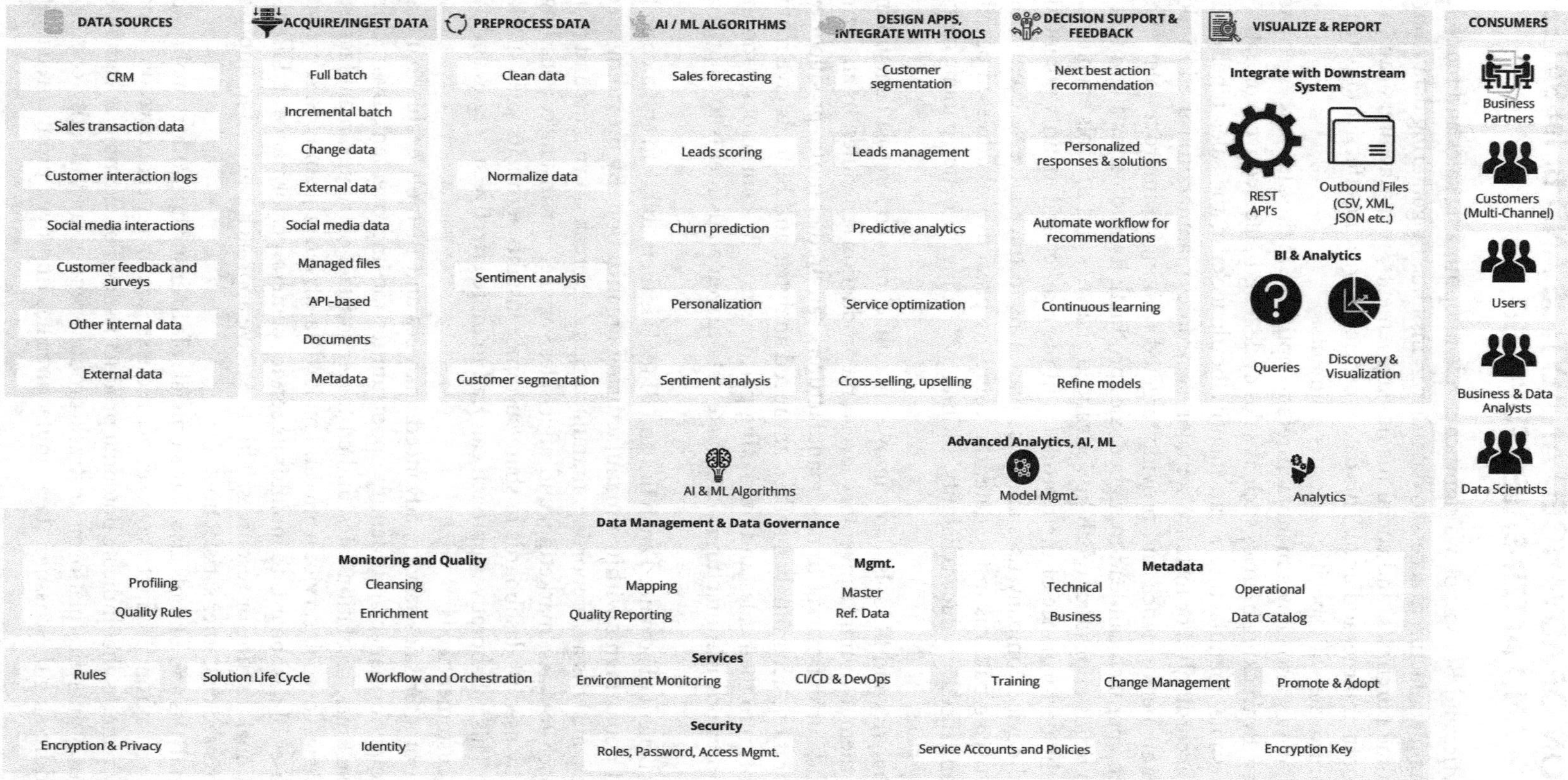

Figure 9.3 Example architecture for sales, marketing, and after-sales service.

APPLICATION OF ADVANCED ANALYTICS AND BI IN SALES & MARKETING AND AFTER-SALES SUPPORT

Advanced analytics and BI tools play a significant role in optimizing sales and marketing strategies by providing manufacturers access to data-driven insights. By gathering data from customer interactions, online purchases, and social media engagements, manufacturers can gain a better understanding of their target audience. This data is crucial in creating effective targeted marketing campaigns that resonate with specific customer segments.

Advanced analytics also allows manufacturers to optimize their sales processes and improve overall efficiency. By tracking metrics such as sales cycle length, lead conversion rates, and average transaction value, manufacturers can identify bottlenecks in the sales pipeline and adjust their strategies accordingly. For instance, a major electronics manufacturer used BI tools to analyze its sales team's performance and provide targeted training to improve sales effectiveness. This data-driven approach can lead to higher sales productivity and better alignment with customer needs. In terms of product reliability, advanced analytics and BI tools enable manufacturers to monitor the performance of their products in the market and identify potential issues. A leading medical device company uses analytics platforms to track adverse events and product recalls. By identifying patterns in the data, the company proactively addressed potential safety concerns and enhanced product reliability, ultimately leading to better patient outcomes and reduced liability risks.

Advanced analytics and BI tools also empower manufacturers to forecast demand more accurately, enabling them to optimize inventory levels and production schedules. By analyzing historical sales data and current market trends, manufacturers can predict future product demand and adjust their operations accordingly. Further, advanced analytics and BI tools give manufacturers real-time insights into their sales and marketing performance. By continuously monitoring key performance indicators, manufacturers can make data-driven adjustments to their strategies on the fly.

Another key benefit of advanced analytics and BI tools is their ability to enhance collaboration across different departments within a manufacturing organization. By providing a centralized platform for data analysis and reporting, manufacturers can break down silos and foster a more cohesive approach to sales, marketing, and after-sales service. For instance, a leading pharmaceutical company used BI dashboards to share customer insights across its sales, marketing, and research teams. This collaborative approach can lead to more effective strategies and a better understanding of customer needs.

In the realm of after-sales service, advanced analytics and BI tools enable manufacturers to monitor service performance and identify opportunities for improvement. By tracking metrics such as response times, resolution rates, and customer satisfaction scores, manufacturers can assess the effectiveness of their service teams and make data-driven adjustments. A major appliance manufacturer uses BI dashboards to monitor its service centers' performance and identify centers that need additional training or resources. This proactive approach can lead to improved service quality and increased customer satisfaction.

Advanced analytics and BI tools also play a crucial role in ensuring product quality and reliability. By analyzing data from product usage, service requests, and maintenance records, manufacturers can identify patterns that indicate potential product issues. For example, a leading electronics manufacturer uses analytics platforms to track product returns and identify common reasons for them. By addressing these issues, the manufacturer has improved product reliability, thereby reducing returns and warranty costs. Further, advanced analytics and BI tools give manufacturers valuable insights into market trends and competitive positioning. By monitoring data from market research, industry reports, and competitor analysis, manufacturers can stay informed about emerging trends and adjust their strategies accordingly.

ML AND AI IN SALES & MARKETING AND AFTER-SALES SUPPORT

AI and ML algorithms have revolutionized the manufacturing sector by helping manufacturers understand their customers and tailor their marketing campaigns. Like advanced analytics and BI, building models using vast amounts of data from various sources such as purchase history, website interactions, and social media engagements, AI/ML can segment customers based on their preferences, behaviors, and purchasing patterns. This facilitates the creation of personalized marketing campaigns that resonate with individual customers and drive engagement and loyalty. AI/ML models also play a crucial role in after-sales service primarily by enabling PdM (although other tools are used as well), which anticipates equipment failures, optimizes maintenance schedules, and enhances equipment reliability. This approach minimizes downtime and maximizes asset lifespan.

AI/ML algorithms for customer segmentation help divide customers into distinct personas based on multiple dimensions. These dimensions include customer demographics, purchasing history, online behavior, and interactions with the brand. These algorithms primarily include clustering algorithms like k-means, hierarchical clustering, and Gaussian mixture models. This data-driven approach enables manufacturers to tailor marketing campaigns that appeal directly to specific personas. For instance, a consumer electronics company used AI algorithms to identify customers who previously purchased high-end products and targeted them with personalized promotions for their latest premium offerings. In the realm of food and beverage manufacturing, AI algorithms can segment customers based on dietary preferences, such as vegan, gluten-free, or organic options. By understanding these preferences, manufacturers can create targeted marketing campaigns highlighting products catering to specific dietary needs, driving customer loyalty and engagement. Similarly, automotive manufacturers leverage AI/ML algorithms to categorize customers based on their driving habits, preferred features, and previous purchases. This enables the creation of targeted campaigns promoting specific vehicle models or features that align with customers' interests, thereby increasing sales and customer satisfaction.

Another AI/ML algorithm commonly used in sales and marketing is the recommender algorithm. These algorithms include collaborative and content-based filtering, providing personalized product recommendations to customers based on their past interactions, preferences, and purchasing history. One notable example of a company that has effectively leveraged recommender systems for customer segmentation is Amazon. Amazon's recommendation engine uses ML algorithms to analyze customers' past purchases, behaviors of similar customers, search history, and browsing behavior. Based on this data, Amazon provides personalized product recommendations, enhancing the shopping experience for customers and increasing the likelihood of additional purchases. The company's targeted marketing campaigns have contributed to its dominance in the e-commerce industry.

As the use of online platforms increases, consumers have become more reliant on social media and online reviews to express their views to manufacturers and, more important, other consumers. NLP algorithms, including sentiment analysis, are widely used to analyze customer reviews, social media interactions, and other textual data to gauge customer sentiment and feedback. Unilever has successfully applied NLP algorithms in its manufacturing operations. It utilizes AI/ML to segment its customers, optimize marketing strategies, and understand what consumers say about its products online. In addition to online data, Unilever uses purchasing data to analyze consumer data from multiple sources. This analysis helps Unilever understand customer preferences and tailor its product offerings and marketing campaigns accordingly. It also allows Unilever to understand customer grievances and respond accordingly to instill customer satisfaction. As a result, the company targets all grievances across specific customer segments with personalized messages, increasing customer engagement and loyalty.[1]

The proliferation of conversational AI (CAI) systems and virtual assistants consisting of NLP, deep learning, and generative AI models enable CAI and virtual assistants to handle customer inquiries, provide product information, and assist with sales and support in real time. In the automotive industry, BMW, along with other luxury car makers, uses CAI and virtual assistants to provide personalized in-car assistance to their customers. BMW uses the data it collects on customer preferences and behaviors from various touchpoints, including online interactions and in-car data. By analyzing this information, in conjunction with CAI and virtual assistants, BMW can provide a personalized driving experience to its customers. It also uses this data to personalize offers for individual customers, enhancing customer loyalty and driving sales.

NLP and deep learning models also enable CAI and virtual assistants to play a key role in customer support. These tools, like IBM's Watson Assistant, help customers with troubleshooting, maintenance guidance, and after-sales support in real time. These models can also refer complex issues to human agents when necessary. A well-designed CAI and virtual assistant will make this exchange seamless, without the customer knowing of the transfer from a machine to a human. These tools can also analyze warranty claims and service requests to identify patterns and trends, helping manufacturers address issues more efficiently and improve product reliability. This also helps reduce warranty fraud and assign responsibility to the correct entity (e.g., manufacturer, supplier, retailer, etc.) in the value chain.

In after-sales service, AI/ML algorithms play a critical role in enabling PdM. Algorithms such as time-series analysis, along with anomaly detection algorithms such as autoregressive integrated moving average (ARIMA), long short-term memory (LSTM), and support vector machines (SVMs), are widely used for after-sales service. They can predict equipment failures and maintenance needs by analyzing sensor data and performance metrics. By using predictive analytics models, manufacturers can optimize maintenance schedules, ensure equipment operates at peak efficiency, and minimize unexpected breakdowns. One example is the application of AI-powered PdM in the aerospace industry. Aircraft manufacturers and airlines employ AI models to monitor engine performance and predict potential maintenance needs. By detecting anomalies in engine data early, airlines can schedule maintenance during planned downtime, reducing flight delays and cancellations while enhancing passenger safety.

Another standard set of algorithms includes classification and clustering, which help manufacturers identify the underlying causes of equipment failures and product defects, enabling them to take corrective actions. Such root cause analysis algorithms are commonly used in aircraft, heavy machinery, and asset-intensive manufacturing, such as paper and steel mills and power plants. Such companies utilize root cause analysis in conjunction with PdM AI/ML algorithms to predict maintenance needs for their assets and products like construction equipment and mining machinery. These models analyze data from sensors embedded in the equipment, detecting wear and tear before it leads to costly breakdowns.

PREDICTIVE ANALYTICS FOR SALES FORECASTING AND SERVICE

Predictive analytics is transforming sales forecasting and service management by leveraging advanced models and algorithms to provide manufacturers with valuable insights. These insights enable manufacturers to forecast sales trends, identify potential opportunities, and anticipate market demand. This, in turn, enables them to optimize inventory management, production planning, and resource allocation, allowing them to effectively meet customer needs and maximize revenue.

In sales forecasting, predictive analytics relies on various statistical and ML models to analyze historical sales data and identify patterns and trends. Some commonly used methods include time series analysis, ARIMA, and exponential smoothing. These methods analyze historical sales data to identify trends and seasonality. They provide a basis for predicting future sales based on past patterns. Beyond this, regression models such as linear and multiple regression models help identify relationships between sales and other various factors such as promotions, seasonality, and economic conditions. These relationships enable manufacturers to predict sales performance more accurately. More sophisticated models using ML and AI are starting to see greater use in the industry. For example, advanced ML models like decision trees, random forest (and its different variants), and SVMs can process large datasets to predict sales trends. These models offer higher accuracy and adaptability, especially when dealing with complex or nonlinear relationships in data. In the AI realm, neural networks (NNs) and deep learning models—particularly recurring NNs (RNNs), LSTM networks, and transformer models—excel at capturing complex patterns and relationships in sales data, thereby obtaining the temporal nature of sales. These models can provide accurate predictions, even in rapidly changing market environments.

In service management, predictive analytics is key in optimizing after-sales service strategies. Manufacturers use various models to predict equipment failures and optimize maintenance schedules. PdM models anticipate equipment failures by analyzing sensor data, maintenance records, and other performance metrics. Algorithms such as SVMs, decision trees, and random forests predict when maintenance is needed, allowing manufacturers to address issues before they cause significant disruptions. In addition to PdM, anomaly detection plays an important role. Anomaly detection algorithms like isolation forests, one-class SVMs, and autoencoders help identify unusual patterns in equipment performance, signaling potential issues that require attention. This proactive approach minimizes downtime and extends asset lifespan. From a service management perspective, it is critical that inventory for essential components is available or can be obtained quickly. Robust inventory optimization algorithms achieve this. Predictive analytics support inventory optimization by forecasting demand for products and parts. Algorithms such as linear programming and dynamic programming help manufacturers determine optimal reorder points and quantities.

When incorporating advanced analytics, BI, AI, and ML into traditional sales, marketing, and after-sales service processes, the components of each step are enhanced with technology. Figure 9.4 depicts the processes involved in modern manufacturing and development. Incorporating advanced technologies transforms each step with components that leverage data and automation to improve efficiency, quality, and decision making throughout the manufacturing process. Unlike Figure 9.2, all processes are semi-automated (depicted by the icon size for computing and human resources) with a much higher contribution from computing resources instead of manual resources. Although semi-automated, most of the *hard* work is done by the models. These models provide guidance to human resources on actions to incorporate in operations. The manual intervention ensures checks and balances are in place by incorporating the *human-in-the-loop*

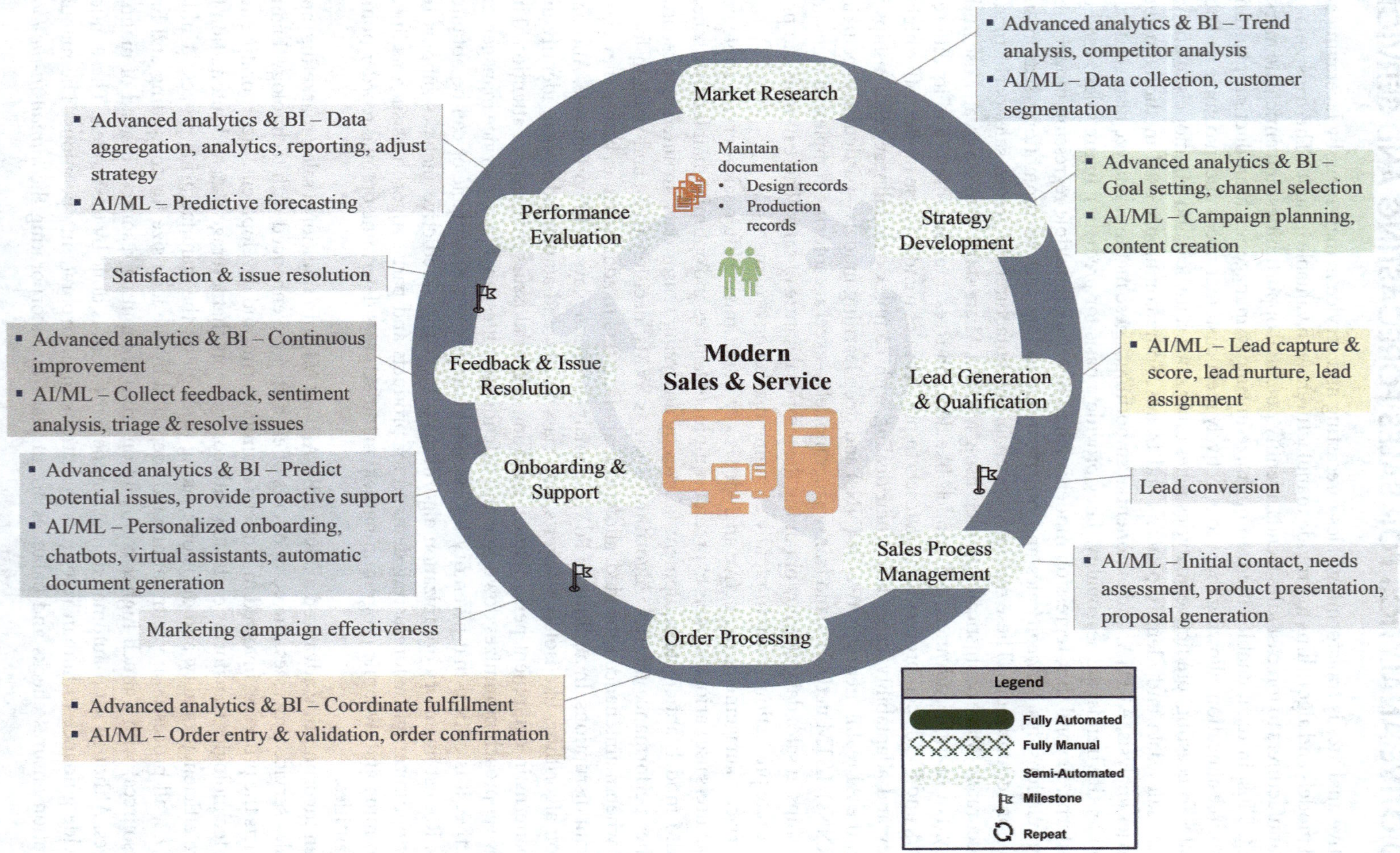

Figure 9.4 Processes within modern sales, marketing, and after-sales service.

framework. This allows humans to quickly react and manually override poor outcomes from the modern tools, preventing them from getting pushed downstream.

General Electric is one example of a company that utilizes these methods. As mentioned in Chapter 2, GE uses SmartSignal for PdM to monitor its industrial equipment, allowing it to identify potential issues before they become major problems.[2] This helps reduce downtime and maintenance costs. Similarly, Siemens uses PdM to monitor and maintain its manufacturing equipment. By using digital twins to analyze data from sensors on production lines, it can predict maintenance needs and potential failures. As described in Chapter 3, Bosch leverages PdM models to monitor its manufacturing machinery and anticipate maintenance needs. This allows Bosch to maintain high levels of equipment reliability and production efficiency while minimizing unexpected breakdowns.

Caterpillar's success with predictive analytics is documented in its service offerings such as Cat® Connect, which leverages data and analytics to enhance performance. Caterpillar also employs its Product Link for remote monitoring and equipment management solutions and uses predictive analytics to monitor the performance of its heavy machinery.[3] By analyzing sensor data and equipment usage patterns, the company can predict when maintenance is needed and schedule service accordingly. Similarly, Ford, Honda, Volkswagen, and Tesla employ predictive analytics to optimize their supply chain and production processes.[4]

Parker-Hannifin is another example; it utilizes ML and predictive analytics algorithms to analyze customer data and sales trends to predict future demand. This helps the company optimize inventory levels and streamline its marketing strategies.[5] The 3M Company uses ML and predictive analytics algorithms to analyze customer data, including purchasing patterns, preferences, and feedback.[6] The company leverages these insights to personalize marketing campaigns and tailor product recommendations to customers. Rolls-Royce Holdings PLC uses PdM to monitor its aircraft engines in real time. By collecting data from sensors on the engines and using advanced algorithms, Rolls-Royce can predict potential issues, anticipate maintenance needs, and schedule proactive repairs. More important, this is done before issues cause downtime or safety risks. Rolls-Royce's use of predictive analytics to improve its service can be seen in its TotalCare service model. TotalCare offers airlines a comprehensive maintenance and support package that includes predictive analytics for proactive engine maintenance.[7]

OPTIMIZING MARKETING CAMPAIGNS AND CUSTOMER SUPPORT WITH AI

AI algorithms are revolutionizing the manufacturing sector by providing powerful tools to optimize marketing campaigns and enhance customer support. In marketing, much like advanced analytics, AI-driven algorithms analyze customer data to segment audiences and build personas based on their preferences, purchasing behaviors, and interactions with products or brands. This segmentation allows manufacturers to develop persona-based targeted marketing strategies and personalized campaigns that resonate with specific customer groups. One popular method for marketing campaigns is customer segmentation that uses clustering algorithms such as k-means or hierarchical clustering. It groups customers into distinct segments based on similar attributes. This enables manufacturers to tailor marketing messages and product offerings to specific segments, improving engagement and conversion rates.

In terms of customer support, conversational AI systems and virtual assistants are transforming the way manufacturers interact with customers. These AI tools can handle routine inquiries, troubleshoot problems, and guide customers through self-service options. This typically frees up human agents by almost 80% and allows them to focus on more complex issues. NLP algorithms enable chatbots to understand and respond to customer queries effectively, improving response times and overall customer satisfaction. Predictive analytics also play a role in customer support by anticipating customer needs and proactively addressing potential issues. For example, AI algorithms can analyze data from product usage patterns and customer interactions to predict when a customer might encounter a problem with a product. This allows manufacturers to proactively engage with customers, offering solutions before the issue escalates, thereby enhancing the customer experience and fostering brand loyalty.

In manufacturing, AI algorithms are critical in optimizing marketing campaigns and customer support, enabling manufacturers to stay competitive in an ever-evolving market. Some ways in which AI supports marketing and customer support include customer journey mapping and personalization. AI algorithms map the entire customer journey, from initial contact to post-purchase support. By analyzing data from various touchpoints, such as website interactions, purchase history, and customer feedback, AI can create detailed profiles of individual customers. This allows manufacturers to create personalized marketing messages and product recommendations to enhance the customer experience and increase sales. In addition, AI-powered programmatic advertising platforms enable manufacturers to optimize their digital marketing campaigns by automating advertisement purchasing and placement based on real-time data. This targeted approach ensures that ads reach the most relevant audience segments, resulting in higher engagement and conversion rates.

AI-driven technologies offer a range of tools and methodologies that enhance marketing campaigns and optimize customer support within after-sales service. By understanding which content formats, tones, channels, and messaging styles generate the most engagement, manufacturers can create more compelling and effective marketing campaigns that align with customer preferences. Using these with AI-driven, targeted advertising platforms allows manufacturers to segment audiences based on preferences, demographics, and behaviors. This enables them to deliver highly personalized advertisements to the appropriate audience at the most advantageous time, in the correct format, and through the proper channel.

CAI and virtual agents supported by NLP are the primary tools used within after-sales support. Well-designed, AI-driven CAI systems can handle a wide range of customer inquiries in real time. These systems can answer common questions, troubleshoot issues, and provide personalized recommendations based on a customer's history with the company. By automating routine tasks, conversational AI improves efficiency and minimizes the involvement of human agents. Virtual assistants leverage NLP algorithms to understand and respond to customer inquiries in a natural, conversational manner and across multiple channels. This technology enables seamless communication between customers and support agents, providing customers with quick and accurate answers to their questions while maintaining a positive brand experience.

Siemens leverages AI algorithms to optimize its digital marketing campaigns.[8] The company uses AI-driven data analytics to understand customer preferences and behavior patterns, enabling them to create personalized content and targeted advertising. As a result, Siemens has seen a significant increase in customer engagement and improved return on investment (ROI) in their marketing efforts.

For AI in customer support, Philips utilizes conversational AI-powered chatbots to enhance its customer support services. By automating routine inquiries and providing quick responses, Philips improves customer satisfaction and reduces the workload on human agents. The company's chatbots use NLP to understand and respond to customer queries effectively.[9]

By harnessing AI-driven technologies, manufacturers can enhance marketing campaigns through predictive analytics, content optimization, and targeted advertising. In after-sales service, CAI, virtual assistants, and NLP algorithms optimize customer support by delivering personalized and efficient service, thereby enriching customer experiences and increasing satisfaction levels. Together, these innovations empower manufacturers to achieve desired marketing outcomes and strengthen their relationships with customers.

SUSTAINABILITY AND ETHICAL CONSIDERATIONS

The increasing adoption of ML and AI algorithms in sales, marketing, and customer support functions within the manufacturing sector offers significant potential for enhancing efficiency and effectiveness. However, these technologies also pose substantial sustainability and ethical considerations that manufacturers must address to ensure responsible use.

One primary concern involves the environmental impact of data centers and the computational resources required to support AI and ML algorithms. AI and ML models, especially generative AI models, often require vast amounts of data and computational power for training and execution, contributing to increased energy consumption and carbon emissions. Manufacturers can address this by choosing energy-efficient data centers, moving their solutions to the flexible and elastic cloud, investing in renewable energy sources, and optimizing algorithms for efficiency. Another ethical concern revolves around data privacy and security. ML and AI rely heavily on customer data, which must be handled with the utmost care to prevent unauthorized access and data breaches. Many companies have established processes to handle internal and external personally identifiable data. Companies must also comply with data protection regulations, such as the General Data Protection Regulation (GDPR) in Europe and the new EU Artificial Intelligence Act, to safeguard customer data and maintain trust. Bias in AI and ML algorithms presents another significant ethical issue. Biases can creep into the models through multiple ways—accidentally or deliberately. The most common biases are due to biased data, biased models, or biased developers. If not appropriately addressed, biases in training data can result in discriminatory marketing and customer support practices. For instance, a manufacturer may inadvertently target or exclude certain demographic groups in marketing campaigns. Companies can mitigate bias by using diverse and representative data sets, regularly auditing models for fairness, and being transparent about using AI and ML in their processes.

Sustainability and ethical considerations are important factors, specifically in extending product life. For instance, manufacturers must ensure that their sales, marketing, and customer support strategies align with responsible sourcing and ethical supply chain practices. AI and ML can aid in monitoring suppliers for ethical compliance and environmental impact. When it comes to specific examples of what manufacturers did correctly, Bosch used AI and ML algorithms to optimize its production processes, improve energy efficiency, and minimize waste. Bosch also prioritizes data privacy and security, complying with relevant regulations. On the other hand, there have been instances where manufacturers failed to manage AI and ML algorithms ethically. For example, there have been cases of algorithmic bias in product recommendations, leading to discrimination and backlash from customers. Such failures highlight the importance of ethical oversight in using AI and ML technologies.

As mentioned in the previous section, Siemens integrated AI and ML into its sales and marketing strategies to offer personalized solutions to its clients. Through predictive analytics, the company anticipates client needs and suggests products and services that best fit their requirements. This level of personalization enhances customer satisfaction and drives long-term relationships.

In addition, Siemens has demonstrated the ethical use of data, maintaining transparency with customers and ensuring compliance with regulations such as GDPR.[10] It is anticipated that Siemens will build a process around the new EU AI Act that went into effect in March 2024. Schneider Electric has made strides in using AI and ML to improve customer support in the manufacturing sector.[11] By implementing conversational AI systems and predictive analytics tools, the company provides customers with quick, efficient support while also reducing its environmental footprint. Schneider Electric has also emphasized the ethical use of data by ensuring that algorithms do not perpetuate biases and that they comply with privacy regulations.[12] Unfortunately, some automotive manufacturers have faced criticism for using ML algorithms in marketing that resulted in targeting vulnerable or potentially exploitable customers. This approach has drawn backlash and regulatory scrutiny, pointing to the need for greater ethical oversight in the use of AI and ML in marketing strategies. In some instances, manufacturers have been accused of using biased data sets in their AI and ML algorithms, leading to discriminatory advertising practices. Manufacturers must remain vigilant in ensuring their algorithms are free from bias and unfair treatment of specific groups.

To navigate the challenges of sustainability and ethics, manufacturers should prioritize transparency and accountability in their use of ML and AI. This includes building an ethical and sustainable framework by regularly conducting AI risk assessments for algorithms in their solutions and checking for fairness, maintaining clear communication with customers about data usage, and adhering to data protection regulations. Manufacturers should also invest in training for employees on the ethical use of AI, fostering a culture of responsibility and respect for customer privacy. Employees should be watching for bias in data, potential bias in developers, and bias in the models. This should be an ongoing effort from the time a company embarks on this journey, not something done after the fact. It becomes more difficult later, as a large U.S.-based global financial institution found out. They had already rolled out over 150 AI models in their business processes when they realized they needed to conduct an AI bias risk assessment. Each model needed a minimum of four weeks of assessment involving multiple resources over this time frame. When a model was found to have an unacceptable level of risk, a lot more effort was invested to redesign, replan, and redeploy an updated solution. It also increased the risk for the financial institution concerning breakdowns in the process, or worse yet, customers getting impacted. Similarly, by embracing sustainable practices and ethical standards, manufacturers can build a stronger relationship with customers and gain a competitive edge in the industry. However, this effort has to begin in the planning process, not later.

Transparency, data privacy, and responsible marketing strategies that align with ethical and environmental standards are crucial to maintaining customer trust and upholding corporate values. Transparency about data usage and clear communication with customers are essential to maintaining trust. AI embedded within responsible marketing strategies enables hyper-targeted advertising, but this can raise concerns about manipulation and overly intrusive marketing. To avoid these issues, manufacturers must employ responsible marketing strategies that respect customer preferences and avoid exploiting vulnerabilities.

In sales and marketing, AI can help manufacturers optimize resource allocation and reduce waste as part of the company's sustainability initiative. For instance, Tetra Pak uses AI to optimize its supply chain, reducing transportation emissions and packaging waste. Similarly, AI can be used to track product life-cycle data and guide customers in responsible disposal methods. Philips uses AI to ensure responsible end-of-life equipment management. It recommends sustainable practices like refurbishment and recycling for its medical equipment. Manufacturers can also leverage AI to optimize maintenance schedules and prolong equipment life, reducing the need for new production and thus mitigating environmental impact.

IMPLEMENTATION CHALLENGES AND BEST PRACTICES

Implementing advanced analytics, BI, ML, and AI in sales and marketing, as well as after-sales service, can provide significant benefits but also poses specific challenges while implementing them. Data quality and integration are among the most common challenges of dealing with inconsistent or poor-quality data from disparate sources. Poor data quality can lead to inaccurate insights and decisions. Unilever addresses this challenge by establishing a data governance framework and implementing data cleansing processes to ensure consistency and accuracy.[13] Once the ongoing processes for data cleansing are in place, manufacturers are using cloud-based data repositories like Snowflake or Databricks to act as the single-source-of-truth for the enterprise.

Another common challenge is cultural resistance within the organization. This resistance to change can hinder the successful adoption of advanced technologies, especially when employees are unfamiliar with the new systems or, more often, concerned about their jobs being replaced by technology. This can be overcome by strong, ongoing change management that begins even before AI projects officially commence. Organizations like Procter & Gamble invest in training programs and workshops to help employees understand and embrace the benefits of AI and analytics.[14] Cultural resistance can also be due to the complexity and scalability of the AI solutions—from data collection to cleansing to using an AI model that is unexplainable. Also, managing the complexity of advanced analytics and AI can be daunting, especially when scaling these technologies across multiple departments. Companies should start small by implementing pilot projects to test and optimize AI solutions before scaling them up.

Integrating advanced analytics and AI with existing after-sales service systems can be challenging due to compatibility issues, but another challenge is the concern for customer privacy. Philips adheres to stringent data protection policies and ensures transparent data practices to build customer trust.[15] Like other AI initiatives, technical skill gaps continue to challenge AI initiatives in after-sales service as well. Implementing advanced technologies requires specialized skills that may be lacking within the organization but should be incorporated from the initiative kick-off. General Electric partners with universities and training institutions to develop talent pipelines and to upskill existing employees.[16]

> *"When deploying AI, whether you focus on top-line growth or bottom-line profitability, start with the customer and work backward."*
>
> —Rob Garf, VP and General Manager at Salesforce Retail

Clearly defining business objectives and aligning them with AI and analytics initiatives helps ensure focused and measurable outcomes. Cross-functional collaboration among departments fosters a holistic approach to implementation, maximizing the potential benefits. Continuous improvement, including constantly monitoring and evaluating the performance of AI and analytics solutions, helps make iterative advancements. Such embedded initiatives within business processes go a long way to optimizing outcomes.

FUTURE TRENDS AND OUTLOOK

Some key trends and technologies shaping the future include AI-driven personalization, in which AI enables hyper-personalization in marketing campaigns. This allows manufacturers to target customers with tailored content and offers. For example, Volkswagen uses AI algorithms to personalize its marketing campaigns, resulting in higher conversion rates and increased customer loyalty.[17] Predictive customer analytics tools help manufacturers anticipate customer needs and behaviors, allowing for more effective marketing strategies. For example, Bayer uses predictive analytics to anticipate customer preferences and optimize product offerings. Chatbots and virtual assistants are also used to engage with customers and provide real-time support in sales and marketing. Siemens uses virtual assistants to provide instant responses to customer inquiries, improving customer satisfaction.

Some trends include PdM for after-sales services. This enables manufacturers to anticipate equipment failures in specific models that have been sold and schedule maintenance proactively to help reduce downtime and improve equipment reliability. ABB uses PdM to predict machine failures and optimize maintenance schedules, reducing downtime by up to 40%.[18] IoT-enabled monitoring systems permit real-time monitoring of products, allowing manufacturers to detect issues early and improve after-sales service. Remote diagnostics and support that uses IoT data within AI and ML models enable manufacturers to diagnose issues remotely and provide virtual support, reducing the need for on-site visits. Kone Corporation uses IBM's Watson products to predict when an elevator may fail, allowing it to proactively dispatch a technician to repair an elevator that seems to be working but may fail soon.[19]

Four future technologies will continue to alter this landscape. Edge computing allows for real-time data processing and analysis at the edge of the network, enabling faster decision making in sales and after-sales service. Processing data closest to where it is generated results in instant feedback that can be acted on quickly. Augmented reality (AR) and virtual reality (VR) technologies can enhance customer experiences by providing immersive product demonstrations and support. AR and VR can also play a key role in after-sales service by guiding technicians to traverse the diagnosis and repair process safely. With use cases such as an assistant for marketing content and promotions planning, a virtual technician assistant supporting after-sales service, and a safety professional for after-sales service, among many others, generative AI is at the forefront of transforming this aspect of the manufacturing value chain.

SUMMARY AND KEY INSIGHTS

> *"If you're a seller and not thinking about AI today, you're going to want to start."*
>
> —Emma McGuigan, Senior Managing Director & Lead, Enterprise & Technology at Accenture

This chapter has explored the transformative impact of advanced analytics, BI, AI, and ML on sales, marketing, and after-sales service in manufacturing. Key insights include the ability of these technologies to provide deep awareness of customer behaviors and preferences, enabling manufacturers to tailor marketing campaigns and personalize customer experiences. Predictive analytics models enable the accurate forecasting of sales trends and market demands, leading to optimized inventory management and resource allocation. AI-driven personalization enhances targeted advertising and content delivery, increasing customer engagement and satisfaction. CAI systems and virtual assistants streamline customer support, expedite issue resolution, improve customer experiences, and foster loyalty. PdM models for after-sales service anticipate equipment failures, reduce downtime, and enhance equipment reliability.

A few of the key insights from this chapter are as follows:

1. Advanced analytics, BI, AI, and ML are poised to shape the future of sales, marketing, and after-sales service in manufacturing. Predictive analytics can optimize inventory, predict maintenance needs, and improve production processes.
2. These technologies include virtual assistants for marketing content, promotions planning, and after-sales service. They empower manufacturers to create personalized customer experiences, optimize marketing campaigns, and enhance after-sales service to help manufacturers improve efficiency and customer satisfaction.
3. Specifically, generative AI is very useful in creating marketing content, promotions planning, and after-sales service.
4. Challenges to implementing these technologies include poor data quality, internal cultural resistance, and customer privacy concerns.

10

UNLEASHING THE POTENTIAL OF ADVANCED ANALYTICS, BI, ML, AND AI IN MANUFACTURING

AI-Powered Transformation Across Stellar Electronics' Manufacturing Value Chain

Stellar Electronics has successfully embedded artificial intelligence (AI) into every stage of its manufacturing value chain, enhancing efficiency, reducing costs, and improving customer satisfaction.

Product Design and Development: Stellar's product design team, led by Rachel Miller and senior manufacturing engineer John Lee, harnessed generative AI to explore rapid design iterations and customer-aligned solutions. By simulating manufacturing processes, the team cut physical prototyping by 25% and reduced the design cycle by 30%. These AI-driven tools enable quicker, data-driven decisions, helping Stellar bring innovative products to market faster.

Procurement and Supplier Management: Laura Gomez and James Yu deployed AI to assess suppliers based on reliability, cost, and performance metrics, achieving a 10% reduction in procurement expenses, a 20% improvement in material availability, and an 18% reduction in inventory holding costs. The AI-driven system allows Stellar to monitor supplier reliability in real time, strengthen supply chain resilience, and minimize the risk of stockouts.

Manufacturing and Production: Tom Willis and Carlos Rivera implemented predictive maintenance (PdM) by analyzing data from Internet of Things (IoT)-enabled sensors. This proactive approach reduced unplanned downtime by 40% and extended machinery lifespan by 25%, resulting in lower maintenance costs. Machine learning (ML) algorithms optimized production schedules, boosting throughput by 15% and ensuring efficient alignment with demand. AI-driven quality control (QC) systems led by Olivia Chen achieved 95% accuracy in defect detection, leading to a 20% reduction in defective units, thereby lowering rework costs and enhancing product quality for Stellar's customers.

Distribution and Logistics: In logistics, COO John Davies and Michael Zhang used AI for dynamic route optimization, resulting in a 12% reduction in delivery times and a 15% improvement in fuel efficiency. Emma Patel led AI-enhanced forecasting and warehouse operations solutions that have helped Stellar reduce stockouts by 15%, excess inventory by 20%, and warehouse equipment downtime by 30%. This has kept customers satisfied with faster and more reliable deliveries.

continued

Sales, Marketing, and After-Sales Service: Julia Sanders and Ryan Cho used AI-driven customer segmentation and predictive sales analytics to achieve a 25% increase in customer engagement through targeted marketing and a 20% improvement in forecast accuracy. An AI-powered chatbot, managed by Linda Nguyen, handles 70% of common inquiries, reducing response times and freeing up support staff to address more complex issues. Proactive maintenance alerts reduced product returns by 15%, strengthening customer satisfaction.

Stellar Electronics' integration of AI across its entire manufacturing value chain has streamlined operations, minimized costs, and elevated customer experiences, positioning the company as a forward-thinking leader in electronics manufacturing.

". . . the use of AI-augmented reality allows the employment of people with limited or no experience, because they are instructed by the software to do certain tasks . . ."

—Yossi Sheffi, Professor at MIT

SUMMARIZING KEY INSIGHTS

This concluding chapter aims to weave together the threads of insights and analyses presented throughout the book. By summarizing the key takeaways from each chapter, it seeks to offer a comprehensive overview of the transformative role that advanced analytics, business intelligence (BI), ML, and AI play in the manufacturing sector. In addition, this chapter will examine the broader implications of these advanced technologies for the industry, including a look at emerging trends and the future industry landscape. Sustainability, ethical and trustworthy AI considerations, and challenges faced by manufacturers in implementing advanced analytics, BI, AI, and ML will continue to play an important role going forward, as well as responsible innovation and adherence to best practices. This chapter also examines the future trajectory of manufacturing, examining how emerging technologies such as edge computing, blockchain, and quantum computing may further revolutionize manufacturing processes and value chains. Additionally, it will explore how these technologies may impact global supply chains and the broader economy, as well as the potential risks and opportunities they present.

Finally, this chapter hopes to inspire and encourage manufacturing practitioners, leaders, and readers to embrace advanced analytics, BI, AI, and ML technologies and integrate them into their manufacturing operations. We hope that through the compelling examples and case studies of successful implementations throughout the book, we illustrated the tangible benefits that can be achieved through the strategic use of these technologies. We also demonstrated some failed initiatives and provided reasons for failure, guiding manufacturers on what they should pay attention to in order to prevent similar failures. The goal is to motivate readers to leverage the insights gained from the book to drive innovation, optimize their operations, and achieve a competitive edge in the evolving landscape of the manufacturing industry.

In the preceding chapters, the book examined the transformative role of advanced analytics, BI, AI, and ML in revolutionizing the manufacturing value chain. The first two chapters set the stage by exploring why manufacturing is transitioning toward a more comprehensive focus on the entire manufacturing value chain. They emphasized the need for holistic optimization and integration and examined the driving forces behind this shift, such as changing consumer demands, advances in technology, and the need for greater competitiveness in the global market. These chapters laid the foundation for understanding the importance of integrating advanced technologies to stay ahead in the industry. The next chapter reviewed advanced analytics and BI within manufacturing. Their potential to provide actionable insights into customer behavior, market trends, and operational performance was highlighted. These insights empower manufacturers to make informed decisions and develop targeted strategies to drive revenue growth and enhance market competitiveness. A detailed discussion of AI and ML, along with common algorithms and tools in AI and ML, followed. This included common AI and ML algorithms, tools, and frameworks that are used (or could be used) in manufacturing. Also included are examples illustrating their applications across different areas of the manufacturing value chain, such as production planning, QC, and supply chain optimization.

The six stages of the manufacturing value chain were also explored within the subsequent chapters, showcasing how advanced analytics, BI, AI, and ML play a role in each stage—from design and development to after-sales service. Each chapter shed light on how data collection, integration, and analysis can optimize production processes, improve QC, and drive innovation. Relevant algorithms were identified that would provide guidance to the user as to which algorithms to use. They also highlighted the specific ways these advanced technologies can be applied at each stage: design and development, sourcing and procurement, production and assembly, distribution and logistics, sales and marketing, and after-sales service.

Throughout these chapters, several significant findings and trends emerged. For instance, AI-driven predictive analytics offers manufacturers the ability to anticipate market demand, optimize inventory management, and streamline supply chains. Best practices were explored, such as the importance of data integration and the ethical use of AI in sales and marketing. Implementation challenges and the best practices used to overcome them were addressed, followed by a discussion of future trends and an outlook on the use of these technologies. In terms of overarching themes, this book underscores the critical role of data-driven decision making in achieving operational efficiency and delivering enhanced customer experiences. Manufacturers are increasingly turning to the advanced technologies referenced in this book to gain insights into customer behavior and preferences, enabling them to tailor products and services to meet specific needs. Figure 10.1 provides an example of an architecture summary across all the stages in the manufacturing value chain. You can use this as a guide to develop your solution. Remember that these are just sample data sources and other components. You have to customize it to your environment and internal standards and work with stakeholders and the architect assigned to manufacturing.

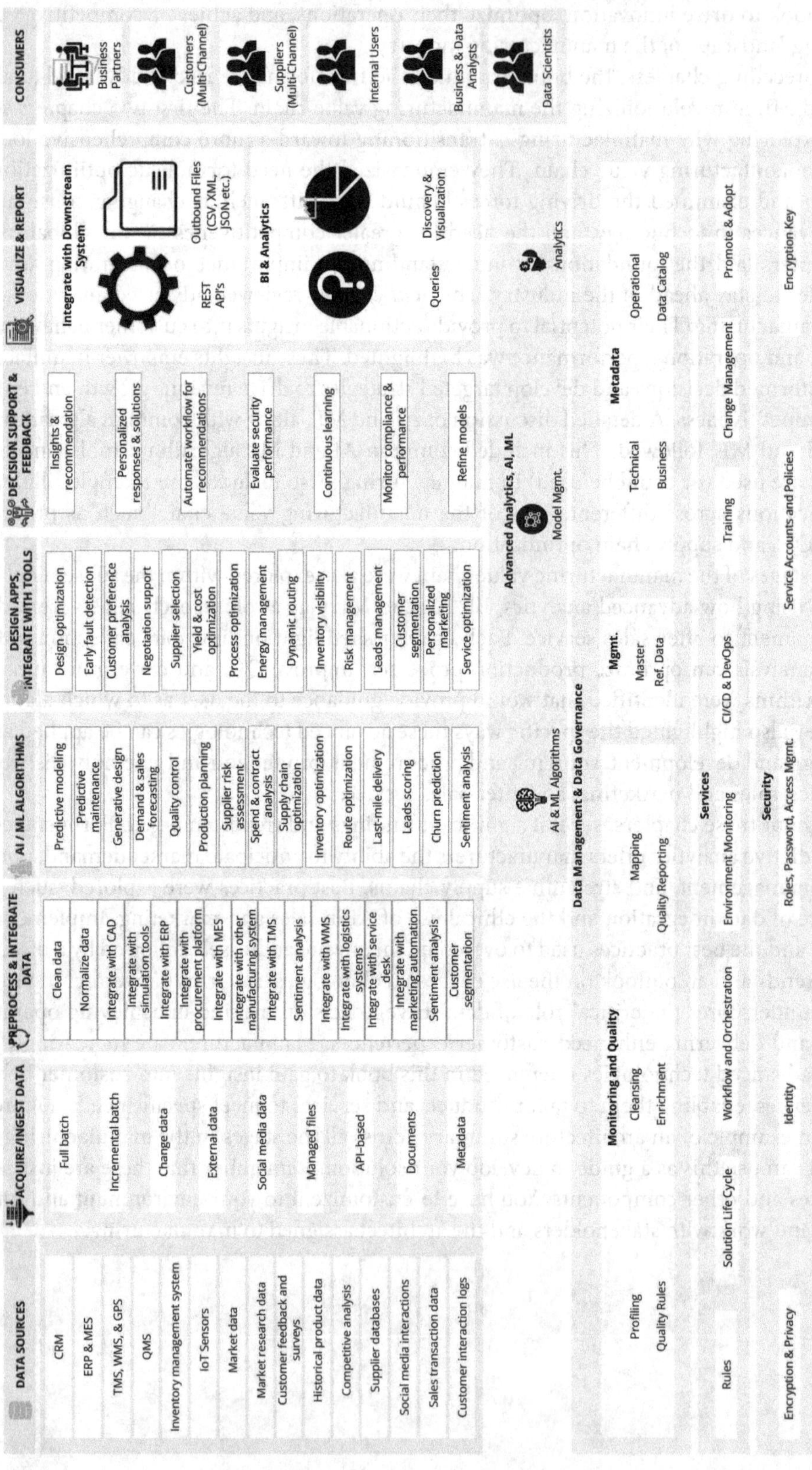

Figure 10.1 Sample architecture summary for the manufacturing value chain.

ALGORITHM USAGE AND TRENDS

Each sector in manufacturing has its own distinct set of challenges and opportunities, which advanced analytics, BI, AI, and ML algorithms can address with varying degrees of success and applicability.

In production, these technologies enable PdM, QC, and process optimization. PdM uses algorithms to analyze sensor data from machinery and equipment to anticipate and prevent breakdowns. This reduces unplanned downtime and maintenance costs while also increasing overall efficiency. QC benefits from computer vision and anomaly detection algorithms that ensure products meet strict quality standards. According to a recent study by Deloitte, AI-enabled PdM can decrease asset breakdowns by 70% and maintenance costs by 30%.[1]

In logistics, these advanced technologies are key in optimizing supply chain management, route planning, and inventory management. AI algorithms can analyze demand patterns to predict and manage inventory levels effectively, reducing overstock or stockouts. In route planning, reinforcement learning (RL) and optimization algorithms enhance delivery routes to minimize costs and delivery times. Recent trends indicate that most manufacturers are exploring advanced analytics, BI, AI, and ML solutions in logistics to improve efficiency and reduce costs.

> *"We don't create tech for tech's sake. There's purpose behind everything we do so that our customers have the tools they need to tackle some of the world's greatest challenges."*
>
> —John May, Chairman & CEO of John Deere

In sales and marketing, advanced analytics, BI, AI, and ML algorithms are widely used for customer segmentation, personalized marketing campaigns, and predictive analytics. Customer segmentation, based on purchasing behaviors and preferences, allows manufacturers to create targeted and effective marketing strategies. Predictive analytics, on the other hand, can anticipate market demand, helping manufacturers optimize their marketing budgets and strategies. According to recent industry reports, many marketing departments in manufacturing companies have adopted these models for various applications. After-sales service is another area where AI and ML technologies have shown great potential. Chatbots and virtual assistants powered by natural language processing (NLP) algorithms can handle customer inquiries and provide timely, personalized support. Moreover, deep learning algorithms can analyze customer feedback and service records to identify areas for improvement in service quality. As a result, manufacturers can enhance customer satisfaction and loyalty.

Emerging technologies, such as deep learning, computer vision, and RL, are expected to play an increasingly significant role in the future. Deep learning, with its ability to process complex data and recognize patterns, can drive additional innovations in QC, PdM, and customer support. Computer vision can further automate inspection processes. For instance, computer vision and deep learning models are utilized in image recognition for QC, enabling real-time monitoring of production lines to detect defects and ensure product consistency. These technologies can aid in streamlining operations and improving production quality. Similarly, RL is gaining traction for its ability to optimize decision making in dynamic environments, such as production lines, supply chains, autonomous robots, and machines that adapt to changing conditions in real time.

> *"The savviest leaders prioritize participation by the rank and file throughout the adoption process."*
>
> —David De Cremer, Professor of Management and Technology at Northeastern University

Ethical and sustainability considerations are also shaping the future of AI and ML in manufacturing. As manufacturers increasingly rely on sophisticated, advanced algorithms, they must prioritize data privacy, transparency, and the responsible use of AI technologies. For example, protecting customer data during after-sales service interactions and ensuring fair treatment of employees in AI-driven environments are ethical imperatives.

Throughout this book, references have been made to several algorithms. To make things clear and to help readers decide how to proceed, Table 10.1 provides a breakdown of the frequency of usage of common algorithms used in manufacturing. Algorithms like NLP, computer vision, and time series forecasting are quickly becoming mainstream, and others will follow soon. Table 10.2 presents the algorithms used in the 15 largest manufacturing companies. As Table 10.2 shows, most of the largest manufacturers are in the automotive sector. The algorithms from Table 10.1 substantiate the usage of algorithms in Table 10.2. Table 10.3 summarizes AI/ML penetration by the size of the manufacturer and the industry (only eight industries are represented here).

Table 10.1 Breakdown of algorithms by usage in manufacturing.

Algorithm	Adoption Prevalence %
Predictive Maintenance	20–25%
Robotics	15–20%
Supply Chain Optimization	10–15%
Process Optimization	8–12%
Quality Control	8–12%
Demand Forecasting	8–12%
Predictive Analytics	5–10%
Automation	5–10%
Energy Optimization	3–7%
Computer Vision	3–7%
Natural Language Processing	2–5%
Ensemble Learning	1–3%
Time Series Forecasting	1–3%
Anomaly Detection	1–3%
Clustering Algorithms	1–3%
Reinforcement Learning	1–2%
Dimensionality Reduction	1–2%
Bayesian Methods	1–2%

Table 10.2 AI/ML algorithms used by the top 15 global manufacturers.

Company Name	Industry	AI/ML Algorithms Used	Success Metrics	Negative Effects of AI/ML Algorithms
Volkswagen AG	Automotive	Predictive maintenance, quality control	Reduced downtime, Improved defect detection accuracy	Increased consumer privacy concerns due to data collection and usage

Continued

Table 10.1 *continued*

Company Name	Industry	AI/ML Algorithms Used	Success Metrics	Negative Effects of AI/ML Algorithms
Toyota Motor Corporation	Automotive	Predictive maintenance, robotics, supply chain optimization	Decreased maintenance costs, Increased production efficiency	Increased vulnerability to cyberattacks due to interconnected systems
Samsung Electronics	Electronics	Robotics, predictive maintenance, process optimization	Reduced defects, Increased equipment uptime	Product recalls due to AI-related errors
Foxconn Technology Group	Electronics	Robotics, process optimization	Increased production throughput, Reduced manufacturing errors	Worker exploitation concerns
Ford Motor Company	Automotive	Robotics, predictive analytics	Increased production throughput, Reduced manufacturing errors	Resistance from labor unions due to AI-driven automation
Honda Motor Co., Ltd.	Automotive	Predictive maintenance, quality control	Decreased downtime, Enhanced defect detection accuracy	Product quality issues due to inaccurate algorithms
General Motors Company	Automotive	Predictive analytics, robotics	Increased production, Reduced maintenance costs	Potential job displacement and workforce resistance
SAIC Motor Corporation Limited	Automotive	Predictive analytics, robotics	Increased production efficiency, Reduced manufacturing errors	Consumer skepticism toward AI-driven products
Fiat Chrysler Automobiles N.V.	Automotive	Predictive maintenance, quality control	Increased equipment uptime, Enhanced product quality	Negative impact on brand reputation due to AI-related incidents
Nestlé SA	Food and Beverages	Supply chain optimization, demand forecasting	Improved demand forecasting accuracy, Reduced logistics costs	Increased risk of market manipulation due to predictive analytics
Hyundai Motor Company	Automotive	Predictive maintenance, quality control	Increased equipment uptime, Enhanced product quality	Consumer backlash over AI-driven recalls
Hitachi, Ltd.	Industrial Goods	Predictive maintenance, process optimization	Decreased maintenance costs, Increased production throughput	Negative impact on employee morale due to job automation
Airbus SE	Aerospace	Predictive maintenance, process optimization	Reduced maintenance downtime, Improved production throughput	Safety concerns over autonomous systems
Bosch GmbH	Industrial Goods	Predictive maintenance, quality control	Enhanced defect detection accuracy, Reduced maintenance costs	Potential bias in decision-making algorithms
Procter & Gamble Company	Consumer Goods	Demand forecasting, supply chain optimization	Improved inventory turnover, Reduced stockouts	Negative environmental impact due to increased production

Table 10.3 Penetration of AI/ML practices by sector and size of manufacturers.

Sector	Small (1–50 employees)	Medium (51–500 employees)	Large (501+ employees)
Automotive	5–15%	10–20%	15%–25%
Electronics	3–12%	7–17%	12–22%
Consumer Goods	4–14%	8–18%	13–23%
Aerospace	2–10%	5–15%	10–20%
Pharmaceutical	2–9%	5–14%	10–19%
Chemical	2–9%	5–14%	10–19%
Food and Beverages	2–8%	4–13%	9–18%
Industrial Goods	3–11%	6–16%	12–22%

By leveraging AI's predictive capabilities, manufacturers can anticipate changes in market demand, enabling them to adjust production schedules and inventory levels accordingly. This agility is crucial in today's fast-paced market, allowing companies to respond swiftly to shifting trends and maintain a competitive edge. As these technologies continue to evolve and become more accessible, manufacturers that embrace them will gain a significant advantage in terms of efficiency, sustainability, and competitiveness. By focusing on the responsible and ethical adoption of AI, manufacturers can unlock the transformative potential of these technologies while maintaining trust and integrity with their customers and stakeholders when explainability is embedded in the solutions. Manufacturers can position themselves at the forefront of industry advancements and remain competitive in an ever-evolving landscape by staying abreast of emerging trends and adopting advanced technologies.

BROADER IMPLICATIONS

In the broader context, these technologies provide manufacturers with the ability to harness data effectively to enhance various aspects of operations, including production planning, QC, and customer service. These technologies enable manufacturers to develop more agile and adaptive processes that respond quickly to market changes and customer demands. However, these technologies have implications beyond business value for manufacturing. They impact society and industry in terms of ethical considerations, regulatory challenges, sustainability issues, and future innovation and transformation. These technologies play a vital role in shaping the industry and guiding it toward a more efficient and responsible future.

Ethical Considerations and Regulatory Challenges

The integration of advanced technologies in manufacturing presents several ethical challenges. These include ensuring transparency in data usage, safeguarding customer and business data, and avoiding biases in AI and ML models. Ethical AI development requires fairness, accountability, and transparency in decision making. For instance, biased algorithms in customer segmentation or hiring practices can perpetuate social inequities. Manufacturers must adhere to data protection laws like the General Data Protection Regulation to ensure compliance and ethical use of data. These regulations will continue to see additions, refinements, and changes. For example, on March 13, 2024, the EU Parliament adopted the Artificial Intelligence Act.[2] The United States is also looking to establish its own AI act, as are other countries. This regulatory challenge is a moving target, and manufacturers must follow these carefully as they embark on their AI journey.

Sustainability Issues

AI and ML technologies offer opportunities to improve sustainability in manufacturing by optimizing resource use, reducing waste, and minimizing energy consumption. For example, PdM can help extend the lifespan of equipment and reduce the need for new manufacturing, thus conserving resources and energy. Additionally, AI-driven supply chain optimization can reduce transportation emissions and streamline production processes. Further, energy optimization can help manufacturers reduce their energy consumption and greenhouse gas emissions and positively impact sustainability. Companies must conduct a critical analysis regarding the use of generative AI since it consumes a significant amount of energy.

Innovation and Transformation

Advanced technologies drive innovation and transformation in manufacturing by enabling automation, smart factories, and PdM. Automation can lead to increased efficiency and productivity, while smart factories integrate IoT devices and data-driven insights for real-time monitoring and control. PdM leverages AI and ML to anticipate equipment failures, allowing for timely repairs and minimizing downtime. Industry 4.0 mandates manufacturers to stay on top of innovation and digital transformation to remain competitive in these times.

The Role of Automation and Smart Factories

The rise of automation and smart factories is one of the most transformative aspects of advanced technologies in manufacturing. Automation helps manufacturers improve efficiency, consistency, and safety in production. Smart factories leverage IoT sensors and AI-driven analytics to create connected, intelligent environments that optimize production and facilitate better decision making. By integrating advanced analytics, BI, and ML models, manufacturers can gain real-time insights into their operations, enabling them to optimize production schedules and reduce downtime.

> *"The strength of the team is each individual member. The strength of each member is the team."*
>
> —Phil Jackson, professional basketball coach

Predictive Maintenance and Its Impact

PdM is another area where AI and ML technologies have a significant impact. Currently, it is the most widely used AI and ML technology within manufacturing (see Table 10.1). By analyzing historical and real-time data, manufacturers can predict potential equipment failures and schedule maintenance proactively. This approach reduces unplanned downtime and maintenance costs, leading to more efficient and reliable operations. However, manufacturers that have adopted this must continue to seek improvements in their technology to improve the performance of their PdM solutions.

Challenges and Opportunities

Earlier chapters have shown that while the potential benefits of advanced technologies in manufacturing are vast, manufacturers must navigate several challenges to realize them fully. These include the need for substantial investments in infrastructure, data management and governance, and employee training. Additionally, addressing ethical considerations and regulatory compliance is essential to build trust with stakeholders and customers. Other challenges manufacturers encounter include managing workforce transformations as automation and AI-driven processes reshape traditional manufacturing roles. Manufacturers must also navigate ethical and regulatory considerations as technology continues to evolve. Finally, it is crucial to include rank-and-file employees to ensure success in digital transformations. Three reasons why AI can alienate employees are: (1) lack of autonomy, (2) lack of understanding, and (3) proliferation of business silos created by AI.[3]

Future Outlook

Despite these challenges, the opportunities presented by advanced analytics, BI, AI, and ML technologies in manufacturing are immense. Manufacturers that embrace these technologies and address challenges effectively can expect improvements in operational efficiency, cost savings, and customer satisfaction. The future of manufacturing is promising due to the continued adoption of advanced technologies. Manufacturers must remain agile and adaptable to emerging trends—such as digital twins, advanced robotics, generative AI, and blockchain for supply chain management. These technologies offer increased productivity, cost savings, and improved customer experiences.

The following quote captures the future outlook for AI and ML: "The integration of AI and ML in manufacturing is not just a trend; it's a necessity for companies aiming to stay competitive in a rapidly changing industry." Manufacturers that successfully adopt advanced technologies and address the associated challenges will be well-positioned to thrive in the manufacturing landscape of the future.

INSPIRING ACTION

To encourage readers to take action and embed advanced analytics, BI, AI, and ML technologies into their manufacturing operations, it is essential to highlight the potential benefits these advanced technologies can offer. By integrating intelligent automation and data-driven decision-making processes, manufacturers can achieve improved productivity, cost savings, and enhanced competitiveness in the industry. For example, by leveraging AI-driven models for PdM, manufacturers can minimize unplanned downtime and maintenance costs. They can also change the mindset of *time-based* maintenance to *just-in-time* maintenance. This will eliminate unnecessary maintenance with its related downtime and costs. These technologies also optimize production schedules and inventory management, leading to more efficient resource utilization. Another key benefit is the ability to personalize customer experiences through data-driven insights. For instance, ML algorithms can analyze customer data to

> *"The only way to win is to learn faster than anyone else."*
>
> —Eric Ries, entrepreneur, author, and creator of the Lean Startup methodology

identify patterns and preferences, allowing manufacturers to tailor marketing campaigns and after-sales services accordingly.

An example of a large global company impacted by the cost of sleeping on AI is Samsung Electronics. It is facing significant challenges in the AI sector, which have contributed to a staggering $122 billion loss in market value.[4] Once seen as poised to benefit from the global AI boom, the company is now struggling to keep up with rivals like SK Hynix and Taiwan Semiconductor Manufacturing Company (TSMC) in AI memory and outsourced chipmaking. Despite an initial surge in profits and a stock price nearing record highs in 2024, Samsung's recent admission of delays in its latest high-bandwidth memory chips has dampened investor optimism. As smaller competitors gain ground and demand for AI-related technologies grows, Samsung's inability to maintain its technological leadership has led to a loss of confidence among investors, prompting many to sell their shares. This situation highlights the critical role of technological leadership in the semiconductor industry, where competitors like Nvidia and TSMC are reaping substantial gains while Samsung grapples with potential management changes and the challenge of regaining its lost position in the rapidly evolving AI landscape. It further emphasizes AI's role in manufacturing, the importance of innovation, and timely execution in the highly competitive AI landscape, where the ability to adapt can determine a company's success or failure.

To begin implementing AI and ML in manufacturing processes, readers should follow these practical steps. Begin by evaluating the current state of data management and governance, analytics, and automation capabilities within the organization. Identify areas where AI and ML technologies can be integrated for maximum impact. Look at what other companies have done (e.g., Table 10.2) and what algorithms (from Table 10.1) might be relevant to your use cases. Then, clear objectives can be set by defining specific, measurable goals for AI and ML implementation. This could include improving production efficiency, reducing waste, or enhancing customer service. Remember to start small with a pilot project in a specific area of manufacturing operations and scale gradually. During the pilot phase, continuously monitor data and model performance. Make necessary model adjustments to improve results. Do not forget to make adjustments to data management and employee training. This includes regularly reviewing data quality, model accuracy, operational outcomes, and employee performance concerning new processes and technologies. Once the pilot proves successful, gradually scale up to other parts of the organization.

Readers must identify and select appropriate AI and ML tools that align with the organization's technology stack, goals, and requirements. Partnering with technology vendors or system integrators can provide additional expertise and support during implementation. As tools and partners are selected, ensure a well-defined working process for tool and partner selection. Readers must also invest in training and talent by equipping employees with the necessary skills and knowledge to work with advanced analytics and AI technologies. This may involve hiring new talent or offering training programs to existing staff. Also, to increase the likelihood of success with AI, involve all employees and implement strong change management.

CALL TO ACTION

The transformative potential of advanced analytics, BI, AI, and ML technologies in the manufacturing sector is undeniable. These technologies offer unprecedented opportunities to optimize operations, improve productivity, enhance customer experiences, and drive revenue growth—all while overcoming challenges and building ethical and trustworthy systems by using algorithms that are explainable with a labor force that is adequately trained. Embracing AI and ML is not just an opportunity but a necessity for manufacturers seeking to excel in a rapidly evolving landscape.

By committing to the adoption and implementation of these technologies now, leaders can catalyze unprecedented efficiency, innovation, and profitability in their organizations. Hence, this is the moment to take bold, strategic action and lead the way into the future of manufacturing. It is crucial for leaders and practitioners to take decisive action and leverage the insights provided in this book to drive transformation within their organizations.

Collaboration with experts in AI and ML, whether academic, industry, or government, is essential to ensure successful implementation and to maximize the benefits of these technologies. Engaging with industry leaders, technology vendors, and consultants can provide valuable support and guidance throughout the transformation journey.

For further learning and development, readers are encouraged to explore resources such as industry associations, research papers, and online courses. There are many wonderful free and paid online courses on most of the algorithms and technologies covered here. These resources offer opportunities to stay updated with the latest advancements and best practices in AI and ML for manufacturing. Remember, by embracing advanced analytics, BI, AI, and ML technologies, manufacturers can position themselves at the forefront of the industry's digital transformation. This will not only lead to improved outcomes and enhanced competitiveness but also contribute to a sustainable and prosperous future for the sector, the country, and the global economy.

CHAPTER REFERENCES

CHAPTER 1

1. *Industries at a Glance, Goods-Producing Industries*, https://www.bls.gov/iag/tgs/iag06.htm.
2. *Standard Industrial Classification (SIC) Manual, Division D: Manufacturing*, https://www.osha.gov/data/sic-manual.
3. *National Institute of Standards and Technology*. (2018), "Manufacturing Extension Partnership (MEP)—What Is Manufacturing?" https://www.nist.gov/manufacturing.
4. *United Nations Industrial Development Organization*, "Report on the World Manufacturing Production," Quarterly report, Quarter 3, 2022, https://www.unido.org/resources-statistics/quarterly-report-manufacturing.
5. *World Economic Forum*, "Advanced Manufacturing Industry," 2024, https://www.weforum.org/communities/advanced-manufacturing/.
6. Toner, P. and Stilwell, F., *Why does Australia need manufacturing industry?* https://search.informit.org/doi/pdf/10.3316/informit.291852359232918.
7. http://www.engineeringnews.co.za/article/manufacturing-sector-outlook-and-challenges-2013-08-06.
8. Share of value added to the gross domestic product of the United States in 2023, by industry, Nov. 2023, https://www.statista.com/statistics/248004/percentage-added-to-the-us-gdp-by-industry/.
9. https://www.nist.gov/el/applied-economics-office/manufacturing/manufacturing-economy/total-us-manufacturing.
10. Business R&D Performance in the United States Tops $600 Billion in 2021, https://ncses.nsf.gov/pubs/nsf23350#:~:text=In%202021%2C%20companies%20in%20manufacturing.
11. Northam, J., *As Overseas Costs Rise, More U.S. Companies are 'Reshoring'*, Jan. 27, 2014, http://www.npr.org/blogs/parallels/2014/01/22/265080779/as-overseas-costs-rise-more-u-s-companies-are-reshoring.
12. Gloeckler, M., *Inside Walmart's $50 Billion Buy American Campaign*, Mar. 2013, https://www.feedbacksystems.com/inside-walmarts-50-billion-buy-american-campaign/.
13. Szal, A., *Reshoring In-Depth: Ford, Caterpillar, GE, GM Among Top Reshoring Companies*, June 2015, https://www.manufacturing.net/supply-chain/news/13099269/reshoring-indepth-ford-caterpillar-ge-gm-among-top-reshoring-companies.
14. *The Multiplier Effect: Which Industries Are the Biggest Job Creators?* Nov. 2021, https://camoinassociates.com/resources/the-multiplier-effect-which-industries-are-the-biggest-job-creators.
15. *Report to the President on Ensuring American Leadership in Advanced Manufacturing*, June 2011, https://obamawhitehouse.archives.gov/sites/default/files/microsites/ostp/pcast-advanced-manufacturing-june2011.pdf.

16. https://www.nist.gov/mep/advanced-manufacturing-technology-servicesindustry-40.
17. Lu, Y., *Industry 4.0: A survey of technologies, applications and open research issues,* Journal of Industrial Information Integration, Vol. 6, June 2017, pp. 1–10, https://www.sciencedirect.com/science/article/abs/pii/S2452414X17300043.
18. *21 Century's Grand Engineering Challenges Unveiled*, Feb. 2008, http://www8.nationalacademies.org/onpinews/newsitem.aspx?RecordID=02152008.
19. National Academy of Engineering Summit for the Grand Challenges in Manufacturing, Research Triangle Park, North Carolina, Oct. 31, 2013–Nov. 1, 2013, http://summit-grand-challenges.pratt.duke.edu/research-triangle-park-2013.html.
20. Lee, J., Bagheri, B., and Kao, H. A., *A Cyber-Physical Systems architecture for Industry 4.0-based manufacturing systems,* Manufacturing Letters, https://www.sciencedirect.com/science/article/abs/pii/S221384631400025X.
21. Oks, S., et al., *Cyber-physical systems in the context of industry 4.0: A review, categorization and outlook, Production: Status Review and Trends*, April 2022, https://link.springer.com/article/10.1007/s10796-022-10252-x.
22. Soori, M., Arezoo, B., and Dastres, R., *Internet of things for smart factories in industry 4.0, a review,* Internet of Things and Cyber-Physical Systems, Vol. 3, 2023, pp 192–204, https://www.sciencedirect.com/science/article/pii/S2667345223000275.

CHAPTER 2

1. https://knowthechain.org.
2. Van der Bosch, T., Schneider, K., and Hemelrijk, C., *The Power of Blockchain in Revolutionising the Food Industry*, https://www2.deloitte.com/nl/nl/pages/risk/articles/the-power-of-blockchain-in-revolutionising-the-food-industry.html.
3. *4D Shoes and Sneakers*, https://www.adidas.com/us/4d-shoes.
4. https://emerj.com/ai-sector-overviews/artificial-intelligence-at-fedex/.
5. https://www.businessinsider.com/ups-using-predictive-analytics-algorithm-2018-7.
6. https://www.ikea.com/global/en/our-business/people-planet/zero-emissions-for-home-deliveries/.
7. https://www.dbschenker.com/ug-en/insights/news-and-stories/press-releases/first-series-production-eactros-1411158.
8. https://stories.starbucks.com/press/2018/starbucks-and-alibaba-announce-partnership-to-transform-coffee-experience/.
9. https://thebigmarketing.com/emerson-electric-marketing-strategy/.
10. https://www.aviso.com/case-study/honeywell.
11. https://news.lockheedmartin.com/2024-02-12-Lockheed-Martin-and-Red-6-Announce-Augmented-Reality-Integration-Progress-for-TF-50.
12. https://www.siemens.com/global/en/company/stories/research-technologies/digitaltwin/virtual-new-normal.html.
13. https://learn.microsoft.com/en-us/hololens/hololens2-options-remote-assist.
14. https://www.youtube.com/watch?v=5V3BP-cf44Y.
15. https://xerox.com/perl-bin/product.pl?mode=recycling.
16. OpenAI, *Robots and Humans Collaborating on a Manufacturing Floor*, Generated using AI by ChatGPT and DALL·E, OpenAI, 2025.

17. Derived from https://www.automate.org/robotics/blogs/cobots-empowering-humans-in-manufacturing.
18. Derived from https://www.bmwgroup.com/en/news/general/2019/open-manufacturing-platform.htmlgh-level.
19. https://www.youtube.com/watch?v=5V3BP-cf44Y—see section II.F.
20. Based on https://fieldserviceusa.wbresearch.com/ge-digital-iiot-technology-strategy-enable-predictive-maintenance-help-oems-ty-u.
21. https://www-lightreading-com.webpkgcache.com/doc/-/s/www.lightreading.com/video-broadcast/boeing-productive-vr-cuts-training-time-by-75-.
22. *Ericsson USA 5G Smart Factory*, https://www.ericsson.com/en/about-us/company-facts/ericsson-worldwide/united-states/5g-smart-factory#:~:text=Smart%20Manufacturing,real%20time%20data%20across%20operations.
23. *How are manufacturers gaining a competitive edge with private 5G networks?* https://www.ericsson.com/en/industries/manufacturing.
24. *Ericsson USA 5G Smart Factory*, https://www.ericsson.com/en/about-us/company-facts/ericsson-worldwide/united-states/5g-smart-factory#:~:text=Smart%20Manufacturing,real%20time%20data%20across%20operations.
25. https://www.we.org/en-US/our-work/we-schools/u-n-sustainable-development-goals.
26. https://www.microsoft.com/en-us/ai/principles-and-approach.
27. https://ai.google/responsibility/principles/#:~:text=AI%20applications%20we%20will%20not%20pursue&text=Technologies%20that%20cause%20or%20are,will%20incorporate%20appropriate%20safety%20constraints.
28. https://initiatives.weforum.org/global-lighthouse-network/home.
29. Based on https://products.xpair.com/innovation/plateforme-batiment-ecostruxure-building.htm.
30. Derived from https://www.capgemini.com/wp-content/uploads/2017/11/brilliant-factories-digital-to-the-core.pdf.
31. Womack, J. P., Jones, D. T., and Roos, D., *The Machine That Changed the World: The Story of Lean Production*, 2007, Free Press.

CHAPTER 3

1. *How Bosch uses AI in manufacturing*, https://www.bosch-presse.de/pressportal/de/en/how-bosch-uses-ai-in-manufacturing-260800.html.
2. *Artificial Intelligence at Procter & Gamble*, https://emerj.com/ai-sector-overviews/artificial-intelligence-at-procter-gamble/.
3. *Ford Motor Company: Supply Chain Strategy Case Study*, https://egor-muretov.mystrikingly.com/blog/ford-motor-company-supply-chain-strategy-case-study.
4. Ray, S., *Samsung Bans ChatGPT Among Employees After Sensitive Code Leak*, https://www.forbes.com/sites/siladityaray/2023/05/02/samsung-bans-chatgpt-and-other-chatbots-for-employees-after-sensitive-code-leak/?sh=367a3ca60781.
5. *Predictive maintenance market: 5 highlights for 2024 and beyond*, https://iot-analytics.com/predictive-maintenance-market/.
6. Kennedy, S., *6 case studies illuminate the value of predictive and prescriptive maintenance*, 2020, https://www.plantservices.com/predictive-maintenance/predictive-maintenance/article/11290707/6-case-studies-illuminate-the-value-of-predictive-and-prescriptive-maintenance.

7. Fuller, H., Tohani, M. and *Alcoa Improves Operational Efficiency and Asset Reliability with Predictive Maintenance*, 2020, https://www.verdantix.com/report/alcoa-improves-operational-efficiency-and-asset-reliability-with-predictive-maintenance.
8. *Optimize your machine tools with Rotating Equipment Performance from SKF*, https://www.skf.com/us/industries/machine-tools/products-and-services/rotating-equipment-performance.
9. *Smart maintenance using artificial intelligence*, 2023, https://www.press.bmwgroup.com/global/article/detail/T0438145EN/smart-maintenance-using-artificial-intelligence?language=en.
10. Thapar, V., *GE Brings AI into Preventive Maintenance to Reduce Jet Engine Failure by One-Third*, Sept. 2019, https://www.sps-aviation.com/story/?id=2646&h=GE-brings-AI-into-preventive-maintenance-to-reduce-jet-engine-failure-by-one-third.
11. *The rise of predictive maintenance is upon us*, 2017, https://venturebeat.com/ai/the-rise-of-predictive-maintenance-is-upon-us/.
12. *Preventing unnecessary elevator maintenance and repair*, 2024, https://www.schindler.com/en/media/news-press-releases/preventing-unnecessary-elevator-maintenance-and-repair.html.
13. *How Railroads Prevent Derailments: Railroad Safety Technology*, 2023, https://www.up.com/customers/track-record/tr032823-train-derailment-safety-technology.htm.
14. *Case Study 16: Nike's 100 Million Dollar Supply Chain "Speed bump,"* https://www.henricodolfing.com/2022/10/case-study-nike-i2-supply-chain-management.html.
15. *Artificial Intelligence at Procter & Gamble*, https://emerj.com/ai-sector-overviews/artificial-intelligence-at-procter-gamble/.
16. *OpenBlue Digital Twin: Designed for Buildings. Infused with AI*, Johnson Controls, https://www.johnsoncontrols.com/openblue/openblue-digital-twin.
17. Ammanath, B., *Trustworthy AI: A Business Guide for Navigating Trust and Ethics in AI*, 2022, John Wiley and Sons, NJ.
18. *IBM SPSS Modeler CRISP-DM Guide*, https://www.ibm.com/docs/it/SS3RA7_18.3.0/pdf/ModelerCRISPDM.pdf.
19. Kotter, J., *Leading Change: Why Transformation Efforts Fail*, 1995, Harvard Business Review, https://hbr.org/1995/05/leading-change-why-transformation-efforts-fail-2.
20. Davenport, T. and Harris, J., *Competing on Analytics: The New Science of Winning*, 2010, Harvard Business Press.
21. *Agencies warn of accelerating attacks on health care by Black Basta ransomware group*, https://www.aha.org/news/headline/2024-05-10-agencies-warn-accelerating-attacks-health-care-black-basta-ransomware-group.
22. Easterly, J., *The Attack on Colonial Pipeline: What We've Learned & What We've Done Over the Past Two Years*, https://www.cisa.gov/news-events/news/attack-colonial-pipeline-what-weve-learned-what-weve-done-over-past-two-years.
23. *AI-Powered Deception to Protect Against Adversarial AI*, https://acalvio.com.
24. *Zero Trust Security—Code to Cloud AppSec + CloudSec*, https://accuknox.com.
25. *Cyber Security and Industrial Automation Systems*, https://www.nrc.gov/docs/ML1319/ML13198A409.pdf.
26. *Edge-Native Kuberbetes for Software-Defined Edge*, https://www.intel.com/content/www/us/en/edge-computing/smart-edge.html.
27. *Industry 4.0: At the Intersection of Readiness and Responsibility*, 2020, Deloitte, https://www2.deloitte.com/ch/en/pages/risk/articles/industry-4-0-intersection-of-readiness-and-responsibility.html.

CHAPTER 4

1. *P&G Levels Up Manufacturing with Machine Learning and AI*, 2022, https://consumergoods.com/pg-levels-manufacturing-machine-learning-and-aigenuer.
2. Genuer, R., Poggi, J. M., and Tuleau-Malot, C., *Variable selection using random forests*, March 2010, Pattern Recognition Letters, Elsevier.
3. Liaw, A. and Wiener, M., *Classification and Regression by Random Forest*, Dec. 2002, R news, https://journal.r-project.org/articles/RN-2002-022/RN-2002-022.pdf.
4. *Overview of clustering methods*, https://scikit-learn.org/stable/modules/clustering.html.
5. *Real-time time-series anomaly detection*, https://github.com/intel-iot-devkit/predictive-maintenance-python/blob/master/README.md.
6. *Hierarchical Cluster Analysis*, https://www.sciencedirect.com/topics/computer-science/hierarchical-cluster-analysis.
7. Goran, Svensson, *Supplier segmentation in the automotive industry: A dyadic approach of a managerial model*, 2004, International Journal of Physical Distribution & Logistics Management, https://www.emerald.com/insight/publication/issn/0960-0035/vol/34/iss/1.
8. Heidarzade, A., Mahdavi, I., and Mahdavi-Amiri, N., *Supplier selection using a clustering method based on a new distance for interval type-2 fuzzy sets: A case study*, 2016, https://www.sciencedirect.com/science/article/abs/pii/S1568494615006018.
9. Kuo, C., *Handbook of Anomaly Detection with Python Outlier Detection—(4) Isolation Forest*, https://towardsdatascience.com/use-the-isolated-forest-with-pyod-3818eea68f08.
10. Phung V. H. and Rhee, E. J., *A High-Accuracy Model Average Ensemble of Convolutional Neural Networks for Classification of Cloud Image Patches on Small Datasets*, https://www.researchgate.net/figure/Schematic-diagram-of-a-basic-convolutional-neural-network-CNN-architecture-26_fig1_336805909.
11. Malekzadeh, T., et al., *Aircraft fuselage defect detection using deep neural networks*, 2017, https://arxiv.org/pdf/1712.09213.
12. Aggarwal, S., *The Ultimate Guide to Building Your Own LSTM Models*, https://www.projectpro.io/article/lstm-model/832.
13. Aggarwal, S., *The Ultimate Guide to Building Your Own LSTM Models*, https://www.projectpro.io/article/lstm-model/832.
14. Radhakrishnan, S., et al., *Creative Intelligence—Automating Car Design Studio with Generative Adversarial Networks (GAN)*, 2019, https://inria.hal.science/hal-02060050/document.
15. Revi, K. Remya, Vidya, K. R., and Wilscy, M., *Detection of Deepfake Images Created Using Generative Adversarial Networks: A Review*, https://www.researchgate.net/figure/General-block-diagram-of-Generative-Adversarial-Network-GAN_fig2_348989084.
16. Vaswani, A., et al., *Attention Is All You Need*, https://arxiv.org/abs/1706.03762.
17. Fei, Y. and Tirumalashetty, V., *How AutoML Vision is helping companies create visual inspection solutions for manufacturing*, https://cloud.google.com/blog/products/ai-machine-learning/ai-and-machine-learning-improve-manufacturing-visual-inspection-process.

CHAPTER 5

1. Bogaisky, J., *GE Says It's Leveraging Artificial Intelligence to Cut Product Design Times in Half*, 2019, Forbes, https://www.forbes.com/sites/jeremybogaisky/2019/03/06/general-electric-ge-artificial-intelligence/?sh=56f74250d881.

2. *Manufacturing Faster, Cheaper and Better, Additive Layer Manufacturing*, Rolls Royce, https://www.rolls-royce.com/products-and-services/defence/digital-innovation/additive_layer_manufacturing.aspx.
3. *Apple's Product Development Process—Inside the World's Greatest Design Organization*, 2023, https://www.interaction-design.org/literature/article/apple-s-product-development-process-inside-the-world-s-greatest-design-organization.
4. Newcomb, A., *Samsung Finally Explains the Galaxy Note 7 Exploding Battery Mess*, 2017, https://www.nbcnews.com/tech/tech-news/samsung-finally-explains-galaxy-note-7-exploding-battery-mess-n710581.
5. *Tide Reinvents Clean on Journey to Decarbonize Laundry with Efforts to Turn Consumers to Cold, Explore Carbon Capture and Reduce Virgin Plastic*, 2021, P&G, https://news.pg.com/news-releases/news-details/2021/Tide-Reinvents-Clean-on-Journey-to-Decarbonize-Laundry-with-Efforts-to-Turn-Consumers-to-Cold-Explore-Carbon-Capture-and-Reduce-Virgin-Plastic/default.aspx.

CHAPTER 6

1. *Steel Manufacturer Reduces Costs and Adds Value through BI and Data Analytics*, https://www.onebridge.tech/case-study/steel-manufacturer-reduces-costs-and-adds-value-through-bi-data-analytics.
2. *Optimizing electronics manufacturing with artificial intelligence and machine learning*, 2023, https://flex.com/resources/optimizing-electronics-manufacturing-with-artificial-intelligence-and-machine-learning.
3. *An Empirical Analysis of the Effect of Supply Chain Disruptions on Long-run Stock Price Performance and Equity Risk of the Firm*, 2009, https://onlinelibrary.wiley.com/doi/abs/10.1111/j.1937-5956.2005.tb00008.x.
4. *Conflict Minerals—Division for Counter Threat Finance and Sanctions*, https://www.state.gov/conflict-minerals/.
5. *Conflict Minerals Regulation: The regulation explained*, https://policy.trade.ec.europa.eu/development-and-sustainability/conflict-minerals-regulation/regulation-explained_en.
6. *Conflict Minerals: 2022 Company Report on Minerals Sources Were Similar to Those Filed in Prior Years*, https://www.gao.gov/products/gao-23-106295.
7. *Demand the Supply: Ranking Consumer Electronics and Jewelry Retail Companies on their Efforts to Develop Conflict-Free Minerals Supply Chains from Congo*, https://enoughproject.org/reports/demand-the-supply.
8. *Looming wave of iron-ore supply threatens $100 price floor*, 2024, https://www.miningweekly.com/article/looming-wave-of-iron-ore-supply-threatens-100-price-floor-2024-07-10.
9. *China's Use of Unofficial Trade Barriers in the U.S.-China Trade War*, 2024, https://sccei.fsi.stanford.edu/china-briefs/chinas-use-unofficial-trade-barriers-us-china-trade-war.
10. *Cost of Data Breach*, https://www.ibm.com/reports/data-breach.
11. *Cost of a Breach by Industry*, https://www.tessian.com/resources/cost-of-a-data-breach-infographic/.
12. *2023 Sustainability Report*, https://s203.q4cdn.com/676879669/files/doc_financials/esg/2023-terex-esg-report.pdf.

13. *2023 Annual Report*, https://s203.q4cdn.com/676879669/files/doc_financials/2023/ar/terex-corp_annual_report_final-2023.pdf.
14. Bradshaw, T. and Srivastava, M., *Ransomware attack on chip supplier causes delays for semiconductor groups*, 2023, https://www.ft.com/content/b8669140-8dde-493e-bb30-f5f1e9830804.
15. *8 Ways How AI Is Changing Contract Management*, https://cloudely.com/artificial-intelligence-and-contract-management/.
16. Sharma, A., *The Dawn of AI in Procurement: Unleashing a New Era of Efficiency and Insight*, 2024, https://www.scmr.com/article/ai-procurement-efficiency-insight#:~:text=40%25%20faster%20transactions%2C%2030%25,are%20driving%20AI%2Dpowered%20procurement&text=In%20the%20fast%2Dpaced%20world,entrance%20into%20the%20procurement%20arena.
17. Dida, V., Mori, L., Noterdaeme, O., and Schmitz, C., *Manufacturing: Analytics unleashes productivity and profitability*, 2017, https://www.mckinsey.com/capabilities/operations/our-insights/manufacturing-analytics-unleashes-productivity-and-profitability.
18. Pollock, D., *Nestlé Expands Use of IBM Food Trust Blockchain to Its Zoégas Coffee Brand*, 2020, https://www.forbes.com/sites/darrynpollock/2020/04/15/nestl-expands-use-of-ibm-food-trust-blockchain-to-its-zogas-coffee-brand/.
19. Williams, L., *AutoVentive & Ford: Transforming Automotive Supply Chain Management*, 2024, https://autoventive.com/av-and-ford-transforming-automotive-supply-chain/.

CHAPTER 7

1. *International Organization of Motor Vehicle Manufacturers: 2023 Production Statistics*, https://www.oica.net/category/production-statistics/2023-statistics/.
2. *The Year Ahead: Mapping the Next Wave of Supply Chain Pressures*, 2023, https://www.businessoffashion.com/articles/retail/the-state-of-fashion-2024-report-bullwhip-effect-supply-chain-volatility-manufacturing-consumer-demand/#:~:text=A%20McKinsey%20survey%20of%20chief,over%20the%20next%20five%20years.
3. Siddharth, Patil, *Predictive maintenance and the smart factory: Connecting machines to reliability professionals*, 2022, https://www2.deloitte.com/us/en/pages/operations/articles/predictive-maintenance-and-the-smart-factory.html.
4. *Transforming advanced manufacturing through Industry 4.0*, 2022, https://www.mckinsey.com/capabilities/operations/our-insights/transforming-advanced-manufacturing-through-industry-4-0.
5. *Addressing the challenges with demand forecasting solutions on AWS*, https://docs.aws.amazon.com/whitepapers/latest/demand-forecasting/addressing-the-challenges-with-demand-forecasting-solutions-on-aws.html.
6. *Identify bad actors and prevent unplanned downtime with comprehensive reliability analysis software*, 2018, https://www.ge.com/digital/applications/asset-performance-management/apm-reliability?utm_source=google_g&utm_medium=paid-search&utm_campaign=PGOG-PG&OG-APM_Reliability-GLOB-Search&utm_content=predictive-preventive&utm_term=predictive%20maintenance%20asset%20management_b&ad=653365695046&placement=&device=c&location=1021225&gad_source=1&gclid=Cj0KCQjw-uK0BhC0ARIsANQtgGN6TLfyUHAyuUXAZBQj6tW0o1iTNh3paCgeiYcBv9oHuqv3kXibAuIaAjfAEALw_wcB.

7. *Chemical industry worldwide—statistics & facts*, https://www.statista.com/topics/6213/chemical-industry-worldwide/#topicOverview.
8. *Revenue of the worldwide pharmaceutical market from 2001 to 2023*, Jan. 2024, https://www.statista.com/statistics/263102/pharmaceutical-market-worldwide-revenue-since-2001/#:~:text=The%20global%20pharmaceutical%20market%20has,around%201.6%20trillion%20U.S.%20dollars.
9. *Digitally enabled reliability: Beyond predictive maintenance*, 2018, https://www.mckinsey.com/capabilities/operations/our-insights/digitally-enabled-reliability-beyond-predictive-maintenance.
10. *Manufacturing: Analytics unleashes productivity and profitability*, 2017, https://www.mckinsey.com/capabilities/operations/our-insights/manufacturing-analytics-unleashes-productivity-and-profitability.
11. Hu, M., *Challenges and Best Practices in Industrial Artificial Intelligence (AI) applications*, 2020, https://www.ge.com/digital/blog/challenges-and-best-practices-industrial-artificial-intelligence-ai-applications.
12. Ammanath, B., *Trustworthy AI: A Business Guide for Navigating Trust and Ethics in AI*, 2022, Wiley.
13. *Rhebo—OT Monitoring Sensor*, 2023, https://developer.community.boschrexroth.com/t5/Store-and-How-to/Rhebo-OT-Monitoring-Sensor/ba-p/18297.

CHAPTER 8

1. *AI-Enhanced Last-Mile: E-Commerce Shipping's Future*, Jun. 2024, (accessed July 12, 2024), https://www.dhl.com/discover/en-sg/ship-with-dhl/start-shipping/ai-enhanced-last-mile-delivery-future-of-ecommerce#:~:text=Using%20AI%2Dpowered%20logistics%20software,like%20volume%20and%20service%20requirements.
2. *UPS to Enhance ORION With Continuous Delivery Route Optimization*, Jan. 2020, (accessed July 12, 2024), https://about.ups.com/us/en/newsroom/press-releases/innovation-driven/ups-to-enhance-orion-with-continuous-delivery-route-optimization.html.
3. *We, Robot: How humans and AI are working together in logistics*, https://www.dhl.com/global-en/delivered/innovation/ai-in-logistics.html#:~:text=We%2C%20Robot%3A%20How%20humans%20and,to%20industries%20rife%20with%20challenges.
4. *UPS to Enhance ORION With Continuous Delivery Route Optimization*, Jan. 2020, (accessed July 12, 2024), https://about.ups.com/us/en/newsroom/press-releases/innovation-driven/ups-to-enhance-orion-with-continuous-delivery-route-optimization.html.
5. Deguchi, H., *P&G Japan to cut delivery trucks 30% using AI demand forecast*, https://asia.nikkei.com/Business/Transportation/P-G-Japan-to-cut-delivery-trucks-30-using-AI-demand-forecast#:~:text=TOKYO%20%2D%2D%20P%26G%20Japan%20seeks,demand%20to%20boost%20logistics%20efficiency.
6. *AI-Enhanced Last-Mile: E-Commerce Shipping's Future*, Jun. 2024, (accessed July 12, 2024), https://www.dhl.com/discover/en-sg/ship-with-dhl/start-shipping/ai-enhanced-last-mile-delivery-future-of-ecommerce#:~:text=Using%20AI%2Dpowered%20logistics%20software,like%20volume%20and%20service%20requirements.
7. *Four cool things Ocado does with AI and ML to improve its robotic workforce*, 2023, https://techinformed.com/four-cool-things-ocado-does-with-ai-and-ml-to-improve-its-robotic-workforce/.

8. *The Future of Warehousing: How AI is Transforming Inventory Management and Order Fulfillment*, 2024, https://rtslabs.com/ai-warehouse-automation-picking-optimization.
9. *UPS to Enhance ORION With Continuous Delivery Route Optimization*, Jan. 2020, (accessed July 12, 2024), https://about.ups.com/us/en/newsroom/press-releases/innovation-driven/ups-to-enhance-orion-with-continuous-delivery-route-optimization.html.
10. Davis, J., *Current Rhine River water levels disrupting shipping routes*, (accessed July 11, 2024), https://www.everstream.ai/articles/current-rhine-river-low-water-levels-disrupt-shipping-routes/.
11. *Army Corps of Engineers announces end of low-water drought in Mississippi Valley Division*, (accessed July 11, 2024), https://www.mvd.usace.army.mil/Media/News-Releases/Article/3665123/army-corps-of-engineers-announces-end-of-low-water-drought-in-mississippi-valle/#:~:text=The%20U.S.%20Army%20Corps%20of,the%20region%20in%20September%202022.
12. *DeepMind AI Reduces Google Data Center Cooling Bill by 40%*, 2016, https://deepmind.google/discover/blog/deepmind-ai-reduces-google-data-centre-cooling-bill-by-40/.
13. Sroka, N., *How AI Is Revolutionizing Solid Waste Management*, 2013, https://swana.org/news/blog/swana-post/swana-blog/2023/12/11/how-ai-is-revolutionizing-solid-waste-management.
14. *How AI can solve manufacturing's waste problem*, 2021, https://www.weforum.org/agenda/2021/04/how-ai-can-cut-waste-in-manufacturing/.
15. *Using artificial intelligence to reduce emissions*, https://new.abb.com/control-systems/industry-specific-solutions/oil-gas-and-petrochemicals/using-artificial-intelligence-to-reduce-environmental-impact.
16. Ahramovich, A., *Computer vision in manufacturing: 9 use cases, examples, and best practices*, 2023, https://www.itransition.com/computer-vision/manufacturing.
17. *Logistics Use Cases: People-Health and Safety*, https://www.dhl.com/us-en/microsites/csi/computer-vision/logistics-use-cases/people-health-and-safety.html#:~:text=AI%2Dpowered%20computer%20vision%20systems,properly%2C%20contributing%20to%20workplace%20safety.
18. *Inside Amazon's robot revolution*, https://www.youtube.com/watch?v=ZWonAz7Kczs.
19. Quinlivan, J., *How Amazon deploys collaborative robots in its operations to benefit employees and customers*, https://www.aboutamazon.com/news/operations/how-amazon-deploys-robots-in-its-operations-facilities.

CHAPTER 9

1. Tapia, A., *Unilever—Designing a Responsible, Self-service Tool for Natural Language Processing*, 2021, https://community.dataiku.com/discussion/18716/unilever-designing-a-responsible-self-service-tool-for-natural-language-processing.
2. *Our Future is Digital*, 2017, https://www.ge.com/digital/blog/our-future-digital.
3. *5 Caterpillar Solutions Transforming Customer's Job Sites*, 2019, https://www.caterpillar.com/en/news/caterpillarNews/customer-dealer-product/5-caterpillar-solutions-transforming-customers—job-sites.html.
4. Leopold, G., *Automakers Embrace Predictive Analytics to Boost Sales*, 2014, https://www.datanami.com/2014/08/20/automakers-embrace-predictive-analytics-boost-sales/.

5. *Parker and Camgian Partner to Deliver AI-Connected Services*, 2022, https://www.parker.com/us/en/about-parker/newsroom/news-release-details/parker-and-camgian-partner-to-deliver-ai-connected-services.html.
6. *Artificial Intelligence*, 2022, https://futures.3m.com/machine-learning.
7. *ALS confirms selection of Rolls-Royce TotalCare for engine servicing*, https://www.rolls-royce.com/media/press-releases/2019/20-11-19-rr-als-confirms-selection-of-rolls-royce-totalcare-for-engine-servicing.aspx.
8. *A Deep Dive into the Marketing Strategies of Siemens*, 2020, https://thebrandhopper.com/2024/05/08/a-deep-dive-into-the-marketing-strategies-of-siemens/#google_vignette.
9. *Philips Lighting Singapore: Empowering Conversational Commerce and Driving Sales via Messenger*, 2020, https://aichat.com/philips-lighting-singapore-ecommerce-chatbot/.
10. Sibik, P. and Fielder, Z., *2024: Polarion vs. Responsible AI*, 2024, https://blogs.sw.siemens.com/polarion/2024-polarion-vs-responsible-ai/.
11. *Our AI-based solutions*, https://www.se.com/us/en/work/solutions/artificial-intelligence/solutions.jsp.
12. Wodecki, B., *Exclusive: Schneider Electric Chief AI Officer on Using Custom ChatGPT*, 2023, https://aibusiness.com/nlp/exclusive-schneider-electric-ai-chief-on-using-custom-chatgpt#close-modal.
13. *Safeguarding data*, 2021, https://www.unilever.com/sustainability/responsible-business/safeguarding-data/.
14. *Procter & Gamble fuels innovation with online learning*, https://www.coursera.org/business/casestudy/pg-fuels-innovation-online-learning.
15. *Privacy at Philips*, 2020, https://www.philips.com/a-w/privacy.html.
16. *Building Talent Pipelines—The GE Way*, 2016, https://www.tmi.org/blogs/building-talent-pipelines-the-ge-way.
17. Hood, J., O'Reilly, L., and Wei, W., *Volkswagen is using AI to speed up and scale marketing, while also integrating ChatGPT into its vehicles, says CMO Susanne Franz*, 2024, https://www.businessinsider.com/volkswagen-is-using-ai-to-speed-up-and-scale-marketing-while-also-integrating-chatgpt-into-its-vehicles-says-cmo-susanne-franz-2024-6.
18. *ABB Ability™ Asset Manager*, https://new.marketplace.ability.abb/s/products/electrification/asset-manager?language=en_US.
19. *IBM Watson—Deathvator*, https://www.youtube.com/watch?v=5V3BP-cf44Y.

CHAPTER 10

1. Cain, D., *AI and Predictive Maintenance: Boosting Efficiency, Slashing Costs*, 2023, https://www.linkedin.com/pulse/ai-predictive-maintenance-boosting-efficiency-slashing-david-cain-ublqc/.
2. *What Is the EU AI Act?* https://artificialintelligenceact.eu/#:~:text=What%20is%20the%20EU%20AI,AI%20to%20three%20risk%20categories.
3. DeCremer, D., *For Success with AI, Bring Everyone on Board*, May–June 2024, https://hbr.org/2024/05/for-success-with-ai-bring-everyone-on-board.
4. Lee, Youkyung and Lee, Yoolim, *Samsung's sudden $122 billion wipeout shows the cost of sleeping on AI*, Oct. 20, 2024, https://www.cnbctv18.com/market/samsung-sudden-wipeout-shows-the-cost-of-sleeping-on-ai-19501223.htm.

INDEX

Note: Page numbers followed by *f* and *t* refer to figures and tables, respectively.

3-D printing. *See* additive manufacturing (3-D printing)
3M Company, 221

ABB, 202–203, 226
access controls, 82, 83
AccuKnox, 83–84, 203
additive manufacturing (3-D printing), 20, 28, 29, 32, 35, 38, 138, 149
Adidas, 20
advanced analytics. *See also* business intelligence (BI)
  cloud-based platforms, 89
  concept, 39–40
  distribution and logistics, 196–197
  implementation, 66–81
  infrastructure, 72
  planning and scheduling, 43
  procurement, 158–159
  product design and development, 144–145
  production, 179–181
  sales, marketing, and after-sales service, 216–217
  tools, 71–72
  training and upskill programs, 78
advanced manufacturing, 3–4
after-sales service, 24–25, 37, 210. *See also* sales and marketing
  advanced analytics and BI tools, 216–217
  AI and ML, 210, 218
  data sources and integration, 212–213
  PdM for, 226
  predictive analytics, 219
  Stellar Electronics, 209
AI. *See* artificial intelligence (AI)
Amazon, 59–60
  AI and robotics, 202, 205–206
  demand forecasting, 180, 181–182
  personalized product recommendations, 217
  predictive analytics, 197
  supply chain operations, 59–60
  targeted marketing campaigns, 217
anomaly detection algorithms, 179
anonymization, 82–83
Apple, 18, 164, 200
  AppleCare program, 24
  data privacy, 30
  demand forecasting, 61, 144
  predictive analytics, 144
  product design and development, 141, 144
  supply chain practices, 19
artificial intelligence (AI), 111–133
  architectures, 112–123, 123*t*
  attention mechanisms, 119–123
  distribution and logistics, 199–201
  generative adversarial networks (GAN), 117–119
  global manufacturers, 234*t*–235*t*
  procurement, 159–162
  product design and development, 141–142, 148–149
  production, 181–182

sales, marketing, and after-sales service, 217–218
sector and size of manufacturers, 236*t*
tools and frameworks, 123–133, 134*t*
transformer architecture, 120–123, 121*f*
Artificial Intelligence Act (EU), 223, 237
attention mechanisms, 119–123
attribute-based access control, 83
augmented reality (AR), 23, 24, 26, 28, 29, 226
Australia, 2
authentication, 82, 83
automation, 35, 36, 237. *See also* artificial intelligence (AI)
AutoML tools, 131–133, 134*t*
AutoVentive, 169

Bac, Nico, 153
bagging and boosting, 132
BASF Chemicals, 43, 44, 57
Bezos, Jeff, 210
BI. *See* business intelligence (BI)
black-box model, 100
blockchain, 19, 24, 36, 87, 165, 169, 206
Blue Digital Twin platform, 63
BMW, 26, 27, 32, 37, 43, 53, 218
Boeing's 737 MAX crisis, 18
Bosch, 43, 51, 53, 185, 221, 223
Brilliant Factory Initiative (GE), 32, 33*f*
business intelligence (BI), 39. *See also* advanced analytics
cloud-based platforms, 89
concept, 40–41
distribution and logistics, 196–197
product design and development, 144–145
production, 179–181
sales, marketing, and after-sales service, 216–217
tools, 45–46

California Consumer Privacy Act (CCPA), 49, 81–82, 160, 165
Caterpillar, 3, 24, 30, 221
CCPA. *See* California Consumer Privacy Act (CCPA)
chatbots, 222, 226, 233
China, 2, 3
CIA (confidentiality, integrity, and availability), 82
CNN. *See* convolutional neural networks (CNN)
Coca-Cola, 60
collaboration, 30
human-AI, 205–206
Colonial Pipeline attack, 82
computer vision (CV), 201, 203
continuous integration/continuous deployment (CI/CD), 76–77, 79*f*
contract compliance, 167
conversational AI (CAI) system, 218, 222, 223, 226
convolutional neural networks (CNN), 113–114, 113*f*
cost, 166–167
Couch, Chris, 138
CRISP-DM framework, 74
cultural resistance, 185, 186, 225
customer centricity, 24
customer feedback, 5, 23, 41, 48, 55, 58, 138, 144, 203, 212, 213
customer segmentation, 23, 217, 221, 233, 236
cybersecurity risks and points of attack, 85*f*

data. *See also* data sources and integration
breaches, 82
classification, 84, 87
collection and preparation, 46–51
integration, 50, 71
life-cycle management, 87
masking, 82
owners/ownership, 84
profiling, 72
stewards/stewardship, 84
data security and governance, 48–49, 81–88
elements and potential solution providers, 86*f*
principles, 82, 83*f*
procurement, 165–166
threats and vulnerabilities, 86*f*
data sources and integration
distribution and logistics, 191, 194–196
procurement, 153–155
production, 175, 177

sales, marketing, and after-sales service, 212–213
data storage, 72
decision trees, 92, 96–99, 97*f*
decomposition models, 45
De Cremer, David, 233
Deloitte, 32, 175
demand forecasting, 61–64
optimization algorithms, 177
predictive analytics, 40
Deming, W. Edwards, 40
DHL, 194, 200, 201, 203
Dierkx, Keith W., 199
digital twins, 88
distribution and logistics, 21–22, 189–207
advanced analytics and BI, 196–197
AI and ML, 199–201
architecture, 195*f*
data sources and integration, 191, 194–196
future trends and outlook, 205–206
implementation challenges, 203–204
importance of, 190
modern processes, 198*f*
optimizing warehouse operations, 201–202
risk mitigation, 202, 203
steps involved, 192*f*
sustainability, 202–203
traditional process, 193*f*
Dow Chemical Company, 43, 44, 57, 203
Drucker, Peter, 138
Dyson, James, 1

EcoStruxure (Schneider Electric), 32, 37–38
encryption, 82, 84
energy optimization algorithms, 179
enterprise asset management (EAM), 76
enterprise resource planning (ERP) systems, 41, 46, 55, 58, 76, 153, 155, 169, 194
environmental and social impact metrics, 167–168
environmental sustainability, 202–203
ERP. *See* enterprise resource planning (ERP) systems
ethical considerations, 236–237
sales and marketing, 223–224
exploratory data analysis (EDA), 72

failure mode and effects analysis (FMEA), 45
FedEx, 21–22, 197
Fiorina, Carly, 40
Ford Motor Company, 3, 43–44, 55, 57, 169

GAN. *See* generative adversarial networks (GAN)
Garf, Rob, 225
GDPR. *See* General Data Protection Regulation (GDPR)
General Data Protection Regulation (GDPR), 36, 49, 81, 83, 160, 165, 223, 224, 236–237
General Electric (GE), 138, 185, 225
APM Reliability, 180
Brilliant Factory Initiative, 32, 33*f*
smart manufacturing practices, 20
SmartSignal predictive maintenance software, 27–28, 221
generative adversarial networks (GAN), 117–119
generative artificial intelligence (AI) tools, 49
Germany, 2, 3
Global Lighthouse Network (World Economic Forum), 30
Gomez, Laura, 151
Good Manufacturing Practice (GMP), 30
Gutzmann, Matthias, 152

Hayes, Samantha, 137
heuristics algorithms, 45
hierarchical clustering, 102, 105–107, 106*f*
human-AI collaboration, 205–206
human-in-the-loop, 146, 182, 199, 221

IBM
Food Trust, 169
predictive analytics, 203
SPSS Modeler, 74
Watson, 24, 28, 226
Watson Assistant, 218
Watson Studio, 200
Industry 4.0, 5, 25–29
infrastructure, 72
innovation, 2–6, 16–17, 237. *See also* Industry 4.0

integration, 50, 71. *See also* data sources and integration
intelligent systems, 121
International Standards Organization (ISO), 49
Internet of Things (IoT), 12, 13, 24, 25, 26, 28–29, 30, 32, 36, 41, 43, 46, 49, 52–55, 88, 89, 124, 165, 169, 185, 196, 197, 200, 205, 212, 226
intrusion detection and prevention systems, 83–84
inventory management, 168
isolation forests, 102, 107–109, 108*f*

Jackson, Phil, 237
Jennings, Jason, 3
Jobs, Steve, 11
Johnson Controls, 63
Johnson & Johnson (J&J), 60
just-in-time maintenance, 238

Keras, 128–130, 134*t*
key performance indicators (KPI), 35
Kimberly-Clark Corporation, 63
k-means clustering, 102–105, 104*f*
Komatsu, 203
Kone Corporation, 28, 226

lead time, 167
Lockheed Martin Corporation, 63–64
logistics. *See* distribution and logistics
Longridge, David, 39
long short-term memory (LSTM), 115–117, 116*f*

machine learning (ML), 91
  distribution and logistics, 199–201
  procurement, 159–162
  product design and development, 141–142, 148–149
  production, 181–182
  reinforcement learning (RL), 109–111, 110*f*
  sales, marketing, and after-sales service, 217–218
  supervised learning algorithms, 92–101
  unsupervised learning algorithms, 102–109
machine learning operations (MLOps), 76, 77
machine vision systems, 48
Mahindra, Anand, 66
Malik, Sarah, 151
malware, 82
manufacturing
  advanced (*See* advanced manufacturing)
  challenges, 4
  definitions, 1–2
  economy, 2, 3
  global, 2
  innovation, 2
  jobs, 3
  offshoring, 2
  production and (*See* production)
  reshoring, 3
  sector, 2
  value chain (*See* value chain)
manufacturing execution systems (MES), 46
marketing. *See* sales and marketing
May, John, 233
McKinsey & Company, 168, 173, 180, 181
memory cell, 116
metadata, 87
Miller, Rachel, 137
ML. *See* machine learning (ML)

Nair, Devadrita, 172
National Academy of Engineering, 4
National Institute of Standards and Technology (NIST), 1–2
natural language processing (NLP), 23, 119, 120, 149, 161, 168, 169, 217–218, 222–223, 233, 234
Nestlé S.A., 62–63
neural networks (NN), 112–113
  convolutional (CNN), 113–114, 113*f*
  recurrent (RNN), 114–117, 115*f*
Ng, Andrew, 92
Nike, 60
North American Industry Classification System Manufacturing (NAICSM), 1
Nvidia, 239

Ocado, 201
Ogilvy, David, 210
omnichannel retailing, 22

optimization algorithms, 45, 59, 219
anomaly detection, 179
demand forecasting, 177
energy, 179
predictive maintenance, 177
procurement decisions, 161, 169
production scheduling, 179, 180–181, 182
quality control, 179
simulation modeling, 45
supply chain, 179, 184
original equipment manufacturers (OEM), 25
ORION (On-Road Integrated Optimization and Navigation), 200, 202

PdM. *See* predictive maintenance (PdM)
PepsiCo, Inc., 62
Petecchi, Daniele, 139
Pfizer, 144, 149
Porter, Michael, 9
predictive analytics, 17, 18, 39, 71, 99, 213. *See also* advanced analytics
after-sales service, 24, 219
demand forecasting, 40, 61–64, 148, 184, 196
distribution and logistics, 196–197, 199, 201, 203, 205
procurement, 152, 158, 159, 162, 164, 168
product design and development, 41, 144–145, 148, 149, 150
production output, 175, 180
production scheduling, 64–66
quality control (QC), 57
sales forecasting and service, 23, 219, 221
supply chain optimization, 59–61, 150
Predictive Emission Monitoring System, 203
predictive maintenance (PdM), 28, 40, 43–44, 51–54, 233
advanced analytics algorithms, 52
after-sales services, 218, 226
attention mechanisms, 120
AutoML, 132
cases of, 43–44, 53–54
decision trees, 98
distribution and logistics, 205
EAM systems and, 76
impact, 237
isolation forests, 107, 108
Keras, 128, 129
k-means clustering, 102, 103–105
market growth, 52
optimization algorithms, 177
procurement, 168
production, 175, 177, 180, 181
PyTorch, 131
random forests, 99, 101
regression analysis for, 92–94
reinforcement learning (RL), 110
scikit-learn, 125
TensorFlow, 127
predictive models/modeling, 41
validation of, 73
prescriptive analytics, 39, 59, 71, 184
Procter & Gamble (P&G), 200, 225
demand forecasting, 44, 62
predictive analytics, 149
product design and development, 144, 149
regression analysis, 93–94
transportation route optimization, 200
procurement, 17–19, 151–170
advanced analytics and BI, 158–159
AI and ML, 159–162
architecture, 157*f*
compliance, 164–165
cycle times, 167
data security and governance, 165–166
data sources and integration, 153–155
future trends and opportunities, 168–169
managing, 166–168
modern processes, 162, 163*f*
risk management, 164–165, 166–167
steps involved, 154*f*
sustainable, 17–19
traditional processes, 156*f*
product design and development, 137–150
advanced analytics, 144–145
AI and ML, 141–142, 148–149
architecture, 143*f*
business intelligence (BI), 145–146, 147*f*
fundamentals, 139–142
process, 140*f*
production, 19–21, 171–187
advanced analytics and BI, 179–181
AI and ML, 181–182
algorithms used, 177, 179

architecture, 178*f*
data sources and integration, 175, 177
future trends and outlook, 186
implementation challenges, 184–185
steps involved, 174*f*
supply chain management, 182, 184
traditional process, 176*f*
production scheduling, 64–66
algorithms, 179
pseudonymization, 83
PyTorch, 130–131, 134*t*

QC. *See* quality control (QC)
quality control (QC), 44, 54–58, 233
algorithms, 179
anomaly detection, 118
AutoML, 132
CNNs, 113
decision trees, 98
hierarchical clustering, 105, 107
isolation forests, 107, 108
Keras, 128, 129
procurement, 160, 168
PyTorch, 131
random forest, 99
regression analysis, 93
scikit-learn, 124, 125
SVMs, 95, 96
TensorFlow, 127
transformer architectures, 120

random forest, 99–101
ransomware, 82
raw material procurement. *See* procurement
real-time insights, 32, 43–44, 57, 64, 88, 180, 181, 202, 212, 216, 237
recurrent neural networks (RNN), 114–117, 115*f*
regression analysis, 45
predictive maintenance (PdM), 92–94
regulatory challenges, 236–237
reinforcement learning (RL), 109–111, 110*f*
decision-making systems, 186
REP Center (SKF), 53
return on investment (ROI), 168
Ries, Eric, 238
risk management, 166–167
RNN. *See* recurrent neural networks (RNN)
role-based access control, 83
Rolls-Royce Holdings PLC, 43, 53, 138, 221
Rometty, Ginni, 112
root cause analysis (RCA), 45
Rosengren, Björn, 13

sales and marketing, 22–24, 209–227. *See also* after-sales service
advanced analytics and BI, 216–217
AI and ML, 217–218
architecture, 215*f*
customer support with AI, 221–223
data sources and integration, 212–213
future trends and outlook, 225–226
implementation challenges, 225
modern processes, 220*f*
optimizing marketing campaigns, 221–223
predictive analytics, 219, 221
steps involved, 211*f*
sustainability and ethical considerations, 223–224
traditional processes, 214*f*
Samsung, 49, 60, 61, 141, 239
Schmidt, Eric, 48
Schneider Electric, 32, 37–38, 224
scikit-learn, 124–126, 134*t*
Sheffi, Yossi, 230
Shingo, Shigeo, 39
Siemens, 20, 30, 57, 223–224
advanced analytics and BI, 43–44, 57
collaborating with third-party service providers, 25
Digital Enterprise Services, 30, 88
Digital Factory, 32
digitalization solutions, 32
digital marketing campaigns, 222
digital twin technology, 88
predictive maintenance (PdM), 221
quality control (QC), 57
rFAT (Remote Factory Acceptance Test) service, 23
virtual assistants, 226
simulation modeling, 45
Sjögren, Anders, 145
skills and talent gap, 50
smart factories, 237
Smart Factory (Deloitte), 32
smart warehousing, 205

Sommers, Kaitlynn, 160
Sondergaard, Peter, 44
South Africa, 2
statistical analysis, 45
Stellar Electronics, 137, 151
supervised learning algorithms, 92–101
  decision trees, 92, 96–99, 97*f*
  random forest, 99–101
  regression analysis, 92–94
  support vector machines (SVM), 92, 95–96, 95*f*
supervisory control and data acquisition (SCADA) systems, 55
supplier relationship management (SRM), 167
supplier selection, 167
supply chain
  data, 46–47
  disruptions, 203
  management, 182, 184
  optimization, 58–61
support vector machines (SVM), 92, 95–96, 95*f*
sustainability, 234, 236
  distribution and logistics, 202–203, 205
  environmental, 202–203
  issues, 237
  procurement, 17–19
  sales, marketing, and after-sales service, 223–224
Sustainable Development Goals (UN), 29
SVM. *See* support vector machines (SVM)

Taiwan Semiconductor Manufacturing Company (TSMC), 239
TensorFlow, 126–128, 134*t*
Tesla, 61
  carbon neutrality, 29
  demand forecasting, 61
  IoT sensors, 24
  Model S, 142
  over-the-air (OTA) software, 24
  quality control (QC), 51
  sustainable value chains, 29–30
Tetra Pak, 224
Thompson, Alex, 137
time-based maintenance, 238
time series analysis, 44–45
tokenization, 82–83
Toyota, 20, 32
  production planning processes, 181
  regression analysis, 93
Toyota Production System (TPS), 20, 32
transformation, 237
transformer architecture, 120–123, 121*f*
transformer models, 186
transportation management systems (TMS), 58, 191

Unilever, 29, 218, 225
UnitedHealth Group, 82
United Nations Industrial Development Organization, 2
United States, 2
United States Department of Labor, 1
unsupervised learning algorithms, 102–109
  hierarchical clustering, 102, 105–107, 106*f*
  isolation forests, 102, 107–109, 108*f*
  k-means clustering, 102–105, 104*f*
UPS, 21–22, 194, 200, 202
Urban, Will, 190

validation of predictive models, 73
value chain, 1, 4–9
  after-sales service, 24–25
  distribution, 21–22
  generic representation, 7*f*
  Industry 4.0, 5, 25–29
  primary activities, 6–7
  procurement, 17–19
  production, 19–21
  support activities, 7–9
virtual assistants, 218, 221, 222, 226, 233
virtual reality (VR), 23, 24, 28, 29, 226

Walmart, 3, 22, 181–182
warehouse layout optimization, 201–202
warehouse management systems (WMS), 58
weighted moving average (WMA) method, 64
World Economic Forum, 2

Yu, James, 151

zero-labor manufacturing, 4
Zero Trust Cloud Security (AccuKnox), 83–84, 203